Lecture Notes in Mathematics

C.I.M.E. Foundation Subseries

Volume 2347

Editors-in-Chief
Jean-Michel Morel, City University of Hong Kong, Kowloon Tong, China
Bernard Teissier, IMJ-PRG, Paris, France

Series Editors
Karin Baur, University of Leeds, Leeds, UK
Michel Brion, UGA, Grenoble, France
Rupert Frank, LMU, Munich, Germany
Annette Huber, Albert Ludwig University, Freiburg, Germany
Davar Khoshnevisan, The University of Utah, Salt Lake City, UT, USA
Ioannis Kontoyiannis, University of Cambridge, Cambridge, UK
Angela Kunoth, University of Cologne, Cologne, Germany
Ariane Mézard, IMJ-PRG, Paris, France
Mark Podolskij, University of Luxembourg, Esch-sur-Alzette, Luxembourg
Mark Policott, Mathematics Institute, University of Warwick, Coventry, UK
László Székelyhidi ⓘ, MPI for Mathematics in the Sciences, Leipzig, Germany
Gabriele Vezzosi, UniFI, Florence, Italy
Anna Wienhard, MPI for Mathematics in the Sciences, Leipzig, Germany

Fondazione C.I.M.E., Firenze

C.I.M.E. stands for *Centro Internazionale Matematico Estivo*, that is, International Mathematical Summer Centre. Conceived in the early fifties, it was born in 1954 in Florence, Italy, and welcomed by the world mathematical community: it continues successfully, year for year, to this day.

Many mathematicians from all over the world have been involved in a way or another in C.I.M.E.'s activities over the years. The main purpose and mode of functioning of the Centre may be summarised as follows: every year, during the summer, sessions on different themes from pure and applied mathematics are offered by application to mathematicians from all countries. A Session is generally based on three or four main courses given by specialists of international renown, plus a certain number of seminars, and is held in an attractive rural location in Italy.

The aim of a C.I.M.E. session is to bring to the attention of younger researchers the origins, development, and perspectives of some very active branch of mathematical research. The topics of the courses are generally of international resonance. The full immersion atmosphere of the courses and the daily exchange among participants are thus an initiation to international collaboration in mathematical research.

C.I.M.E. Director (2002 – 2014)
Pietro Zecca
Dipartimento di Energetica "S. Stecco"
Università di Firenze
Via S. Marta, 3
50139 Florence
Italy
e-mail: zecca@unifi.it

C.I.M.E. Director (2023 –)
Paolo Salani
Dipartimento di Matematica "U. Dini"
Università di Firenze
viale G.B. Morgagni 67/A
50134 Florence
Italy
e-mail: salani@math.unifi.it

C.I.M.E. Secretary
Daniele Angella
Dipartimento di Matematica "U. Dini"
Università di Firenze
viale G.B. Morgagni 67/A
50134 Florence
Italy
e-mail: daniele.angella@unifi.it

CIME activity is carried out with the collaboration and financial support of INdAM (Istituto Nazionale di Alta Matematica)

For more information see CIME's homepage: **http://www.cime.unifi.it**

Manfred Einsiedler • Giovanni Forni •
Vadim Kaloshin • Jasmin Raissy

Modern Aspects of Dynamical Systems

Cetraro, Italy 2021

Claudio Bonanno • Alfonso Sorrentino • Corinna Ulcigrai
Editors

Authors
Manfred Einsiedler
Department of Mathematics
ETH Zurich
Zürich, Switzerland

Vadim Kaloshin
Department of Mathematics
University of Maryland
College Park, MD, USA

Giovanni Forni
Department of Mathematics
University of Maryland
College Park, MD, USA

Jasmin Raissy
Institut de Mathématiques de Bordeaux
University of Bordeaux
Talence, France

Editors
Claudio Bonanno
Department of Mathematics
University of Pisa
Pisa, Italy

Corinna Ulcigrai
Institute of Mathematics
University of Zurich
Zurich, Switzerland

Alfonso Sorrentino
Department of Mathematics
University of Rome 'Tor Vergata'
Rome, Italy

ISSN 0075-8434 ISSN 1617-9692 (electronic)
Lecture Notes in Mathematics
ISSN 2946-1812 ISSN 2946-1820 (electronic)
C.I.M.E. Foundation Subseries
ISBN 978-3-031-62013-3 ISBN 978-3-031-62014-0 (eBook)
https://doi.org/10.1007/978-3-031-62014-0

Mathematics Subject Classification: 37-XX, 37-02, 37A17, 11J13, 11J25, 22D40, 22E40, 37A45, 37D40, 35A01, 65L10, 65L12, 65L20, 65L70

© The Editor(s) (if applicable) and The Author(s), under exclusive license to Springer Nature Switzerland AG 2024

This work is subject to copyright. All rights are solely and exclusively licensed by the Publisher, whether the whole or part of the material is concerned, specifically the rights of translation, reprinting, reuse of illustrations, recitation, broadcasting, reproduction on microfilms or in any other physical way, and transmission or information storage and retrieval, electronic adaptation, computer software, or by similar or dissimilar methodology now known or hereafter developed.
The use of general descriptive names, registered names, trademarks, service marks, etc. in this publication does not imply, even in the absence of a specific statement, that such names are exempt from the relevant protective laws and regulations and therefore free for general use.
The publisher, the authors and the editors are safe to assume that the advice and information in this book are believed to be true and accurate at the date of publication. Neither the publisher nor the authors or the editors give a warranty, expressed or implied, with respect to the material contained herein or for any errors or omissions that may have been made. The publisher remains neutral with regard to jurisdictional claims in published maps and institutional affiliations.

This Springer imprint is published by the registered company Springer Nature Switzerland AG
The registered company address is: Gewerbestrasse 11, 6330 Cham, Switzerland

If disposing of this product, please recycle the paper.

Preface

From August 2 to 6, 2021, Fondazione C.I.M.E. (*Centro Internazionale Matematico Estivo*[1]) organized a school in Cetraro, Italy, titled *Modern Aspects of Dynamical Systems*. The theory of dynamical systems is among the most transversal and applicable in the whole of mathematics, and it shares ideas and tools with a large number of mathematical fields. Due to its interdisciplinary nature, research in this area has contributed to the development of new ideas and techniques, extending well beyond the specific problem at hand. This has had ripple effects in unforeseen fields of research, both theoretical and applied. This might be one of the reasons behind the increasing prominence that dynamical systems have in the current mathematical panorama.

The C.I.M.E. school aimed to provide an overview of recent advances in the theory of dynamical systems, with a particular emphasis on their connections to other areas of mathematical research, including number theory, geometry, mathematical physics, complex analysis, and celestial mechanics. The following courses, taught by some of the world's leading experts in this field, were offered:

- *Homogeneous Dynamics and Diophantine Problems*
  Lecturer: Manfred Einsiedler (ETH Zürich, Switzerland)
- *Effective Ergodic Theory for Translation Flows*
  Lecturer: Giovanni Forni (University of Maryland, USA)
- *Integrability and Rigidity for Convex Billiards*
  Lecturer: Vadim Kaloshin (Institute of Science and Technology Austria)
- *Holomorphic Dynamics*
  Lecturer: Jasmin Raissy (Université de Bordeaux, France)
- *Exponentially Small Phenomena and Its Role in the Dynamics*
  Lecturer: Tere Martinez-Seara (Universitat Politècnica de Catalunya, Spain)

This book compiles the lecture notes from some of the contributions presented at the school.

---

[1] International Mathematical Summer Center.

Let us begin by introducing some of the topics and their underlying motivations. We will also provide a brief overview of the content covered in the collected lecture notes.

The modern theory of dynamical systems can be traced back to the work of Henri Poincaré, who focused on the qualitative theory of solutions to differential equations and their applications, particularly in the context of the three-body problem in celestial mechanics. Models stemming from mathematical physics have continued to inspire many classical examples within the field of dynamical systems. One class of such models are *mathematical billiards*, which appear naturally in many areas of mathematical physics, from mechanics, to celestial mechanics, to optics.

A (planar) mathematical billiard is a dynamical system describing the inertial motion of a point (the massless ball) inside a domain $\Omega \subset \mathbb{R}^2$ with a smooth or piecewise smooth boundary. The ball moves with unit velocity and without friction following a rectilinear path; when it hits the boundary it reflects elastically according to the standard reflection law: the angle of reflection is equal to the angle of incidence.

This conceptually simple, yet dynamically rich model, has been proposed by G. D. Birkhoff as a mathematical playground where "[...]the formal side, usually so formidable in dynamics, almost completely disappears, and only interesting qualitative questions need to be considered." Despite their simple local dynamics, indeed, the qualitative dynamical properties of mathematical billiards are extremely non-local and profoundly intertwined with the geometry (e.g., the shape) of the domain.

The course "Integrability and Rigidity for Convex Billiards," taught by Vadim Kaloshin, focused on the the study of rigidity properties in what we now refer to as *Birkhoff billiards*. These billiards are played inside a convex domain $\Omega \subset \mathbb{R}^2$ and their analysis serves as a model for *Hamiltonian dynamics* and integrable systems in *mathematical physics*. One of the longstanding conjectures in this field is the *Birkhoff conjecture* regarding the shape of the domain of an integrable billiard. Kaloshin's course aimed to highlight recent significant advances toward resolving this conjecture. These advances have drawn attention to compelling rigidity phenomena, which continue to inspire numerous intriguing questions and conjectures and highlight connections to other mathematical disciplines, including spectral analysis, symplectic geometry, and Riemannian geometry.

In the lecture notes by Vadim Kaloshin, co-authored with Corentin Fierobe and Alfonso Sorrentino, the authors provide an introduction to the dynamics of Birkhoff billiards, with a particular emphasis on exploring the existence and significance of periodic orbits, invariant curves (which are related to so-called *caustics*), and action-minimizing sets. The authors specifically explore the connections between these objects, Birkhoff conjecture, and the spectral rigidity properties of billiards, namely, with the question: "Can you hear the shape of a billiard?"

Coming back to the general theory of dynamical systems, basic questions about the dynamics revolve around finding periodic orbits, understanding the behavior of orbits near these points, and characterizing the long-term behavior of the system based on initial conditions. The study of these questions depends crucially on the topological properties of the space in which the dynamics act and from the properties of the map which defines the dynamics in the case of discrete dynamical systems.

When the space is the complex plane $\mathbb{C}$ or its compactification (the Riemann sphere), the study of bounded orbits under the dynamics of the corresponding *holomorphic* maps, more specifically, complex polynomials in the case of $\mathbb{C}$ or quotient of two complex polynomials (i.e., rational maps) in the case of the Riemann sphere, typically produces fractal sets, like the *Julia sets*, whose beautiful and often intricated patterns one can nowadays frequently see plotted and reproduced. The investigation of the properties of Julia sets and of their complements, the so-called *Fatou sets*, began in the early twentieth century with the works by Pierre Fatou and Gaston Julia.

The course "Holomorphic Dynamics" by Jasmin Raissy and the related lecture notes co-authored with Xavier Buff, focus on the classification of periodic points and their associated Fatou components. These components describe the common qualitative behavior of orbits in the vicinity of such points. The investigation begins with the classical case of rational maps on the Riemann sphere and then extends to the case of polynomial maps in complex dimension two, which remains to date an active area of research. The fundamental definitions and preliminary proofs in holomorphic dynamics, which are introduced at the beginning of the notes, make use of results and tools from *complex analysis*.

Whereas in one-dimensional dynamics, the classification of Fatou components (such as attracting, super-attracting, and parabolic domains, Siegel disks and Herman rings) summarized in the lecture notes is by now considered a classical result, in the two (or more) variable setting, the classification remains an open problem and is far less complete even for holomorphic endomorphisms $F : \mathbb{C}^2 \to \mathbb{C}^2$ in complex dimension two.

In the second part of the notes, the authors discuss recent progress in understanding the classification of Fatou components in the special case of *skew-products* in $\mathbb{C}^2$, which are maps of the form $F(z, w) = (f(z, w), g(w))$ where f and g are complex polynomials in two and one variable, respectively. These skew-products provide a more manageable example due to the presence of a one-dimensional fiber. After describing known type of Fatou components in this setting, they present a recent result, jointly proved by the two authors together with Matthieu Astorg, Romain Dujardin, and Han Peters. Their work reveals a new phenomenon which appears only in higher dimensions, by demonstrating the existence of *wandering Fatou components*, namely, Fatou components such that all their iterates under the dynamics are disjoint. The notes introduce some of the key ideas behind the *parabolic implosion* technique and highlight its central role in this novel construction.

One of the areas of pure mathematics where ergodic theory (the part of dynamical systems that deals with measure-preserving actions) has found the most striking applications is perhaps *number theory*. The historical roots of this connection can be dated back to the work of Michael Artin in 1924, connecting the study of the geodesic flow of the modular surface with the properties of continued fractions (later generalized by Caroline Series and Shrikrishna G. Dani, among others, and nowadays known as the Artin-Dani correspondence).

Deep connections between various problems in number theory, such as the Oppenheim conjecture or questions in Diophantine approximation on the one hand and *homogeneous dynamics*[2] on the other, have been discovered starting from the late 1970s. This development began in particular with seminal works by Hillel Furstenberg, addressing equidistribution of fractional parts of polynomials, as well as Gregory Margulis' celebrated proof of the Oppenheim conjecture.

Pushing this philosophy further, new advances and powerful tools in homogeneous dynamics later led to surprising and profound applications in the field of number theory. Examples include Marina Ratner's celebrated rigidity theorems for unipotent flows, which led to the resolution of Raghunathan's conjectures, the approach to metric Diophantine Approximation questions such as Mahler conjecture via homogeneous flows developed by Dmitry Kleinbock and Gregory Margulis, and the result on the exceptions to the Littlewood conjecture obtained as application of advances on measure rigidity by Manfred Einsiedler, Anatole Katok, and Elon Lindenstrauss.

The lecture notes for the course "Homogeneous Dynamics and Diophantine Problems", jointly written by Manfred Einsiedler with Thomas Ward, provide an introduction to this fruitful interplay. They start with the classical Artin-Dani connection and background material on the space of lattices and ergodic theory. After presenting classical results, such as Ratner's rigidity theorems and Margulis-Dani's quantitative non-divergence results for unipotent dynamics, the lecture notes focus on recent applications in Diophantine approximation. These include the study of (very) well and badly approximable vectors, as well as other fine questions related to Dirichlet approximable vectors and their prevalence within a given one-parameter family (which can be deduced from Nimish Shah's work on equidistribution of curves in the space of lattices).

Finally, in the last 40 years, one of the flourishing areas of research in modern dynamics which has seen contributions from several Fields medalists such as Jean-Christophe Yoccoz, Curtis McMullen, and Maryam Mirzakhani, lies at the intersection of dynamics and geometry and is known as *Teichmüller dynamics*. The dynamical systems which can be studied with tools from Teichmüller dynamics (in particular studying dynamics on the moduli space of surfaces) include a class of mathematical billiards known as *(rational) polygonal billiards*. These are planar billiards whose domain Ω is a polygon with angles that are rational multiples of π. These billiards can be reduced to the study of a linear flow on so-called *translation surfaces*, which are compact, orientable surfaces carrying a locally Euclidean metric, except for a finite number of conical singular points.

A successful approach to studying the fine equidistribution properties of linear flows on translation surfaces and polygonal billiards is based on *renormalization*

---

[2] The study of dynamics on algebraic, *homogeneous* spaces, i.e., quotients of the form G/Γ, where G is a Lie group and Γ a discrete subgroup, one of the prime examples being $SL(n, \mathbb{R})/SL(n, \mathbb{Z})$.

techniques in the parameter space of these geometric structures (namely, the moduli space of translation surfaces), exploiting the *Teichmüller flow* to deform the geometry. The conjectures on the fine ergodic properties of linear flows proved exploiting this *Teichmüller renormalization* idea include the seminal proof of Keane's unique ergodicity conjecture by Howard Masur and William A. Veech in the 1980s, the works by Giovanni Forni, Artur Avila, and Marcelo Viana on the Kontsevich-Zorich conjecture, and the proof of typical weak mixing by Artur Avila and Giovanni Forni in the first decade of this century.

In the course "Effective Ergodic Theory for Translation Flows" and the related lecture notes, Giovanni Forni first introduces fundamental concepts such as translation surfaces and their moduli space, along with basic ergodic theory definitions. He then presents the Hodge theory approach to both classical and recent results. This approach, originating from ideas of Maxim Kontsevich and Anton Zorich and based on the study of the action on (co)homology of renormalization (through the so-called *Kontsevich-Zorich cocycle*), is a distinctive perspective developed by Giovanni Forni himself in several of his works. Developments of this point of view in different directions have also been pursued more recently by Martin Moeller and Simion Filip, albeit with different motivations.

After presenting an original reinterpretation (using the cohomological approach) of a classical bound by Anatole Katok concerning the number of (transverse) invariant measures of a translation flow, the lecture notes discuss two types of quantitative (also referred to as *effective*) results: polynomial estimates for ergodic averages (which give effective unique ergodicity) and polynomial weak mixing estimates (which provide an effective form of weak mixing). The lecture notes emphasize the similarity between the proof of the first result, accomplished by Giovanni Forni in the early 2000s studying variations of Hodge norms, and the very recent results on quantitative weak mixing, which can be viewed as a version of the former for a *twisted* cocycle on a twisted cohomological bundle.

Through the courses just described, the school provided all participants with a captivating insight of the intricate interplay between dynamical systems and various other mathematical domains, in particular mathematical physics, complex analysis, number theory, and geometry. All of this was set against the backdrop of the stunning Cetraro coastline and sea.

As organizers, we acknowledge *Fondazione C.I.M.E.* for affording us this fantastic opportunity and extend our acknowledgments to *Grand Hotel San Michele*, where the school took place, for the hospitality. We are also grateful to the other sponsors that made it possible to organize the school, namely, the MIUR PRIN2017 project *"Regular and stochastic behaviour in dynamical systems,"* the Department of Excellence grant 2018–2022 awarded to the Department of Mathematics of University of Rome Tor Vergata, the *Zurich Graduate School in Mathematics*, cofinanced by the Department of Mathematics at ETH Zürich and the Institute of Mathematics at the University of Zurich, and the *Laboratoire Ypatia des Sciences Mathématiques* (LYSM).

We would also like to express once again our heartfelt gratitude and our deep appreciation to the lecturers, as well as to the collaborators who contributed to writing these lecture notes and to thank more broadly all participants of the School. It is our hope that these lecture notes will serve in the future as valuable reference material and as inspiration for many more researchers in dynamical systems, as well as in some of the many fields in which dynamics has found applications.

Pisa, Italy
Rome, Italy
Zurich, Switzerland
November 12, 2023

Claudio Bonanno
Alfonso Sorrentino
Corinna Ulcigrai

Contents

1 **Lecture Notes on Birkhoff Billiards: Dynamics, Integrability and Spectral Rigidity** 1
   Corentin Fierobe, Vadim Kaloshin, and Alfonso Sorrentino

2 **Introduction to Fatou Components in Holomorphic Dynamics** 59
   Xavier Buff and Jasmin Raissy

3 **Homogeneous Dynamics and its Connection to Diophantine Approximation** 105
   Manfred Einsiedler and Thomas Ward

4 **Effective Unique Ergodicity and Weak Mixing of Translation Flows** 161
   Giovanni Forni

Chapter 1
Lecture Notes on Birkhoff Billiards: Dynamics, Integrability and Spectral Rigidity

Corentin Fierobe, Vadim Kaloshin, and Alfonso Sorrentino

Abstract A mathematical billiard is a system describing the inertial motion of a point mass inside a domain, with elastic reflections at the boundary. The study of the associated dynamics is profoundly intertwined with the geometric properties of the domain (e.g., the shape of the billiard table). While it is evident how the shape determines the dynamics, a more subtle and difficult question is to which extent the knowledge of the dynamics allows one to reconstruct the shape of the domain. This translates into many intriguing unanswered questions and difficult conjectures that have been the focus of active research over the last decades. In these lectures note, we shall describe the main dynamical properties of so-called Birkhoff billiards, with particular emphasis on the problem of classifying integrable billiards (also known as Birkhoff conjecture) and the possibility of inferring dynamical information on the billiard map from its Length Spectrum (i.e., the lengths of its periodic orbits) and related spectral rigidity phenomena.

1.1 Lecture I: Billiard Dynamics

In these lecture notes we would like to introduce and investigate an interesting class of dynamical systems, known as *mathematical billiards*. Billiard is a generic term to refer to a very wide range of dynamical models with impacts; we refer the interested reader to [68, 94, 100, 101] for a more exhaustive presentation. Billiard-like models have been capturing the attention of researchers in various areas of mathematics

---

C. Fierobe
Institute of Science and Technology Austria, Klosterneuburg, Austria

V. Kaloshin (✉)
Institute of Science and Technology Austria, Klosterneuburg, Austria
University of Maryland, College Park, MD, USA

A. Sorrentino
Department of Mathematics, University of Rome Tor Vergata, Rome, Italy
e-mail: sorrentino@mat.uniroma2.it

Fig. 1.1 Two consecutive reflections of a billiard trajectory in a strictly convex planar domain Ω with smooth boundary $\partial\Omega$

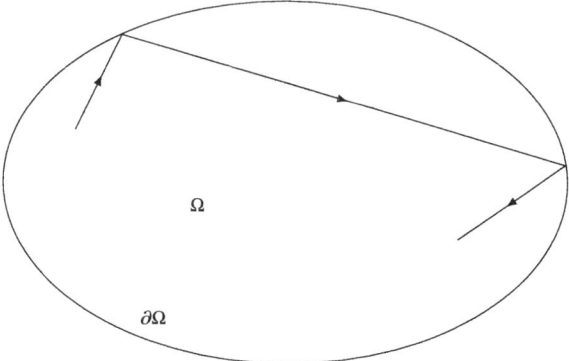

for many years: not only their law of motion is very physical and intuitive, but billiard-type dynamics is ubiquitous. Mathematically, according to the shape of the billiard, they offer models in every subclass of dynamical systems (integrable, regular, chaotic, etc.). Moreover, thanks to their manifold nature, they provide a fruitful laboratory where different ideas and expertises (from dynamical systems, analysis, geometry, etc...), have the possibilty to interface and beneficially integrate.

In these notes, we will be mostly interested in the study of the so-called *Birkhoff billiards* (see Definition 1.1). Such billiards are designed to describe the trajectory of a ray of light evolving inside a homogeneous convex cavity bounded by perfectly reflecting mirrors. From a mechanical point of view, if we consider a convex bounded domain $\Omega \subset \mathbb{R}^d$, where $d \geq 2$, with a C^1-smooth boundary $\partial\Omega$, we can think of a massless ball moving with unit velocity and without friction following a geodesic path: when the ball hits the boundary, it reflects elastically according to the standard *reflection law*, namely the angle of reflection is equal to the angle of incidence;[1] see Fig. 1.1.

The dynamics of the ball (or of the ray of light) evolving in Ω can be described by two approaches, a time-continuous one (the *billiard flow*), and a time-discrete one (the *billiard map*).

- *The billiard flow*: a *billiard trajectory* consists of the path followed by a ray of light evolving inside Ω, and bouncing on its boundary $\partial\Omega$ according to the law of reflection of optics, that is *angle of incidence = angle of reflection*. This first continuous approach can be translated in the existence of a quantity $B^t(p, v)$, where p is a point in the domain Ω, v is a non-zero vector and $t \in \mathbb{R}$ represents a

---

[1] One could also assume that the boundary $\partial\Omega$ has some point of non-differentiability; in this case, when the ball hits the boundary at one of these points (a sort of "holes"), then the motion stops.

time variable, giving the position at time t of the ray of light emitted from p with speed v. The map $(p, v, t) \mapsto B^t(p, v)$ is called the *billiard flow*.[2]

- *The billiard map*: complementary to the billiard flow, which provides a continuous model for the billiard, there is a discrete way to describe the billiard dynamics, namely using the so-called *billiard map*. Roughly speaking, we keep track only of bouncing points on the boundary and their bouncing direction. More specifically, a ray of light can be modeled by an oriented line intersecting Ω. The set of all such oriented lines defines the *phase space* $\mathcal{L}$ of the billiard map. The reflection of an oriented line is given by a transformation $T : \mathcal{L} \to \mathcal{L}$ which associates to an oriented line $\ell \in \mathcal{L}$, the oriented line $\ell' = T(\ell)$ which corresponds to the trajectory of the light after being reflected from ℓ. The map T is called the *billiard map* associated to Ω.

Remark 1.1 In the above models, one can easily replace $\mathbb{R}^d$ by any manifold with a complete Riemannian metric g and oriented lines by oriented geodesics of g. This allows one to define a notion of billiard in other geometrical contexts. The reader can consult [101] for more details.

1.1.1 Birkhoff Billiards

In the following, we will be interested in the time-discrete point of view. Moreover, we assume that the billiard domain satisfies some (strict) convexity assumption. Although in these lecture notes we will focus on planar billiards, we present some of the first properties for billiards in any dimension.

Definition 1.1 A *Birkhoff billiard* is a triple $(\Omega, \mathcal{L}, T)$, where Ω is a strictly convex domain of $\mathbb{R}^d$, and $\mathcal{L}$ and T are the associated phase-space and billiard map. A *billiard orbit* is a sequence $(T^n(\ell))_{n \in \mathbb{Z}}$ for a given $\ell \in \mathcal{L}$.

1.1.2 The Unit-Bundle Model

A way to define the phase space of a Birkhoff billiard Ω is to consider the set Φ of pairs (p, v) where $p \in \partial\Omega$ lies on the boundary of Ω and v is a unit vector pointing inside Ω when we choose p as its origin. Indeed, any oriented line intersecting Ω is uniquely defined by such a pair.

---

[2] The billiard flow might not be defined for any time $t > 0$, as it was shown by Halpern [50]: there exist Birkhoff billiards of C^2-smooth boundary such that the billiard flow is not defined for all times. However, when the boundary is C^3-smooth this phenomenon cannot occur.

The billiard map can be then defined as the map $F : \Phi \to \Phi$ associating to a pair $(p, v) \in \Phi$ a new pair $F(p, v) = (p', v') \in \Phi$ such that the ray of light emitted from (p, v) is reflected by the boundary as a ray emitted from (p', v').

Proposition 1.1 *Suppose that the boundary of Ω is C^r-smooth, $r \geq 2$. Then Φ can be endowed with a structure of C^{r-1}-smooth manifold which is $2(d - 1)$-dimensional, and $F : \Phi \to \Phi$ is a C^{r-1}-smooth diffeomorphism.*

Proof (See also [64, Theorem 4.2 in Part V]) Fix $p \in \partial\Omega$. Since the boundary is C^r-smooth, one can find a C^r-smooth local isometry $j : U \to V \subset \partial\Omega$ between an open set $U \subset \mathbb{R}^{d-1}$ and an open set $V \subset \partial\Omega$ containing p. For any point $q \in V$, one can define

- a *unit normal vector* $N(q)$ pointing inside Ω when we fix its origin at q;
- the *tangent space* of $\partial\Omega$ at $q = j(x)$ by $T_q\Omega = dj(x)\left(\mathbb{R}^{d-1}\right)$.

Consider the set

$$B(0, 1) = \{r \in \mathbb{R}^{d-1} \mid \|r\| < 1\}$$

of vectors of $\mathbb{R}^{d-1}$ with Euclidean norm less than 1. Given $v \in B(0, 1)$ and $q = j(x) \in V$, one can define the unit vector of $\mathbb{R}^d$

$$V(x, r) := dj(x) \cdot r + \sqrt{1 - \|dj(x) \cdot r\|^2} N(x).$$

The reader can check that $V(x, r)$ is the unit vector pointing inside $\partial\Omega$ when we choose its origin at q, and projecting onto $T_q\partial\Omega$ with the component $dj(x) \cdot r$. Note that $dj(x)$ is a linear isometry between $B(0, 1)$ and the unit open ball of $T_q\partial\Omega$. Hence, the C^{r-1}-smooth, injective map given by

$$(x, r) \in U \times B(0, 1) \mapsto (j(x), V(x, r))$$

defines a chart on Φ.

The fact that the billiard map is C^{r-1}-smooth comes from the implicit function theorem: consider two distinct points $p, p' \in \partial\Omega$ and a vector $v \in T_p\Omega$ of norm < 1. The map G defined by

$$G(p, v, p') = \frac{p' - p}{\|p' - p\|} - \left\langle \frac{p' - p}{\|p' - p\|} \mid N(p) \right\rangle N(p) - v$$

is C^{r-1}-smooth and vanishes if and only if p' is the point of impact of a ray emitted from p with unit vector given by $V_{v,v(p)}$. Indeed, the two first terms in G define the projection of the unit vector $\frac{p'-p}{\|p'-p\|}$ on the tangent space $T_p\partial\Omega$. Hence it is enough to show that locally, there is a C^{r-1} smooth map g in the variables (p, v) such that

$$G(p, v, p') = 0 \quad \Leftrightarrow \quad p' = g(p, v).$$

By the implicit function theorem, it is enough to check that $\tilde{G} : q \in \partial\Omega \mapsto G(p, v, q) \in T_p\partial\Omega$ is a local diffeomorphism at $q = p'$. We let the reader check this match by computing the differential of $\tilde{G}$ at p'.

Hence we have shown that if the billiard map writes as $(p, v) \mapsto (p', v')$, then $p' = p'(p, v)$ is C^{r-1} smooth. Now $v' = v'(p, v)$ is also C^{r-1}-smooth since it is obtained by projecting $\frac{p'-p}{\|p'-p\|}$ orthogonally to $T_p\partial\Omega$. □

Exercise 1.1 Let $i : \Phi \to \mathcal{L}$ be the map associating to a pair (p, v) the oriented line ℓ containing p and directed by v. Show that i is a bijection satisfying $i \circ F = T \circ i$.

1.1.3 The Cylinder Model in Dimension 2

From now we shall consider planar Bikrhoff billiards, namely we assume that $d = 2$, and $\Omega \subset \mathbb{R}^2$ is a strictly convex bounded domain with a C^r-smooth boundary, $r \geq 3$. The boundary $\partial\Omega$ is a closed curve which can be parametrized by an arc-length coordinate s, viewed modulo the perimeter of the boundary $L = |\partial\Omega|$. In this case any pair $(p, v) \in \Phi$ in the unit bundle model can be encoded by a unique pair $(s, \varphi) \in \mathbb{R}/L\mathbb{Z} \times (0, \pi)$ where s is the arc-length coordinate of p and φ is the angle made by v with the tangent line of $\partial\Omega$ at p oriented according to the parametrization of the boundary. The phase-space in this case is the annulus (or cylinder)

$$\mathbb{A}_L := \mathbb{R}/L\mathbb{Z} \times (0, \pi).$$

The billiard map is the one which naturally arises from this construction, acting therefore on $\mathbb{A}_L$; see Fig. 1.2.

Let us start with an example.

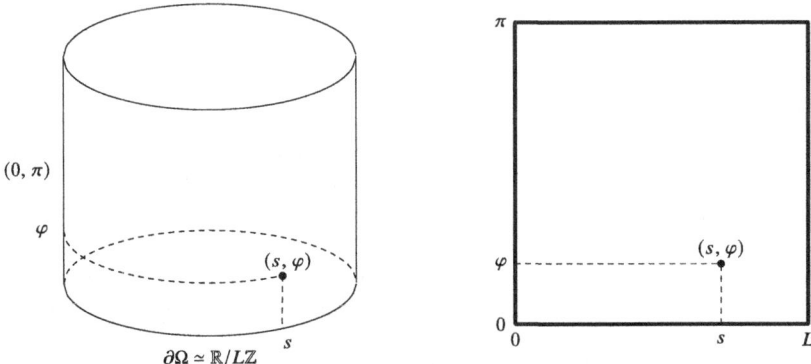

Fig. 1.2 The phase space $\mathbb{A}_L$ in dimension 2. On the left, its representation as a cylinder; on the right, its representation as an affine chart on $\mathbb{R}/L\mathbb{Z} \times (0, \pi)$

Fig. 1.3 The billiard reflection in a disk: two successive impact points p and p', such that the line pp' makes an angle φ with the boundary at p

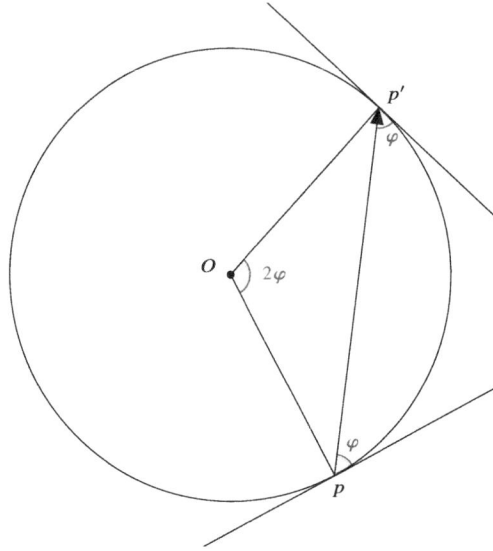

1.1.4 Example: Billiard in the Unit Disk

We consider the billiard in the unit disk $\mathbb{D}^1 := \{(x, y) \in \mathbb{R}^2 \mid x^2 + y^2 \leq 1\}$, whose boundary has perimeter $L = 2\pi$ and we assume to be parametrized by arclength. We denote by θ the length of the arc joining a point p_θ from a fixed origin $p_0 \in \partial \mathbb{D}^1 = \mathbb{S}^1$ in a given orientation (we count θ modulo 2π). Given an oriented line ℓ intersecting $\partial \mathbb{D}^1$ at a point p, we denote by φ the oriented angle between the tangent line $T_p \partial \mathbb{D}^1$ and ℓ. See Fig. 1.3.

Proposition 1.2 *In (θ, φ)-coordinates, the billiard map in $\mathbb{D}^1$ is the map $F : \mathbb{A}_{2\pi} \to \mathbb{A}_{2\pi}$ given by*

$$F(\theta, \varphi) = (\theta + 2\varphi, \varphi). \tag{1.1}$$

Proof Let us write $(\theta', \varphi') := F(\theta, \varphi)$. We consider the following geometric configuration. Denote by O the center of $\mathbb{D}^1$ and consider an oriented line intersecting $\partial \mathbb{D}^1$ at a point $p = p_\theta$ and such that the oriented angle between the tangent line $T_p \partial \mathbb{D}^1$ and ℓ is $\varphi \in (0, \pi)$. Let $p' = p_{\theta'}$ be the second point of intersection of ℓ with $\partial \mathbb{D}^1$. See Fig. 1.3.

By definition of $\mathbb{D}^1$, the triangle Opp' is isosceles in O and the lines Op and Op' intersect $\mathbb{D}^1$ orthogonally. This allows to deduce the following statements:

- The oriented angle between the tangent line $T_{p'} \partial \mathbb{D}^1$ and ℓ is $-\varphi$.
- The oriented angle between Op and Op' is 2φ.

Hence $\theta' = \theta + 2\varphi$ and ℓ is reflected at p' into an oriented line making an angle φ with the tangent line $T_{p'} \partial \mathbb{D}^1$, i.e., $\varphi' = \varphi$. □

1.1.4.1 Periodic and Dense Orbits

In particular, Proposition 1.2 implies that φ stays constant along the orbit (i.e., it represents what is called an *integral of motion* for the map); hence, the property of the orbits are determined by the corresponding angle $\varphi = \pi\omega$, with $\omega \in (0, 1)$; ω is called the *rotation number* of the orbit. Then:

- If $\omega = \frac{m}{n} \in (0, 1) \cap \mathbb{Q}$, for two coprime integers $m, n > 0$, then for any $\theta \in \mathbb{R}$, by induction F satisfies

$$F^n(\theta, \varphi) = (\theta + 2n\varphi, \varphi) = (\theta + 2\pi m, \varphi) = (\theta, \varphi) \quad \text{mod. } 2\pi \mathbb{Z}. \qquad (1.2)$$

Hence, the orbit is periodic with minimal period n. We say that (θ, φ) is a periodic point of rotation number m/n. In fact it corresponds to a polygonal periodic trajectory (if $m = 1$) or a star-shaped one (if $m > 1$), with n denoting its period, and m the number of times that the trajectory winds around the boundary $\partial \mathbb{D}^1$ before closing (see Fig. 1.4).
- If $\omega \in (0, 1) \setminus \mathbb{Q}$, then the orbit is not periodic and it hits the boundary $\partial \mathbb{D}^1$ on a dense set of points (by Kroenecker's theorem).

1.1.4.2 Invariant Curves

By Eq. (1.1), we observe that given a fixed $\varphi_0 \in (0, \pi)$, the curve $C_{\varphi_0} := \{(\theta, \varphi) \in \mathbb{A}_{2\pi} \mid \varphi = \varphi_0\}$ is invariant by F, that is $F(C_{\varphi_0}) = C_{\varphi_0}$. Moreover, the set of all possible C_{φ_0}, for $\varphi_0 \in (0, \pi)$, foliates the annulus $\mathbb{A}_{2\pi}$ (see Fig. 1.5).

Remark 1.2 Equation (1.2) implies that all points in $C_{\frac{m}{n}\pi}$ are periodic of rotation number m/n (and in fact, all such points lie on $C_{\frac{m}{n}\pi}$). We say that $\mathbb{D}^1$ has *1-parameter families* (i.e., curves in the phase space) of periodic points. In fact, it was proven (see [83]) that this phenomenon is degenerate in the sense that for a generic domain (i.e., an intersection of open and dense set of domains), the set of

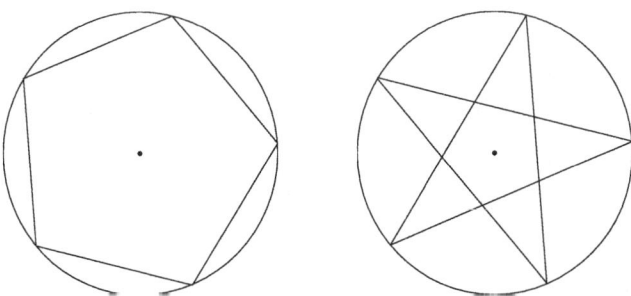

Fig. 1.4 Two periodic trajectories of period 5 in a disk, with rotation numbers 1/5 (on the left) and 2/5 (on the right)

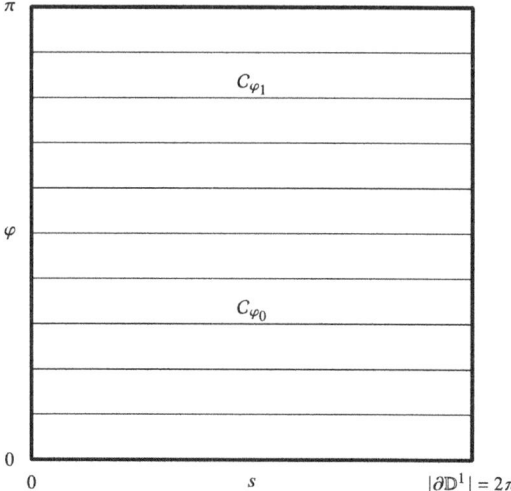

Fig. 1.5 The phase space $\mathbb{A}_{2\pi}$ of the billiard map in a disk with its foliation by horizontal invariant curves C_φ, where $\varphi \in (0, \pi)$

periodic orbits in this domain of any given period $n \geq 2$ is finite (see also [4] for some related results for general maps).

1.1.5 Generating Function

Consider $\gamma : \mathbb{R} \to \mathbb{R}^2$ be a C^r-smooth L-periodic parametrization of the boundary $\partial \Omega$. Define the map $H : \mathbb{R}^2 \to \mathbb{R}^+$ given by

$$H(s, s') = -\text{dist}(\gamma(s), \gamma(s')) \qquad \forall s, s' \in \mathbb{R}.$$

Definition 1.2 The map H is called the *generating function* of the billiard in Ω.

Remark 1.3 The negative sign in the definition of H is just a convention, that will make sense a bit later when we will consider minimizers of H, which otherwise would be maximizers of the distance.

Exercise 1.2 Compute H when Ω is a unit disk.

The function H is called generating because it contains the dynamics of the billiard (see Fig. 1.6), as stated in the following proposition.

Proposition 1.3 *The map H is C^r-smooth outside the set $\Delta = \{(s, s) \mid s \in \mathbb{R}\}$. Moreover, if $(s, \varphi), (s', \varphi') \in \mathbb{A}_L$ are such that $s \neq s'$, then the following statements are equivalent:*

(i) $F(s, \varphi) = (s', \varphi')$;
(ii) $\partial_1 H(s, s') = -\cos \varphi$ *and* $\partial_2 H(s, s') = \cos \varphi'$, *where ∂_k denotes the derivative with respect to the k-th variable ($k = 1, 2$).*

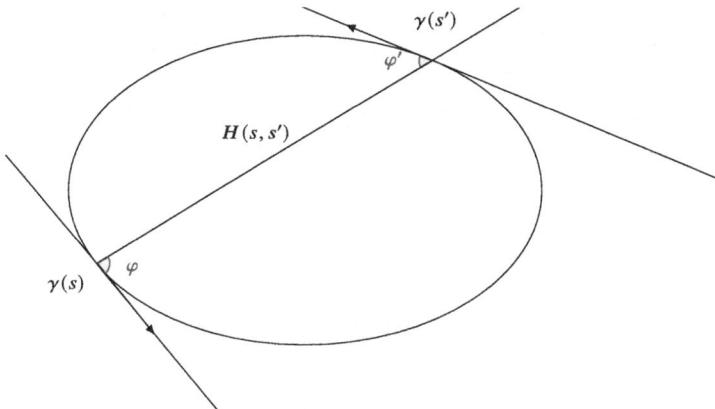

Fig. 1.6 The generating function encodes the dynamics of the billiard

Proof The smoothness condition is obvious since $\partial\Omega$ is C^r-smooth. Given $s \neq \tilde{s}$, let us compute $\partial_1 H(s, \tilde{s})$. Differentiating the Euclidean norm, we obtain

$$\partial_1 H(s, \tilde{s}) = -\left\langle \frac{\gamma(\tilde{s}) - \gamma(s)}{\|\gamma(\tilde{s}) - \gamma(s)\|} \,\Big|\, \gamma'(s) \right\rangle.$$

Here $u := \frac{\gamma(\tilde{s}) - \gamma(s)}{\|\gamma(\tilde{s}) - \gamma(s)\|}$ is a unit vector from $p = \gamma(s)$ to $\tilde{p} = \gamma(\tilde{s})$ and $\gamma'(s)$ has unit norm (since γ parametrizes $\partial\Omega$ with arc-length). It follows that $\partial_1 H(s, \tilde{s})$ is $-\cos\varphi$ where φ is the angle between the line $p\tilde{p}$ and the tangent line to $\partial\Omega$ at p. Similarly, $\partial_2 H(s, \tilde{s}) = \cos\tilde{\varphi}$ where $\tilde{\varphi}$ is the angle between the line $p\tilde{p}$ and the tangent line to $\partial\Omega$ at $\tilde{p}$. From this we obtain $F(s, \varphi) = (\tilde{s}, \tilde{\varphi})$ and hence $F(s, \varphi) = (s', \varphi')$ if and only if $\tilde{s} = s'$ and $\tilde{\varphi} = \varphi'$, which is tantamount to item (ii). □

Exercise 1.3 Let Ω be a Birkhoff billiard with C^2-smooth boundary. Let $\kappa(s)$ be the radius of curvature of $\partial\Omega$ at a point of arc-length coordinate s.

(i) Show that (see [64, Theorem 4.2 in Part V]) :

$$\begin{cases} \partial_s s' = \dfrac{\kappa(s) d(s, s') - \sin\varphi}{\sin\varphi'} & \partial_\varphi s' = \dfrac{d(s, s')}{\sin\varphi'} \\[2ex] \partial_s \varphi' = \dfrac{\kappa(s)\kappa(s') d(s, s') - \kappa(s)\sin\varphi' - \kappa(s')\sin\varphi}{\sin\varphi'} & \partial_\varphi \varphi' = \dfrac{\kappa(s') d(s, s') - \sin\varphi'}{\sin\varphi'} \end{cases}$$

where $d(s, s')$ denotes the distance in the plane between the points on the boundary of the billiards corresponding to s and s'.

(ii) Assume that Ω is a Birkhoff billiard with C^4-smooth boundary. Show that when $(s' - s) \to 0$, the generating function H associated to Ω can be written as

$$H(s, s') = -|s' - s|\left(1 - \frac{\kappa(s)^2}{24}(s' - s)^2 - \frac{\kappa(s)\kappa'(s)}{24}(s' - s)^3\right.$$
$$\left. + O\left((s' - s)^4\right)\right).$$

Moreover, for small φ, the differential of the billiard map has the following form

$$Df(s, \varphi) = L(s) + \varphi A(s) + O(\varphi^2)$$

where

$$L(s) := \begin{pmatrix} 1 & \dfrac{2}{\kappa(s)} \\ 0 & 1 \end{pmatrix}$$

and

$$A(s) := \begin{pmatrix} -2\dfrac{\kappa'(s)}{\kappa^2(s)} & -\dfrac{8}{3}\dfrac{\kappa'(s)}{\kappa^3(s)} \\ 0 & \dfrac{4}{3}\dfrac{\kappa'(s)}{\kappa^2(s)} \end{pmatrix}.$$

1.1.6 Area-Preservation

Consider the 1-form λ on $\mathbb{A}_L$ defined by $\lambda = \cos \varphi \, ds$. The exterior differential of λ is the area form

$$\omega = d\lambda = \sin \varphi \, ds \wedge d\varphi.$$

Proposition 1.3 has the following consequence. We recall that given a k-form α on a manifold M, and a smooth map $F : M \to M$, we define α's pullback by F as the k-form $\beta := F^*\alpha$ such that

$$\beta_p(v_1, \ldots, v_k) = \alpha_{F(p)}(dF_p(v_1), \ldots, dF_p(v_k)) \qquad \forall\, p \in M \quad \forall\, v_1, \ldots, v_k.$$

Exercise 1.4 (Exact Symplecticity) Prove that

$$F^*\lambda - \lambda = dH$$

(F is a so-called *exact-symplectic map*). Deduce that F preserves ω, i.e., $F^*\omega = \omega$.

This suggests the following change of coordinates:

$$\begin{cases} x := s \\ y := -\cos\varphi. \end{cases}$$

Notice that

$$\omega = ds \wedge d(-\cos\varphi) = dx \wedge dy.$$

The billiard map F in these new coordinates is given by

$$f : \mathbb{R}/L\mathbb{Z} \times (-1, 1) \to \mathbb{R}/L\mathbb{Z} \times (-1, 1),$$

defined by the condition $f(s, -\cos\varphi) = (s', -\cos(\varphi'))$ where $(s', \varphi') = F(s, \varphi)$ for all $(s, \varphi) \in \mathbb{A}_L$; by by construction f is conjugated to F. From Exercise 1.4 we can deduce the following proposition.

Proposition 1.4 *The billiard map f is area-preserving, i.e. it preserves the area form $dx \wedge dy$.*

Billiard maps are examples of so-called *exact-symplectic twist maps*; we refer to [43, 94] for more details about these maps. See also Sect. 1.4 for their relation with the so-called Aubry-Mather theory.

1.2 Lecture II: Variational Principle and Periodic Orbits

Let $\Omega \subset \mathbb{R}^2$ be a strictly convex bounded domain with C^3-smooth boundary and assume that its boundary $\partial\Omega$ is parametrized by an arc-length coordinate s, viewed modulo the perimeter of the boundary that, without loss of generality, we assume to be equal to 1. We denote by $\gamma : \mathbb{R} \to \mathbb{R}^2$ this parametrization.

Let us denote as before $\mathbb{A} := \mathbb{R}/\mathbb{Z} \times (0, \pi)$ and consider the billiard map $F : \mathbb{A} \longrightarrow \mathbb{A}$ and a lift of F to the universal cover $\mathbb{R} \times (0, \pi)$ of $\mathbb{A}$:

$$\widetilde{F} : \mathbb{R} \times (0, \pi) \longrightarrow \mathbb{R} \times (0, \pi)$$
$$(s, \varphi) \mapsto (s', \varphi').$$

Let us also denote by $H : \mathbb{R} \times \mathbb{R} \to \mathbb{R}$ its generating function

$$H(s, s') = -\text{dist}(\gamma(s), \gamma(s')) \quad \forall s, s' \in \mathbb{R}.$$

Definition 1.3 An *orbit* of the map $\widetilde{F}$ is a sequence $\underline{p} = (s_k, \varphi_k)_{k \in \mathbb{Z}}$ such that for any $k \in \mathbb{Z}$, $\widetilde{F}(s_k, \varphi_k) = (s_{k+1}, \varphi_{k+1})$. The *rotation number* of the orbit, if it exists, is defined as the quantity

$$\lim_{k \to +\infty} \frac{s_k - s_0}{k}.$$

An orbit is said to be *periodic* if there exist an integer $n > 0$ and $m \in \mathbb{Z}$, such that for any $k \in \mathbb{Z}$ $s_{k+n} = s_k + m$. The period of the orbit is the minimal n satisfying this condition.

Exercise 1.5 Check that in the case of a periodic orbit, rotation number exists and it is given by the ratio m/n, where n is the period and m the corresponding integer (see definition above).

Remark 1.4 The rotation number m/n has a simple geometric interpretation in the case of billiard maps: n is the number of time an orbit bounces on the billiard boundary before repeating itself—or its *period*, and m is the number of time it winds around the boundary.

Definition 1.4 A *stationary configuration* for $\widetilde{F}$ is a sequence $\underline{s} = (s_k)_{k \in \mathbb{Z}} \subset \mathbb{R}^{\mathbb{Z}}$ such that for any $k \in \mathbb{Z}$

$$\partial_2 H(s_{k-1}, s_k) + \partial_1 H(s_k, s_{k+1}) = 0.$$

From Proposition 1.3, we can easily deduce a correspondance between stationary configurations and orbits. More specifically:

Proposition 1.5 *The map associating to an orbit $(s_k, \varphi_k)_{k \in \mathbb{Z}}$ the stationary configuration $(s_k)_{k \in \mathbb{Z}}$ is a bijection between orbits and stationary configurations.*

Given a Birkhoff billiard, we can ask about if it has periodic orbits. The answer is in fact yes, and it is given by the famous theorem of Birkhoff (see [17] and also the presentations in [43, 94]).

Theorem 1.1 (Birkhoff) *Let Ω be a Birkhoff billiard with C^2-smooth boundary. Then for any $m/n \in (0, 1) \cap \mathbb{Q}$ there exist at least two geometrically distinct periodic orbits of rotation number m/n.*

Proof The proof of Theorem 1.1 relies on the variational principle given by Proposition 1.5. Fix coprime integers $m, n > 0$ such that $m/n \in (0, 1) \cap \mathbb{Q}$.

1 Lecture Notes on Birkhoff Billiards

The first periodic orbit of rotation number m/n is given by minimizing the functionnal $h_{m,n} : K_{m,n} \to \mathbb{R}$ defined on the compact set

$$K_{m,n} = \{(s_0, \ldots, s_n) \in \mathbb{R}^{n+1} \mid s_n = s_0 + m, \ s_0 \in [0,1] \text{ and } s_k \leq s_{k+1} \ \forall k\}$$

by

$$h_{m,n}(s_0, \ldots, s_n) := \sum_{k=0}^{n-1} H(s_k, s_{k+1}).$$

The function $h_{m,n}$ is continuous hence reaches its minimal value at a certain $\underline{s} = (s_0, \ldots, s_n)$. In fact, one can show that $\underline{s}$ can be taken in the interior of $K_{m,n}$. Indeed, if for some k we have $s_{k-1} = s_k$, but $s_k < s_{k+1}$ then for a fixed $t \in (s_k, s_{k+1})$ if we denote by

$$\underline{\tilde{s}} = (s_0, \ldots, s_{k-1}, t, s_{k+1}, \ldots, s_n)$$

we have by construction $h_{m,n}(\underline{\tilde{s}}) < h_{m,n}(\underline{s})$ because the sum of distances of the chords from s_{k-1} to t and from t to s_{k+1} is strictly greater than the distance of the chord from $s_{k-1} = s_k$ to s_{k+1} hence

$$H(s_{k-1}, t) + H(t, s_{k+1}) < H(s_{k-1}, s_k) + H(s_k, s_{k+1}).$$

This contradicts the minimality of $\underline{s}$. Moreover, adding or substracting 1 to the components of $\underline{s}$ does not change the minimality, so we can assume $s_0 \in [0, 1)$. Now, since $\underline{s}$ is in the interior of $K_{m,n}$, it is a critical point of $h_{m,n}$: in particular for any $k \in \{1, \ldots, n-1\}$,

$$0 = \partial_k h_{m,n}(\underline{s}) = \partial_2 H(s_{k-1}, s_k) + \partial_1 H(s_k, s_{k+1})$$

and by Proposition 1.5, this corresponds to a periodic orbit of the billiard map of rotation number m/n.

The other one is obtain by minmax techniques (see figure below for an idea of the so-called mountain pass method) and one needs to show that also this critical configuration must lie in the interior of the set, hence correspond to an orbit of the map (Fig. 1.7). □

Remark 1.5 The lower bound on the number of periodic orbits with a given rational rotation number is optimal. For example, if one takes an ellipse with eccentricity $0 < e < 1$, then there are exactly two periodic orbits corresponding, respectively, to the minor and major axis.

Fig. 1.7 Mountainpass critical points

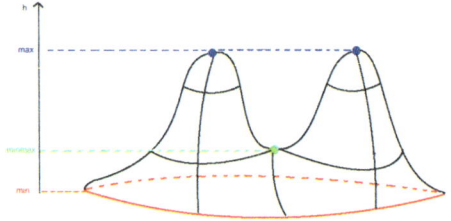

Exercise 1.6 Let $\varepsilon > 0$ and consider the domain $\Omega_\varepsilon \subset \mathbb{R}^2$ whose boundary is given in polar coordinates (φ, r) by

$$r = 1 + \varepsilon \cos 3\varphi.$$

(1) Show that there exists $\varepsilon_0 > 0$ such that for any $0 \leq \varepsilon \leq \varepsilon_0$ the domain Ω_ε is strictly convex.
(2) Consider the points p_1, p_2, p_3 on $\partial \Omega_\varepsilon$ corresponding to $\varphi = 0, \frac{2\pi}{3}, \frac{4\pi}{3}$. Show that (p_1, p_2, p_3) is a 3-periodic orbit. Does this orbit maximize the length functional?
(3) Same question as in (2) with the points q_1, q_2, q_3 on $\partial \Omega_\varepsilon$ corresponding to $\varphi = \frac{\pi}{3}, \pi, \frac{5\pi}{3}$.

1.2.1 On the Quantity of Periodic Orbits and Ivrii's Conjecture

Describing the set of periodic orbits of a dynamical system like a billiard allows to understand it better. The question of their existence is given by Birkhoff's theorem (Theorem 1.1) which states the existence of periodic orbits of any rotation number. A further question one can ask is about their *quantity*. The following statement was conjectured by Ivrii in [59].

Ivrii's Conjecture *Given a strictly convex billiard in $\mathbb{R}^d$, the set of its periodic orbits has zero measure.*

The conjecture can be understood as follows: the billiard map inside a strictly convex domain $\Omega \subset \mathbb{R}^d$ acts on the pairs (p, v) where p lies on the boundary of Ω and v is a unit vector pointing inside Ω from p. It can hence be viewed as a diffeomorphism $F : B(T\partial\Omega) \to B(T\partial\Omega)$ of pairs $(p, v) \in T\partial\Omega$ such that v is tangent to Ω at p and has norm ≤ 1. In this setting, Ivrii's conjecture states that the set of periodic points of F, namely

$$\{(p, v) \in B(T\partial\Omega) \mid \exists n \in \mathbb{Z}^{>0} \ F^n(p, v) = (p, v)\}$$

has zero measure.

This conjecture is related to the so-called Weyl law related to the Dirichlet eigenvalues of the Laplace-Beltrami operator Δ in a bounded convex domain $\Omega \subset \mathbb{R}^d$. For $\lambda \geq 0$, define $N(\lambda)$ the number of eigenvalues (with multiplicities) of Δ for which the corresponding eigenvector vanish on the boundary $\partial \Omega$. Weyl [106] proved the following asymptotics

$$N(\lambda) \sim a_d \operatorname{vol}(\Omega) \lambda^{d/2}$$

where $a_d > 0$ is a constant depending only on d and conjectured the second order term, so that $N(\lambda)$ should follow the following asymptotics:

$$N(\lambda) = a_d \operatorname{vol}(\Omega) \lambda^{d/2} + a_{d-1} \operatorname{area}(\partial \Omega) \lambda^{\frac{d-1}{2}} + o\left(\lambda^{\frac{d-1}{2}}\right) \qquad (1.3)$$

where $a_{d-1} \in \mathbb{R}^*$ only depends on d. Ivrii showed that if the set of periodic orbits inside a domain Ω has measure zero, then the asymptotics (1.3) holds.

Ivrii's conjecture remains open until now; however partial answers have been given. A generic positive answer has been given in [83], where it is proven using a transversality theorem that for a generic domain the set of periodic orbits of any given period is finite. Note that in fact Ivrii's conjecture holds if and only if it is true for the set of periodic orbits of any given period. Following this idea, [90] and later [98] proved the conjecture for 3-periodic orbits. There result was extended to any dimension in [105]. See also [20, 107] for different proofs of this result which also apply for other geometries. These results were also extended to 4-periodic orbits, see [41, 42], but also [37–39]. The latter result also applies to billiards in other geometries or with different reflection laws, see for example [20, 34, 35].

This conjecture was proven by Vasiliev [103] under the assumption that the boundary of the domain is analytic. Let us sketch the proof of this result.

Theorem 1.2 ([103]) *Let $\Omega \subset \mathbb{R}^d$ be a strictly convex domain with analytic boundary. Then the set of periodic orbit for the billiard inside Ω has zero measure.*

Proof Let us sketch the proof of this result for $d = 2$. In this setting, the billiard map F inside Ω is an analytic diffeomorphism of the cylinder $\mathbb{A}$. If the set of periodic points of F has striclty positive measure, then there should exists an $n \geq 2$ such that the set of periodic points of F of period n has strictly positive measure. Hence the identity $F^n = \operatorname{id}$ is satisfied on a set of non zero measure, and by analytic continuation it is satisfied everywhere. But this contradicts the existence of periodic orbits of rotation number $1/q$ where q is bigger than n. □

Remark 1.6 The proof of Theorem 1.2 shows that in the analytic setting the existence of a set of non-zero measure of periodic orbits implies the existence of an open set of periodic orbits (here the whole phase space).

Remark 1.7 Consider a billiard F in a strictly convex domains. Let $x_0 = F^q(x_0)$ be a periodic of some period $q > 1$. Call x_0 *absolutely periodic (resp. absolutely periodic of order n)* if the differential of F^q is the identity and all second order and higher partial derivatives of F^q (resp. of order up to n) at x_0 are 0. In a recent preprint [24], K. Callis showed that for any natural number n, the set of domains containing absolutely periodic orbits of order n are dense in the set of bounded strictly convex domains with C^∞ smooth boundary. This result is a step toward disproving the following conjecture by Safarov-Vassiliev in [91], namely, that no absolutely periodic billiard orbits of infinite order exist in Euclidean billiards. This is also an indication that Ivrii's Conjecture about the measure set of periodic orbits might not be true.

Let us conclude this section by recalling that the set of perimeters of periodic orbits in a strictly convex billiard has zero measure. This is a simple consequence of Sard's lemma and the fact that periodic orbits are critical points of the length functional. This however does not allow to answer Ivrii's conjecture.

Theorem 1.3 (Birkhoff [17]) *Let $\mathcal{L}(\Omega) \subset \mathbb{R}$ be the set of perimeters of periodic orbits inside a Birkhoff's billiard. Then $\mathcal{L}$ has zero Lebesgue measure.*

Proof It is enough to show this result for the set $\mathcal{L}_{m/n}(\Omega)$ of periodic orbits of a fixed rotation number $m/n \in (0, 1)$. As in the proof of Theorem 1.1, define the map $h_{m,n} : V_{m,n} \to \mathbb{R}$ on the open set

$$V_{m,n} = \{(s_0, \ldots, s_n) \in \mathbb{R}^{n+1} \mid s_n = s_0 + m, \forall k \quad s_k < s_{k+1}\}$$

by

$$h_{m,n}(s_0, \ldots, s_n) = \sum_{k=0}^{n-1} H(s_k, s_{k+1}).$$

As we saw, the set $\text{Crit}(h_{m,n})$ corresponds to the stationary configurations, and hence to periodic orbits of rotation number m/n of the billiard map. Moreover, the perimeter of the orbit obtained from $\underline{s} \in V_{m,n}$ corresponds to the quantity $h_{m,n}(\underline{s})$. Hence we just showed that

$$\mathcal{L}_{m/n}(\Omega) = h_{m,n}(\text{Crit}(h_{m,n}))$$

is the so-called set of critical values of $h_{m,n}$. By Sard's Lemma, this set has zero measure. □

Exercise 1.7 Construct a C^∞ domain with the Length spectrum of positive Hausdorff dimension.

1.2.2 Laplace Spectrum and Length Spectrum

We define the *length spectrum* of Ω as the set

$$\mathcal{L}_\Omega := \mathbb{N}^+ \cdot \{\text{lengths of periodic orbits in } \Omega\} \cup \mathbb{N}^+ \cdot |\partial\Omega|,$$

where $|\partial\Omega|$ denotes the length of the boundary, namely the set of multiples of the lengths of all periodic orbits and multiples of the perimeter of Ω.

As we have already pointed out before (see Ivrii's conjecture), a remarkable relation exists between the length spectrum of a billiard in a convex domain Ω and the spectrum of the Laplace operator in Ω with Dirichlet boundary condition:

$$\begin{cases} \Delta f + \lambda^2 f = 0 & \text{in } \Omega \\ f|_{\partial\Omega} = 0. \end{cases}$$

From the physical point of view, the eigenvalues are the eigenfrequencies of the membrane Ω with a fixed boundary. Denote by $\mathrm{Spec}_\Delta(\Omega) = \{0 < \lambda_1 \leq \lambda_2 \leq \ldots\}$ the Laplace spectrum of eigenvalues solving this problem.

The famous question of M. Kac in its original version asks *if one can recover the domain from the Laplace spectrum*. For general manifolds there are counterexamples (see [44]).

K. Anderson and R. Melrose [2] proved the following relation between the Laplace spectrum and the length spectrum (see also [47, 84, 91]):

Theorem (Anderson-Melrose) *The wave trace*

$$w(t) := \mathrm{Re}\left(\sum_{\lambda_n \in \mathrm{Spec}_\Delta(\Omega)} e^{i\lambda_n t} \right)$$

is well-defined as a distribution and smooth away from the length spectrum:

$$\mathrm{sing\,supp.}\big(w(t)\big) \subseteq \pm\mathcal{L}(\Omega) \cup \{0\}. \tag{1.4}$$

Namely, if $\xi > 0$ belongs to the singular support of this distribution, then there exists either a closed billiard trajectory of length ξ, or a closed geodesic of length ξ in the boundary of the billiard table. Generically, equality holds in (1.4).

Exercise 1.8 An example to convince about this relation is the following. Consider $\Omega = (0, \pi) \times (0, \pi)$; then, its Laplace spectrum is given by

$$\mathrm{Spec}(\Omega) = \{n^2 + m^2 : (n, m) \in \mathbb{N} \times \mathbb{N} \setminus \{(0, 0)\}\}.$$

Discuss its relation with the lengths of periodic orbits in $\overline{\Omega}$.

Remark 1.8

(*i*) The above inclusion holds for non-convex C^∞ domains in arbitrary dimension (see [84, Theorem 5.4.6]).

(*ii*) Observe that there are not known examples of domains in which the singular support of the wave trace is strictly included inside the Length spectrum: the equivalence between these sets is strictly related to the problem whether Laplace spectral rigidity implies Length spectral rigidity. Observe that it may be possible that the singular support of the wave trace is strictly included inside the Length spectrum. If they are the same it is closely related to the problem whether Laplace spectral rigidity implies Length spectral rigidity. In an unpublished manuscript Hezari and Zelditch constructed an example of an analytic domain where the equality fails. In a recent preprint Kaloshin-Koval-Vig [61], it was show that for a dense set of eccentricities $e \in (0, 1)$, there is a small perturbation Ω of an ellipse $\mathcal{E}$ of eccentricity e such that the equality of the singular support of $w_\Omega(t)$ and the length spectrum (1.4) fails.

A very interesting result in this direction has been recently provided in [53], where the authors prove that ellipses of sufficiently small eccentricities are Laplace spectrally unique (up to isometry) among all smooth domains (without any assumption on symmetry, convexity, or closeness to other ellipses). A key result in their proof consists in showing that for nearly circular domains, the lengths of periodic orbits of rotation number $1/q$ is contained in the singular support of the wave trace [53, Theorem 1.4]. This observation was used in a recent work of Koval [67] about local Laplace spectrum rigidity of ellipses.

1.2.3 Laplace Spectral Rigidity

Given a class $\mathcal{M}$ of domains and a domain $\Omega \in \mathcal{M}$, we say that Ω is *spectrally determined in* $\mathcal{M}$ if it is the unique element (modulo isometries) of $\mathcal{M}$ with its Laplace Spectrum: if $\Omega, \Omega' \in \mathcal{M}$ are *isospectral*, i.e., $\mathrm{Spec}_\Delta(\Omega') = \mathrm{Spec}_\Delta(\Omega)$, then Ω' is the image of Ω by an isometry (i.e., a composition of translations and rotations).

The question of Kac can be thus formulated as follows, assuming we have fixed a class of domains $\mathcal{M}$: *Is every $\Omega \in \mathcal{M}$ spectrally determined?*

If $\mathcal{M}$ is the space of all planar domains, the answer is well known to be negative (see e.g., [44], which generalizes some results previously obtained for compact manifolds without boundary (see [99, 104])).[3] However, all known examples of domains that are not spectrally determined are not convex, moreover, they are bounded by curves that are only piecewise analytic (e.g. plane domains with corners). On the other hand, Zelditch proved in [109] that the inverse spectral

---

[3] Remarkably, Sunada (see [99]) exhibits isospectral sets (i.e., sets of isospectral manifolds) of arbitrarily large cardinality.

problem has a positive answer when $\mathcal{M}$ is a generic class of analytic $\mathbb{Z}_2$-symmetric convex domains (i.e., symmetric with respect to reflection about a given axis).

More recently, as we have already mentioned, Hezari and Zelditch [53] proved that ellipses of sufficiently small eccentricities are Laplace spectrally unique (up to isometry) among all smooth domains (withouth any assumption on symmetry, convexity, or closeness to other ellipses).

The problem for non-analytic domains is substantially more challenging. In the C^∞ category, Osgood et al. [79–81] showed that isospectral sets are necessarily compact in the C^∞ topology. Sarnak (see [93]) also conjectured that an isospectral set consists of isolated domains. In other words, C^∞-close to a C^∞ domain there should be no isospectral domains, except those that can be obtained by an isometry.

A weaker version of this conjecture can be stated as follows: a domain Ω is said to be *spectrally rigid in* $\mathcal{M}$ if any C^1-smooth one-parameter isospectral family $(\Omega_\tau)_{|\tau|\leq 1} \subset \mathcal{M}$ with $\Omega_0 = \Omega$ is necessarily an isometric family. We can then ask: "*Are all C^∞ domains spectrally rigid?*"

The problem of spectral rigidity is in principle much simpler than the inverse spectral problem; yet it turns out to be extremely challenging. Hezari–Zelditch (see [52]) provided a result in the affirmative direction: let Ω_0 be bounded by an ellipse $\mathcal{E}$, then any one-parameter isospectral C^∞-deformation $(\Omega_\tau)_{|\tau|\leq 1}$ which additionally preserves the $\mathbb{Z}_2 \times \mathbb{Z}_2$ symmetry group of the ellipse is necessarily flat (i.e., all derivatives have to vanish for $\tau = 0$).[4] Popov–Topalov [86] recently extended these results (see also [87]).

Further historical remarks on the inverse spectral problem can also be found in [52] and in the surveys [108] and [110].

In the case of Riemannian manifolds, we mention that Guillemin–Kazhdan in [46] showed that any negatively curved surface is spectrally rigid among negatively curved surfaces. This result has been later extended to compact manifolds of negative curvature in [27].

1.2.4 Length Spectral Rigidity

The relation between the Laplace Spectrum and the Length Spectrum, immediately raises the following question:
Does the knowledge of the lengths of periodic orbits determine the shape of the billiard domain?

All counterexamples to the inverse spectral problem mentioned earlier also constitute a negative answer to this question. Likewise, at present, there is no known counterexample realized by either convex domains or domains with a C^∞ smooth boundary. Moreover, the above mentioned result by Zelditch (in [109]) also holds

---

[4] Results of this kind are usually referred to as *infinitesimal spectral rigidity*.

in the dynamical context. In the case of sufficiently smooth convex domain, the problem is open and presents the same challenges as the inverse spectral problem.

In [29], the following dynamical problem corresponding to spectral rigidity has been investigated: we say that a domain $\Omega_0 \in \mathcal{M}$ is *dynamically spectrally rigid in* $\mathcal{M}$ if any C^1-smooth one-parameter dynamically isospectral family $(\Omega_\tau)_{|\tau|\leq 1} \subset \mathcal{M}$ is necessarily an isometric family. More specifically, the authors proved the following theorem:

Theorem 1.4 (De Simoi et al. [29]) *Let $\mathcal{M}$ be the set of strictly convex domains with sufficiently (finitely) smooth boundary and axial symmetry and that are sufficiently close to a circle. Then, $\Omega \in \mathcal{M}$ is dynamically spectrally rigid in $\mathcal{M}$.*

Remark 1.9 This work leaves several natural open problems:

- Remove axial symmetry. Similar symmetry assumption appears in a work of Zelditch [109] and double symmetry assumption appears in Colin de Verdière [31].
- Nonlocal dynamically spectrally rigid in $\mathcal{M}$ is another exciting open problem.
- A closely related setting is dynamically spectrally rigid for standard maps of a cylinder $(x, y) \mapsto (x + y + V'(x), y + V'(x))$, where the x-component is taken $(mod. 1)$ and $V(x)$ is a smooth period function, i.e. $V(x + 1) \equiv V(x)$.
- Questions of dynamically spectrally rigid for geodesic flows on 2-torus seems a closely related open problem. Even when geodesic flow is close to integrable.

Let us point out that the above-mentioned results are concerned with spectral rigidity for smooth domains. Some results in the analytic category, yet for non-Birkhoff billiards, are contained in:

- De Simoi et al. [30], where under suitable symmetry and genericity assumptions, it is proved that the Marked Length Spectrum determines the geometry of billiard tables obtained by removing from the plane finitely many strictly convex analytic obstacles satisfying the so-called non-eclipse condition;
- Bunimovich stadia and squash-type stadia are beautiful examples of chaotic billiards. A Bunimovich stadium is a convex domain whose boundary is formed by four segments: two parallel segments forming a rectangle and two strictly convex segments connecting the end points of segments. For example, one can take semicircles. Bunimovich showed that such billiards are chaotic. One can consider segments not to be parallel and connect them by two strictly convex segments forming a squash. In [25] it is established the dynamical spectral rigidity for piecewise analytic Bunimovich stadia and squash-type stadia satisfying an additional "symmetry" assumption.

1.2.5 Some Ideas on the Proof of Deformational Spectral Rigidity (Theorem 1.4)

Here we introduce the key elements of the proof of Theorem 1.4. Let $(\Omega_\tau)_{|\tau|\le 1} \subset \mathcal{M}$ be a isospectral family of domains.

The first step is to establish existence of a countable family of maximal periodic orbits given by q-gons for all $q \ge 2$.

Lemma 1.1 (See Lemma 4.3 [29]) *Let $\Omega \in \mathcal{M}$; for any $q \ge 2$, there exists a periodic orbit of rotation number $1/q$ passing through the marked point of $\partial\Omega$ and having maximal length among other periodic orbits passing through the marked point. We call such an orbit marked symmetric maximal periodic orbit and denote it by $S^q(\Omega)$.*

Let $S^q = (s_q^k, \varphi_q^k)_{k=0}^{q-1}$ be the maximal symmetric periodic orbit. Associate to S^q and a continuous function $v : \mathbb{T} \to \mathbb{R}$ a linear functional

$$\ell_{\Omega,q}(v) = \sum_{k=0}^{q-1} v(s_q^k) \sin \varphi_q^k.$$

Given a parameterization γ of a family $(\Omega_\tau)_{|\tau|\le 1}$ in $\mathcal{M}$, we define the infinitesimal deformation function:

$$n_\gamma(\tau, \xi) = \langle \partial_\tau \gamma(\tau, \xi), N_\tau(\tau, \xi) \rangle,$$

where $\langle \cdot, \cdot \rangle$ is the usual scalar product in $\mathbb{R}^2$ and $N_\tau(\tau, \xi)$ is the outgoing unit normal vector to $\partial\Omega_\tau$ at the point $\gamma(\tau, \xi)$. Observe that n_γ is continuous in τ and $n_\gamma(\tau, \cdot)$ is smooth for any $|\tau| \le 1$. By the normalization condition of $(\Omega_\tau)_{|\tau|\le 1}$, we conclude that $n_\gamma(\tau, \cdot)$ is an even function, i.e., $n_\gamma(\tau, \xi) = n_\gamma(\tau, -\xi)$, and $n_\gamma(\tau, 0) = 0$ for any $|\tau| \le 1$. Naturally the space of perturbations can be identified with the space of smooth even functions on the circle denoted C_{sym}.

Proposition 1.6 (See [29, Proposition 4.6]) *Let $(\Omega_\tau)_{|\tau|\le 1}$ be an isospectral family, then for any $|\tau| \le 1$, $q \ge 2$ and having fixed arbitrarily $\bar{S}_\tau^q$ a maximal marked symmetric periodic orbit for Ω_τ, we have*

$$\ell_{\Omega_\tau, q}(n(\tau, \cdot)) = 0.$$

For any domain Ω (parameterized by the length s) with the radius of curvature ρ, we define the linear functional

$$\ell_{\Omega,0}(v) := \int_0^1 \frac{v(s)}{\rho(s)} ds$$

As it is shown in (4.3), [29] if $(\Omega_\tau)_{|\tau|\leq 1}$ be an isospectral family, then for any $|\tau| \leq 1$ we have $\ell_{\Omega,0}(n(\tau,\cdot)) = 0$.

Define the following **key notion**. Call the linearized isospectral operator $\mathcal{T}_\Omega : C_{sym} \to \mathbb{R}^{\mathbb{N}}$:

$$\mathcal{T}_\Omega v = (\ell_{\Omega,0}(n(\tau,\cdot)), \ell_{\Omega,1}(n(\tau,\cdot)), \ldots \ell_{\Omega,q}(n(\tau,\cdot)), \ldots).$$

In fact, $\mathcal{T}_\Omega$ has range in ℓ^∞, by definition of the functionals $\ell_{\Omega,q}$, since by Avila et al. [8, Lemma 8] there exists some $C > 0$ so that for any $q \geq 2$ we have $\sin \varphi_q^k \leq \frac{C}{q}$.

The linearized isospectral operator bears a strong analogy with the X-transform (see [45, Section 2.2]).

Theorem 1.5 ([29, Theorem 4.9]) *In the space of sufficiently smooth axis symmetric domains there is a neighborhood of the circular domain such that the operator $\mathcal{T}_\Omega : C_{sym} \to \ell^\infty$ is injective.*

This Theorem implies the rigidity Theorem above. In the case of the domain Ω_0 being the circle the linearized isospectral operator $\mathcal{T}_{\Omega_0}$ is easy to compute. For $j \geq 1$ and $q \geq 2$

$$\ell_q(e_j) = \delta_{q|j},$$

where $\delta_{q|j} = 1$ is j is divisible by q and zero otherwise. For the circle $\mathcal{T}_{\Omega_0}$ is clearly indective. In [29, Lemma B.1] we compute a perturbative expression for $\ell_{\Omega,q}(e_j)$ when a domain Ω is close to the circle. In a proper sense perturbation of $\mathcal{T}_{\Omega_0}$ is also injective.

1.2.5.1 Related Prior Results

The problem of isospectral deformations of manifolds without boundary were considered in some early works on variations of the spectral functions and wave invariants.

Let (M, g) be a compact boundaryless Riemannian manifold. A family $(g_\tau)_{|\tau|\leq 1}$ of Riemannian metrics on M depending smoothly on the parameter $|\tau| \leq 1$ is called a *deformation of the metric* g if $g_0 = g$. A deformation is called *trivial* if there exists a one-parameter family of diffeomorphisms $\varphi_\tau : M \to M$ such that $\varphi_0 = \text{Id}$, and $g_\tau = (\varphi_\tau)^* g_0$. For each homotopy class of closed curves in M, consider the infimum of g-lengths of curves belonging to the given homotopy class. The *Length Spectrum* $\mathcal{L}(M, g)$ is defined as the union of these lengths over all homotopy classes. The *inverse spectral problem* in this setting is to show that two metrics with the same Length Spectrum are isometric.

Likewise, a deformation $(g_\tau)_{|\tau|\leq 1}$ is said to be *isospectral* if $\mathcal{L}(M, g_\tau) = \mathcal{L}(M, g)$. We say that a Riemannian manifold (M, g) is *length spectrally rigid* if it does not admit non-trivial isospectral deformations.

It is worth mentioning that for there is a partial solution of the inverse spectral problem due independently to Croke [26] and Otal [82] which can be stated as follows: any negatively curved manifold is uniquely determined by its *Marked Length Spectrum* (see Sect. 1.2.6 for the corresponding billiard problem).[5]

Recently, Guillamou and Lefeuvre [45] proved that in all dimensions, the marked length spectrum of a Riemannian manifold (M, g) with Anosov geodesic flow and non-positive curvature, locally determines the metric in the sense that two close enough metrics with the same marked length spectrum are isometric.

Another example of deformational spectral rigidity appears in De la Llave et al. [28]. Recall that one can associate to a symplectic map a generating function. Then, for each periodic orbit, one can define the corresponding *action* by summing the generating function along the orbit. This value of the action is invariant under symplectic coordinate changes. The union of the values all these actions over all periodic orbits is called the action spectrum of the symplectic map. In [28, Theorem 1.3], it is proved that there are no non-trivial deformations of exact symplectic mappings B_τ, $\tau \in [-1, 1]$, leaving the action spectrum fixed, when B_τ are Anosov's mappings on a symplectic manifold. One of the reasons for symplectic rigidity in [28] is that all periodic points of B_τ are hyperbolic and form a dense set.

1.2.6 Marked Length Spectral Rigidity

One of the difficulties in working with the length spectrum is that all of this information on the periodic orbits come in a non-formatted way. For example, we lose track of the rotation numbers corresponding to each length. A way to overcome this difficulty is to "organize" this set of information in a more systematic way, for instance by labelling each length with corresponding rotation number. This new set is called the *Marked Length Spectrum* of Ω and denoted by $\mathcal{ML}_\Omega$:

$$\mathcal{ML}_\Omega := \{(\text{length}(\gamma), \text{rot}(\gamma)) : \gamma \text{ periodic orbit of the billiard in } \Omega\},$$

where $\text{rot}(\gamma)$ denotes the rotation number of γ.

One could also reduce this set of information by considering not the lengths of all orbits, but selecting some of them. More precisely, for each rotation number p/q in lowest terms, one could consider the maximal length among those having rotation number p/q. We call this map the *Maximal Marked Length Spectrum* of Ω, namely $\mathcal{ML}_\Omega^{\max} : \mathbb{Q} \cap [0, 1/2] \to \mathbb{R}$ given by:

$$\mathcal{ML}_\Omega^{\max}(p/q) = \max \left\{ \text{lengths of periodic orbits with rot. number } p/q \right\}.$$

---

[5] The *Marked Length Spectrum* in the case of negatively curved surfaces without boundary consists of the set of pairs of homotopy classes and length of the shortest geodesic in that homotopy class.

Marked Spectral Rigidity Question *Let Ω_1 and Ω_2 be two strictly convex planar domains with smooth boundaries and assume that they are isospectral, i.e., $M\mathcal{L}_{\Omega_1} \equiv M\mathcal{L}_{\Omega_2}$. Is it true that Ω_1 and Ω_2 are isometric?*
Similarly, one could ask whether this same question has an affirmative answer by asking only that $M\mathcal{L}_{\Omega_1}^{\max} \equiv M\mathcal{L}_{\Omega_2}^{\max}$.

Remark 1.10 (i) The above question could be reformulated—and it remains still meaningful and interesting—by asking that they two domains are only isospectral near the boundary, i.e., $M\mathcal{L}_{\Omega_1}(p/q) = M\mathcal{L}_{\Omega_2}(p/q)$ for all $p/q \in \mathbb{Q} \cap [0, \varepsilon)$, for some $0 < \varepsilon \leq 1/2$.

See Sect. 1.4 for a reformulation of this question in terms of the so-called Mather's minimal average action (or β-function) and for some partial answers to the Marked Length rigidity question related to the proof of the perturbative Birkhoff conjecture (see Sect. 1.4.3).

1.3 Lecture III: Caustics, Invariant Curves and Integrability

1.3.1 Caustics and Invariant Curves

In this section, we would like to recall the concept of *caustic* of a billiard and discuss its relations with invariant curves for the billiard map. Let us first start to introduce the concepts of caustic and integrability, starting with some motivating means of two examples; the definition of caustic will be given in Sect. 1.3.4.

1.3.2 Example 1: Circular Billiards

As we saw, the billiard in the unit disk defines a map $f : \mathbb{A}_{2\pi} \to \mathbb{A}_{2\pi}$ for which the sets

$$C_\varphi := \{(\theta, \varphi') \in \mathbb{A}_{2\pi} \mid \varphi' = \varphi\}$$

are invariants by f, that is if $(\theta, \varphi) \in C_\varphi$ then $f(\theta, \varphi) = (\theta + 2\varphi, \varphi) \in C_\varphi$. In particular, as we already notices, φ stays constant along the orbit and it represents an *integral of motion* for the map, and the property of the orbits are determined by the corresponding angle $\varphi = \pi\omega$, with $\omega \in (0, 1)$ (see Sect. 1.3.4).

Moreover, this billiard enjoys the peculiar property that all orbits with $\varphi = \pi\omega$ are tangent to the same concentric circle of radius $R \cos \pi\omega$ (see Fig. 1.8); this concentric circle is an example of *caustics* (see Definition 1.6) and it is related to the existence of a homotopically non-trivial invariant curve for the corresponding billiard map, namely the $C_\omega = \mathbb{R}/2\pi R\mathbb{Z} \times \{\pi\omega\}$ (this relation between caustics

Fig. 1.8 A billiard trajectory in the disk remains tangent to the same concentric circle after successive reflections

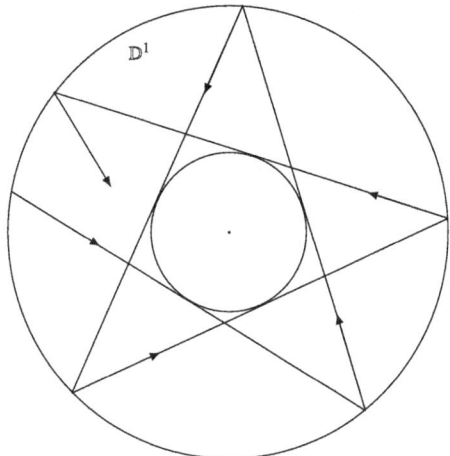

and invariant curves is more subtle, see Remark 1.15). Observe that the whole phase space of the circular billiard map—which is topologically a cylinder—is completely by these C_ω and, looking at the billiard table, this is completely foliated by caustics (this foliation is a singular foliation, due to the special role of the center of the disc): in this regard, circular billiards are example of *integrable billiards*; see Fig. 1.9.

1.3.3 Example 2: Elliptic Billiard

Consider the billiard in an ellipse $\mathcal{E}$ given, for $0 < b < a$ by

$$\mathcal{E} = \left\{ (x, y) \in \mathbb{R}^2 \;\middle|\; \frac{x^2}{a^2} + \frac{y^2}{b^2} = 1 \right\}.$$

Since $\mathcal{E}$ is not a circle, it has two distinct foci F_1 and F_2 lying on the x-axis. The billiard in the ellipse can be described geometrically as follows (see Fig. 1.11):

Proposition 1.7 *Let ℓ be an oriented line intersecting the ellipse $\mathcal{E}$ transversally, and consider the billiard flow induced by ℓ in $\mathcal{E}$. Then one of the following situations is satisfied:*

1. *ℓ contains one of the foci; in this case, the successive reflections of ℓ contain a focus and the latter differs from the focus of previous reflection. Moreover, the succesive reflected lines will converges to the x-axis.*
2. *ℓ crosses the x-axis between the foci; in this case, the successive reflections of ℓ always cross the x-axis between the foci. Moreover, they are supported by lines which are tangent to one and the same confocal hyperbola.*

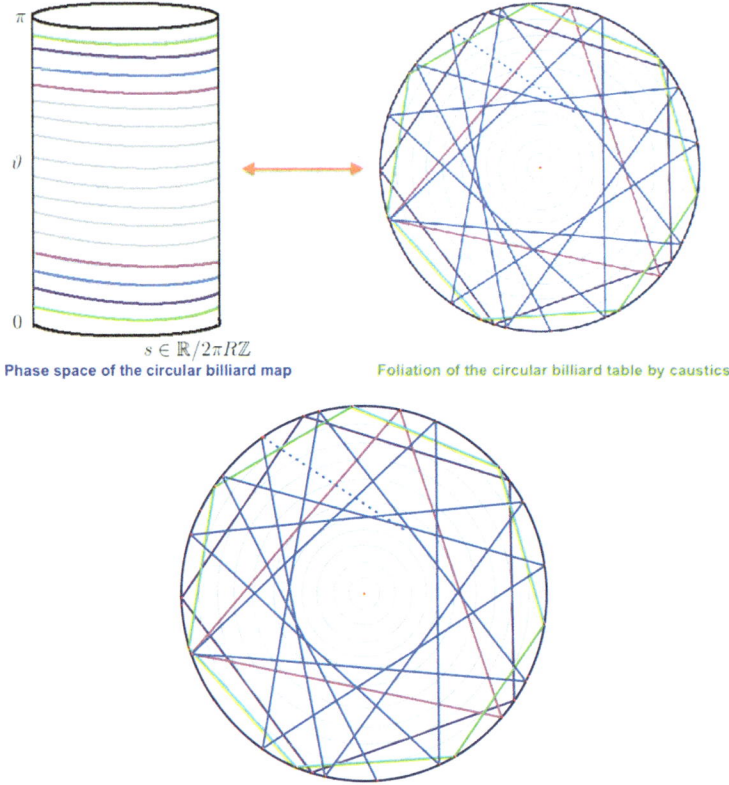

Fig. 1.9 Circular billiard and its phase space

3. ℓ crosses the x-axis outside the foci; in this case, the successive reflections of ℓ always cross the x-axis outside the foci. Moreover, they remain tangent to one and the same confocal ellipse.

This proposition tells us that the trajectory of a billiard in an ellipse are tangent to curves, which are confocal conics to $\mathcal{E}$, including hyperbolae.

Exercise 1.9 Using the description of an ellipse as the data of two distinct points F_1, F_2 and a constant $r > 0$ such that

$$\mathcal{E} = \{M \in \mathbb{R}^2 \mid F_1 M + F_2 M = r\},$$

show item 1 of Proposition 1.7.

The proof of Proposition 1.7 can be shown using the so-called *Joachimsthal invariant*:

Proposition 1.8 *Let $p = (x, y) \in \mathcal{E}$ and $v = (v_x, v_y)$ be a unit vector starting at p and pointing inside of $\mathcal{E}$. Then the quantity $J(p, v)$ defined by*

$$J(p, v) = \frac{x v_x}{a} + \frac{y v_y}{b}$$

is invariant by the billiard map inside $\mathcal{E}$. It is called the Joachimsthal invariant of the ellipse.

A proof of Proposition 1.8 can be found in [33, 101].

Exercise 1.10 Let $0 < b < a$ and consider the pencil of confocal conics C_λ given for any $0 < \lambda < a$ and $\lambda \neq b$ by the equations

$$C_\lambda : \quad \frac{x^2}{a^2 - \lambda} + \frac{y^2}{b^2 - \lambda} = 1.$$

Show that given a trajectory which does not contain the foci, it remains tangent to the conics C_{λ_0} where

$$\lambda_0 = (ab J(p, v))^2$$

and $J(p, v)$ is Joachimsthal invariant associated to the trajectory (Fig. 1.10).

1.3.3.1 Description of the Phase Space of the Billiard in an Ellipse

Optical properties of conics (an alternative way to consider the billiard ball motion inside a conic) were already well known to ancient Greeks. We refer to [101] for a more detailed discussion (see also [94]). Proposition 1.7 leads to a nice description of the phase space of the billiard in $\mathcal{E}$. Consider an arc-length coordinate s on $\mathcal{E}$ defined modulo the perimeter L of the elllipse, and φ the usual angle of reflection. The phase space in $\mathcal{E}$ we consider is

$$\mathbb{A}_L = \{(s, \varphi) \mid s \in \mathbb{R}/L\mathbb{Z}, \varphi \in (0, \pi)\}.$$

$\mathbb{A}_L$ is the disjoint union of the following objects (see Fig. 1.11):

1. *Hyperbolic 2-periodic points.* The two points $O_1 = (0, \pi/2)$ and $O_2 = (L/2, \pi/2)$, corresponding to the major semi-axis, are called *hyperbolic* and corresponds to the billiard flow induced by the x-axis. The billiard map permutes O_1 and O_2.
1'. *Elliptic 2-periodic points.* The two points $O'_1 = (L/4, \pi/2)$ and $O'_2 = (3L/4, \pi/2)$, corresponding to the minor semi-axis, are called *elliptic* and

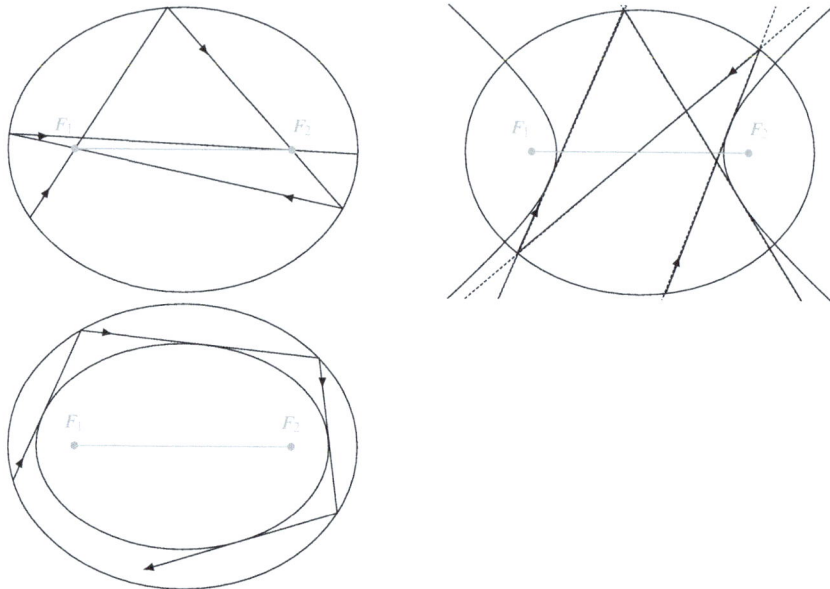

Fig. 1.10 The three different possibilities for a billiard trajectory in a ellipse. Left: the trajectory always alternatively passes through one of the two foci. Center: the trajectory always crosses the segment between the foci, and remains tangent to a confocal hyperbola. Right: the trajectory never crosses the segment joining the foci and remain tangent to a confocal ellipse

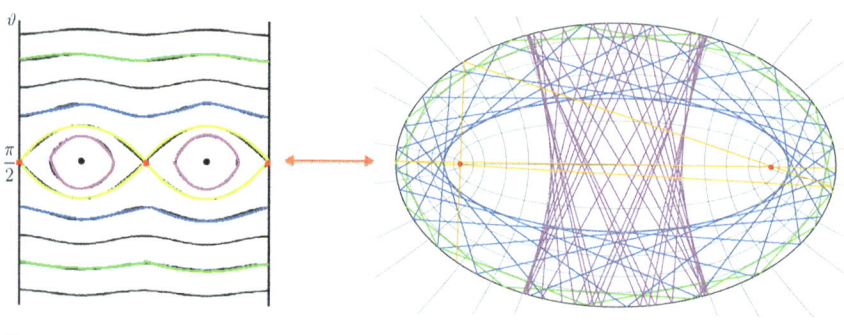

Phase space of an elliptic billiard map   Dynamics inside an elliptic billiard and caustics

Fig. 1.11 Elliptic billiard and its phase space

corresponds to the billiard flow induced by the y-axis. The billiard map permutes O_1' and O_2'.

2. *Stable and unstable manifolds.* Two graphs over s, Γ_1, Γ_2, intersecting at O_1 and O_2 and forming two *eyes* in the phase space. The curve Γ_i consists of the point (s, φ) such that the line intersecting $\mathcal{E}$ at s with an angle φ contains the focus F_i. They correspond to the *stable and unstable manifolds* of O_1 and O_2 and satisfies the following properties: the billiard map f permutes them, they

are invariant under f^2 and the points in Γ_i converge to O_{1-i} under iteration of f^2. The graph Γ_2 is called stable manifolds of O_1 since the points on it converges to O_1 under positive iteration of f^2; Γ_1 is called unstable manifold of O_1 since the points on it converges to O_1 under negative iteration of f^2. The same remarks hold for O_2 by permuting the roles of Γ_1 and Γ_2.

3. *Homotopically trivial invariant curves.* Given a confocal hyperbola H, consider the set of pairs (s, φ) such that the line intersecting $\mathcal{E}$ at s with angle φ is tangent to H. It consists of two closed invariant curves (depending on the orientation of the tangency points of ℓ with γ—one is given by the positive tangencies, the other one by negative tangencies) located in the eyes described at point 2. These curves are contractible in the cylinder $\mathbb{A}_L$.

4. *Homotopically non-trivial (or essential) invariant curves.* Given a confocal ellipse γ nested in $\mathcal{E}$, consider the set of pairs (s, φ) such that the line intersecting $\mathcal{E}$ at s with angle φ is tangent to γ. This consists of two closed invariants graphs over s (depending on the tangency orientation, as in 3.) located outside the eyes described at point 2. These curves are homotopically non-trivial in the cylinder $\mathbb{A}_L$.

Remark 1.11 Confocal ellipses are therefore examples of caustics and they foliate everything but the closed segment between the two foci (see Fig. 1.11). Hence, this could be also considered as an example of integrable billiard; see Fig. 1.11. Observe that also hyperbolae can be considered examples of caustics, although, differently from concentric circles or confocal ellipses, they are not connected, closed or convex; see Sect. 1.3.4 for a more precise discussion.

1.3.4 Caustics

Let us introduce the concept of *convex caustic*[6] and its relation with invariant curves for the billiard map. We refer to [49] for a more detailed (and extended) presentation of these topics. We also discuss some results and questions about their existence.

Let us start by recalling the definition of *invariant circle* (or *homotopically non-trivial invariant curve*, or *essential invariant curve*) for a billiard map.

Definition 1.5 We say that a curve $\gamma \subset \mathbb{A}_L$ is an *invariant circle* for the billiard map $f : \mathbb{A}_L \to \mathbb{A}_L$ if γ is isotopic to a boundary component of $\mathbb{A}_L$ and $f(\gamma) = \gamma$.

---

[6] *Caustic* comes from the greek word καυστικός (kaustikós), meaning "burning"; this terminology is related to optics and refers to the envelope of reflected or refracted rays of light, namely concentration of lights that can potentially lead to burns.

Fig. 1.12 Caustic and Lazutkin invariant (figure credits [94, Fig. 3.6])

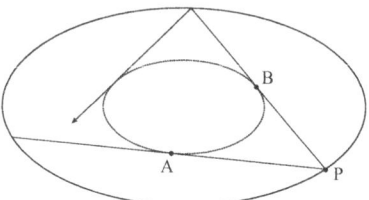

Remark 1.12

(i) Observe that both boundary components of $\mathbb{A}_L$ are trivial invariant circles. It follows from Birkhoff's theorem that invariant circles must be Lipschitz graphs (see [19] and also [94, Theorem 1.3.3]).
(ii) Clearly, a billiard map may possess invariant curves that are not invariant circles: see for example the billiard map in an ellipse (see discussion above) and its homotopically trivial (disconnected) invariant curves, corresponding to orbits intersecting the segment between the foci.

In the spirit of what we have seen in the examples of circular and elliptic billiards, let us give the following definition (Fig. 1.12).

Definition 1.6 A C^1 simple closed curve Γ in the interior of Ω is called a *convex caustic* for the billiard map f, if γ bounds a convex set D_Γ and any supporting line to D_Γ remains a supporting line to D_Γ after the billiard reflection in Ω. In other words, every time a trajectory is tangent to Γ, then it remains tangent after every each reflection.

In our discussion, we will focus on convex caustics, however one could consider a more general notion of caustic that does not require the properties of bounding a convex region, of being closed (see for example, confocal hyperbola for elliptic billiards, see discussion above), nor to be necessarily C^1. Since this will not be the object of our investigation, we refer to the discussion in [5, 49, 66]. See Fig. 1.13 for some examples.

Remark 1.13 An interesting example of billiard maps with invariant circles are billiards whose boundary is a curve of *constant width*, i.e., namely a curve that

Fig. 1.13 Examples of non-convex caustics in billiards of constant width (figure credits [66, Fig. 6])

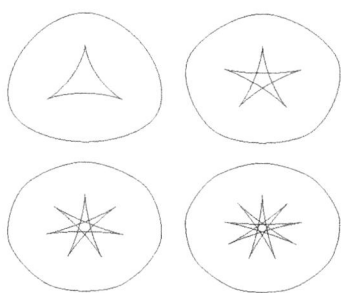

bounds a convex planar region whose width (defined as the perpendicular distance between two distinct parallel lines each having at least one point in common with the region's boundary but none with its interior) is the same regardless of the orientation of the curve (to construct such curves, see, for example, [66, Section 4] and [101, Exercise 3.13]. The corresponding billiard map has an invariant circle consisting of 2-periodic orbits. These curves corresponds to caustics that, in general, may have cusps; see [66, Section 4 and Fig. 6].

Billiard tables (other than ellipses) with a 1-parameter family of 3-periodic trajectories have been constructed by Innami in [57].

As in the case of billiard in a disk, convex caustics and invariant curves are related (see Fig. 1.14). One can prove the following.

Proposition 1.9 *Let Γ be a caustic of a Birkhoff billiard Ω with a chosen orientation. Consider the set γ of pairs (p, φ) where p is any point on $\partial\Omega$ and v is a unit inward pointing vector supporting the line containing p and tangent to Γ with the same orientation at the tangency point. Then γ is a graph in s which is invariant by the billiard map associated to Ω.*

Observe that to every convex caustic has a well-defined rotation number. In fact, the dynamics tangent to it, induces a circle homeomorphism from the boundary to itself; the rotation number of the caustic corresponds to the Poincaré rotation number of this circle homeomorphism (Fig. 1.15).

Exercise 1.11 In the case of the billiard in the unit disk, show that the invariant curve $C_\varphi := \{(\theta, \varphi') \in \mathbb{A}_{2\pi} \mid \varphi' = \varphi\}$ has rotation number 2φ.

Remark 1.14

(i) The notion of caustics is often connected to the so-called *whispering gallery*, a phenomenon that can be detected under some particular domes, in which whispers can be clearly transmitted and received from distant parts of the gallery, as long as the talker/listener are close the wall.
(ii) If Γ_ω is a convex caustic with rotation number $\omega \in (0, 1/2]$, then one can associate to it an invariant, the so-called *Lazutkin invariant* $Q(\Gamma_\omega)$. More precisely

$$Q(\Gamma_\omega) = |A - P| + |B - P| - |\widehat{AB}| \tag{1.5}$$

where $|\cdot|$ denotes the Euclidean length and $|\widehat{AB}|$ the length of the arc on the caustic joining A to B (see Fig. 1.16). This quantity is connected to the value of Mather's α-function, as it will be discussed in Sect. 1.4.

Remark 1.15 One could wonder about the relation between caustics for the billiard in Ω and invariant circles for the corresponding billiard map f. While one can show that to a convex caustic in Ω (not necessarily C^1) corresponds an invariant circle for the billiard map (see Proposition 1.9), however, the converse is however is not

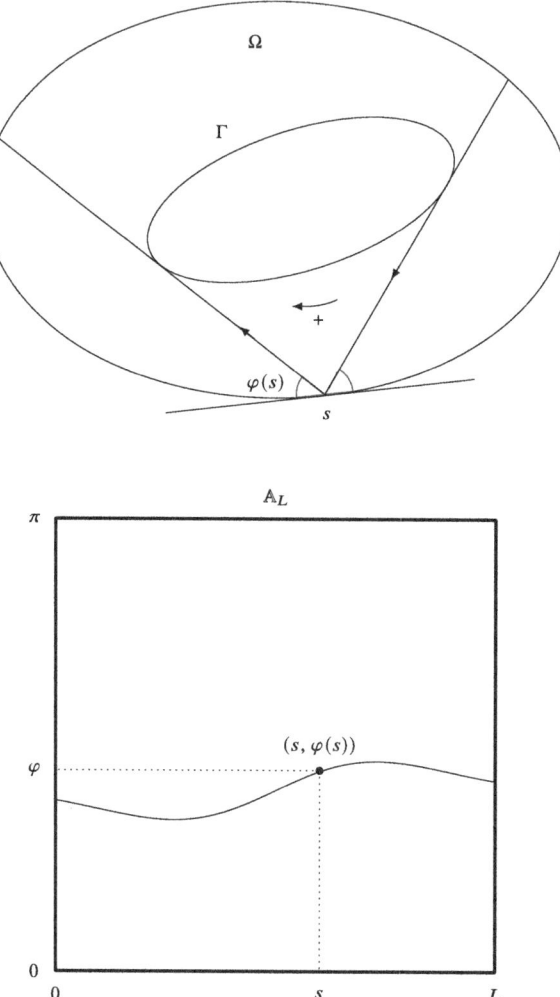

Fig. 1.14 *Left:* a Birkhoff billiard Ω having a caustic Γ; once we fix a positive orientation, for each point p of arc-length parameter s there is only one possible oriented line emitted from p and tangent to Γ, we denote by $\varphi(s)$ its angle at p with $\partial\Omega$. *Right:* The invariant graph $s \mapsto (s, \varphi(s))$ associated to the caustics Γ and drawn in the phase space $\mathbb{A}_L$ where L is the perimeter of Ω

always true: given an invariant curve γ of pairs (p, v) where p is a point on the bounary and v is a unit inward pointing vector, one can consider the envelope of the lines passing through p and directed by v; this curve might be not convex nor smooth. We refer to [5, 49, 66, 101] for more details.

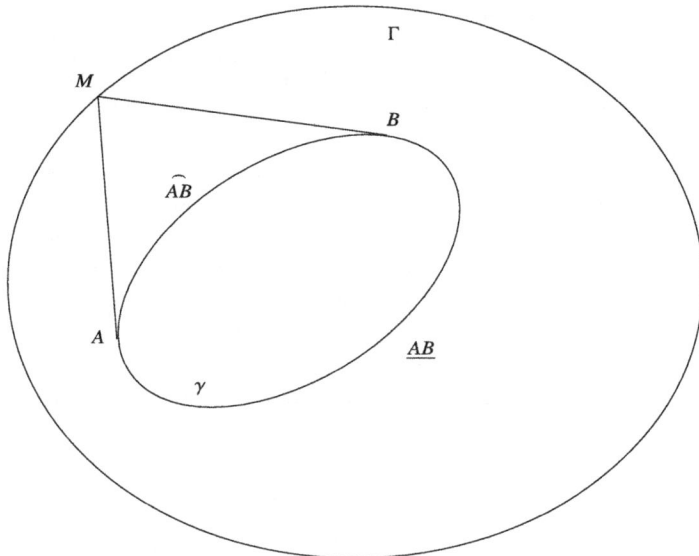

Fig. 1.15 The string construction around a fixed curve γ: an inextensible string is wrapped around γ and stretched so that the quantity $MA + MB + \underline{AB}$ remains the same. This is the same as saying that Lazutkin's invariant, namely $Q(M) = |MA| + |MB| - |\widehat{AB}|$, is constant

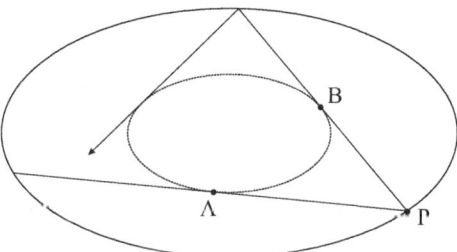

Fig. 1.16 Lazutkin invariant

1.3.5 Existence of (Convex) Caustics

A natural question that one could wonder is whether the existence of (convex) caustics is a common or a rare phenomenon. As we have seen before, circular and elliptic billiards possess many convex caustics.

Questions *Are there other Birkhoff billiards with (convex) caustics?* And in case of an affirmative answer: *How many of them is reasonable to expect?*

Note In the following we will often write caustic in place of convex caustic (unless differently specified). However, most of these questions can be addressed for more general notions of caustics.

1.3.5.1 String Construction

Constructing a Birkhoff billiard with **at least one** caustic is easy: it is enough to perform the so-called *string construction*, similarly to the well-known one to draw a circle as the set of points equidistant from a fixed center, or to construct an ellipse as the locus of points whose distances from two fixed points have a constant sum. Pictoriallky, (see for example [101, Chapter 5] for a more precise construction), given $\gamma \subset \mathbb{R}^2$ a smooth closed convex curve of length L_0, let $L > L_0$ and Γ be the closed curve obtained as follows: consider a closed inextensible string of length L wrapped around γ, pull it tight at a point and move this point around γ: the curve that one obtains, corresponds to a billiard domain that has γ as a caustic.

More precisely, Γ can be defined as the set of point M outside γ such that, if A and B are the two points of tangencies of the lines tangent to γ and containing M, then the length

$$|AM| + |BM| + |\widehat{AB}|$$

is constant equal to L, where $|\widehat{AB}|$ is the length of the arc $\widehat{AB}$ of the curve γ between A and B and located the furthest from M. Note that the level sets of this quantity are the same as the one of

$$Q(M) := AM + BM - |\widehat{AB}|$$

where $|\widehat{AB}|$ is the arc of the curve γ between A and B the closest from M (compare this quantity with *Lazutkin invariant* of γ, see Remark 1.14).

1.3.5.2 Lazutkin's KAM Caustics

Are there other billiards with **infinitely many** caustics? Quite surprisingly, the answer is affirmative: all (sufficiently smooth) Birkhoff billiards have infinitely many smooth convex caustics that accumulate to the boundary of the billiard domain. In [69], in fact, V. Lazutkin introduced a very special change of coordinates that reduces the billiard map f to a very simple form (as usual, $L = |\partial \Omega|$). Let $L_\Omega : \mathbb{R}/L\mathbb{Z} \times [0, \pi] \to \mathbb{R}/\mathbb{Z} \times [0, \delta]$ with small $\delta > 0$ be given by

$$L_\Omega(s, \varphi) := \left(x = C_\Omega^{-1} \int_0^s \rho^{-2/3}(s) ds, \quad y = 4 C_\Omega^{-1} \rho^{1/3}(s) \sin \varphi/2 \right),$$

where ρ denotes the radius of curvature of $\partial \Omega$, and $C_\Omega := \int_0^\ell \rho^{-2/3}(s) ds$ (sometimes called the *Lazutkin perimeter*). In these new coordinates the billiard map has a more simple expression:

$$B(x, y) = \left(x + y + O(y^3), y + O(y^4) \right).$$

In particular, near the boundary $\{y = 0\}$, this map can be seen as a small perturbation of the integrable map $(x, y) \longmapsto (x + y, y)$, and hence, under suitable regularity assumptions, KAM theorem can be applied (it is sufficient, for example, that $\partial\Omega$ is C^6, so that the map is at least C^5). Hence, there exists a positive measure Cantor set of smooth invariant circles for the map which accumulates on $\{y = 0\}$ and on which the motion is smoothly conjugate to a rigid rotation with Diophantine rotation number (see [69] and also [85] for a refined version); this translates into the existence of a positive measure set of caustics, accumulating to the boundary of the billiard table.

1.3.5.3 Non-existence of Caustics

Observe that in this context it is extremely important that Ω is strictly convex. In [76], in fact, Mather proved the non-existence of caustics if the curvature of the boundary vanishes at one point. An alternative proof of this result has been provided by Gutkin and Katok in [49], where the authors also investigate how the shape of the domain determines the location of caustics, establishing the existence of open regions which are free of caustics and estimating (from below) the size of these regions. More specifically, given a caustic Γ with Lazutkin invariant $L = L(\Gamma)$, if we denote by $\delta_{\max}(\Gamma, \partial\Omega)$ the maximum distance of Γ from the boundary $\partial\Omega$, they proved the following estimates (see [49, Propositions 1.2-3]):

$$\frac{\delta_{\max}^2(\Gamma, \partial\Omega)}{d} \leq L \leq \min\{2d^3\kappa^2, 2/K\},$$

where $d = d(\Omega)$ denotes the diameter of Ω, while κ and K are respectively the minimum and the maximum of the curvature of $\partial\Omega$.
It follows from this that if $\kappa = 0$ at some point, then caustics cannot exist.

1.3.6 Integrability and Birkhoff Conjecture

Next step then consists in asking in which cases these caustics **foliate** the whole billiard table or an open dense subset of it, as it happens in the circular and elliptic cases. In other words: *are there other examples of integrable billiards?*
This appearantly naïve question turns out to be much more difficult to extricate, and it has given rise to one of the most famous (and somehow impenetrable) open problem in dynamical systems: the so-called *Birkhoff conjecture*.

Conjecture (Birkhoff) *Circular and elliptic billiards are the only examples of integrable Birkhoff billiards.*

Remark 1.16 Although some vague indications of this question can be found in [17], to the best of our knowledge, its first appearance as a conjecture was in a paper by Poritsky [88], where the author attributes it to Birkhoff himself.[7] Thereafter, references to this conjecture (either as *Birkhoff conjecture* or *Birkhoff-Poritsky conjecture*) repeatedly appeared in the literature: see, for example, Gutkin [48, Section 1], Moser [78, Appendix A], Tabachnikov [100, Section 2.4], etc.

This conjecture assumes very different connotations and levels of complexity, according to the notion of integrability that one takes into account. Despite its long history and the amount of attention that it has captured over the last decades, many interesting formulations of this conjecture still remain unanswered.

We shall see in the Sect. 1.4 how also this conjecture/question can be rephrased as a regularity question for Mather's minimal average action (or *β-function*).

1.3.6.1 Global Integrability

In [9], Bialy proved the following result under the assumption of full global integrability.

Theorem (Bialy) *If the phase space of the billiard ball map is fully foliated by continuous invariant circles, then it is a circular billiard.*

Remark 1.17 An integral-geometric approach to prove Bialy's result was proposed by Wojtkowski in [107], by means of the so-called mirror formula. This approach was later exploited by Bialy [10] for billiards on the sphere and the hyperbolic plane, as well as for magnetic billiards.

Observe that Bialy and Wojtkowski's result is not in contrast with what we have discussed in the case of elliptic billiards. In fact, in that case the family of convex caustics represented by confocal ellipses do not foliate the whole domain (the segment between the two foci is left out) neither the set of homotopically non-trivial invariant curves (invariant circles) have full ω-measure in the phase space: the homotopically trivial invariant curves corresponding to orbits tangent to confocal hyperbolae, foliate a positive ω-measure set (in the phase portrait—see Fig. 1.11— this set corresponds to the area below the separatrix, i.e., the stable/unstable manifold of the hyperbolic 2-periodic orbit corresponding to the major semi-axis of the ellipse).

What about other notions of integrability? In the study of integrable systems, in fact, in most of the cases integrals of motion are non-degenerate not everywhere, but either on an open-dense subset of the phase space (we shall refer to this as *global integrability*) or just a proper (non-trivial) open subset (we shall refer to this as *local integrability*).

---

[7] Poritsky was a National Research Fellow in Mathematics at Harvard University, presumably under the supervision of Birkhoff and refers to Birkhoff and stated that he wrote that paper while in Harvard. However, [88] was published several years after Birkhoff's death.

1 Lecture Notes on Birkhoff Billiards

Remark 1.18

(i) An interesting result by Innami [58] shows that the existence of convex caustics with rotation numbers accumulating to $1/2$ implies that the billiard must be an ellipse. This regime of integrability is somehow opposite to the one we are interested in, which is concerned with caustics near the boundary of the billiard table, i.e., with small rotation numbers. Innami's proof is based on Aubry-Mather theory; a simpler and more geometric proof of Innami's result has been recently given in [5]. Observe that in this result it is decisive that the caustics are convex.

(ii) In this regard, Treschev in [102] gave numerical indication that there might exist analytic billiards, different from ellipses, for which the dynamics in a neighborhood of the elliptic period-2 orbit is conjugate to a rigid rotation. These billiards could be seen as an instance of *local integrability*; however, as we have already remarked above, this regime is somehow complementary to the one usually considered for Birkhoff conjecture, since it is concerned with integribilility a neighborhood of an elliptic periodic orbit of period 2. Very interestingly, this fact—if verified—would provide an intriguing indication that these regimes of integrability are significantly different.

1.3.6.2 Perturbative Birkhoff Conjecture

Instead of considering all possible Birkhoff billiards, one could restrict the analysis to what happens for domains that are sufficiently close to ellipses and try to study the Birkhoff conjecture in this class of domains, which can be considered as *perturbations* of ellipses. More specifically, we can state the following perturbative version of Birkhoff conjecture.

Birkhoff Conjecture (Perturbative Version) *A smooth strictly convex domain that is sufficiently close (w.r.t. some topology) to an ellipse and whose corresponding billiard map is integrable, is necessarily an ellipse.*

First results in this direction were obtained:

- Levallois [71] and Levallois–Tabanov [72]: Non-integrability of certain algebraic perturbations of elliptic billiards.
- Delshams and Ramírez-Ros [32]: Non-integrability of entire symmetric perturbations of ellipses (these perturbations break integrability near the homoclinic solutions.

More recently, Avila, De Simoi and Kaloshin proved in [8] that the claim of the perturbative version of Birkhoff conjecture is true, for domains that are sufficiently close to a circular billiard. The complete proof for domains sufficiently close to an ellipse of any eccentricity, has been provided in [60].

Let us describe this result more precisely, starting with the following definition.

Definition 1.7 Let Ω be a strictly convex domain.

(i) We say Γ is an *integrable rational caustic* for the billiard map in Ω, if the corresponding invariant circle Γ consists of periodic points; in particular, the corresponding rotation number is rational.
(ii) Let $q_0 \geq 2$ be a positive integer. If the billiard map inside Ω admits integrable rational caustics for all rotation numbers $0 < \frac{p}{q} < \frac{1}{q_0}$, we say that Ω is q_0-*rationally integrable*.

The main result proved in [60] is the following.

Theorem 1.6 (Kaloshin–Sorrentino [60]) *For any eccentricity $0 \leq e_0 < 1$ outside of locally finite set in $[0, 1)$ the following holds. Let $\mathcal{E}_0$ be an ellipse of eccentricity e_0 and semi-focal distance c; let $k \geq 39$. For every $K > 0$, there exists $\varepsilon = \varepsilon(e_0, c, K)$ such that the following holds: if Ω is a 2-rationally integrable C^k-smooth domain, whose boundary $\partial\Omega$ is*

- *K-close to $\mathcal{E}_0$, with respect to the C^k-norm,*
- *ε-close to $\mathcal{E}_0$, with respect to the C^1-norm,*

then Ω is an ellipse.

Remark 1.19 Actually, it is sufficient to ask only the existence of integrable rational caustics of rotation number $\frac{1}{q}$, for all $q \geq 3$.

1.3.6.3 Local Integrability and Birkhoff Conjecture

What can be said for *locally integrable* Birkhoff billiards? As we have noticed in Remark 1.18, the correct regime that one should consider seems to be integrability in a neighborhood of the boundary of the billiard table, i.e., for small rotation numbers.

Let us denote with $\mathcal{E}_{e,c} \subset \mathbb{R}^2$ an ellipse of eccentricity e and semifocal distance c. We state the following local version of Birkhoff conjecture.

Local Birkhoff Conjecture *For any integer $q_0 \geq 3$, there exist $e_0 = e_0(q_0) \in (0, 1)$, $m_0 = m_0(q_0)$, $n_0 = n_0(q_0) \in \mathbb{N}$ such that the following holds. For each $0 < e \leq e_0$ and $c \geq 0$, there exists $\varepsilon = \varepsilon(e, c, q_0) > 0$ such that the following holds.*
If $\mathcal{E}_{e,c}$ is an ellipse of eccentricity e and semi-focal distance c, and Ω is a q_0-rationally integrable C^{m_0}-smooth domain, whose boundary $\partial\Omega$ is ε-close to $\mathcal{E}_0$, with respect to the C^{n_0}-norm, then Ω must be an ellipse.

This conjecture has been first studied in [55]. More precisely, the following results have been proved (see also Sect. 1.3.8 for more recent advances by Bialy and Mironov [14], Kaloshin et al. [62], Koval [67]).

Theorem 1.7 (Huang et al. [55])

(i) *The Local Birkhoff Conjecture holds true for $q_0 = 2, 3, 4, 5$, with $m_0 = 40q_0$ and $n_0 = 3q_0$.*

(ii) *The Local Birkhoff Conjecture holds true for $q_0 > 5$ with $m_0 = 40q_0$ and $n_0 = 3q_0$, subject to checking that $q_0 - 2$ matrices (which are explicitely described) are invertible.*

Remark 1.20

(i) Case $q_0 = 2$ was proven in [8] (see also [58, 60]).
(ii) Smoothness exponents are probably not optimal.
(iii) Notice that in the proof we actually need only the existence of rationally integrable caustics of rotation numbers, less than $1/q_0$, of the form j/q for $j = 1, 2, 3$.
(iv) The invertibility condition on finitely many matrices, to which the claim of part (ii) of Theorem 1.7 is subject, is explicit and computable. In [56] it is described how to implement an algorithm to verify it by means of symbolic computations. The coefficients of these matrices are completely determined by the e-expansion of the action-angle parametrisation of the ellipse, which, in turn, is explicitly given by elliptic integrals; it turns out that the entries of these matrices are either 0, 1 or of the form $\xi \cos^{-2j}(w\pi)e^{2j}$, where $\xi \in \mathbb{Q}, j \in \mathbb{N}$, $w \in \{\frac{1}{2k+1}, \frac{2}{2k+1}, \frac{1}{2k}, \frac{3}{2k} : k > j\}$.

Recently a complete proof of this conjecture for ellipses of almost all eccentricities has been given in [67].

1.3.7 Some Ideas on the Proofs of Perturbative Birkhoff Conjecture and its Local Version (Theorems 1.6 and 1.7)

1.3.7.1 Perturbative Birkhoff Conjecture (Theorem 1.6)

Let us provide a description of the strategy that we adopted in [60] to prove Theorem 1.6.

For small eccentricities, Theorem 1.6 was proven in [8]. Let us start by describing the simplified setting of integrable infinitesimal deformations of a circle. This provides an insight into the strategy of the proof in the general case.

Let Ω_0 be a circle centered at the origin and radius $\rho_0 > 0$. Let Ω_ε be a one-parameter family of smooth deformations given in the polar coordinates (ρ, φ) by

$$\partial\Omega_\varepsilon = \{(\rho, \varphi) = (\rho_0 + \varepsilon\rho(\varphi) + O(\varepsilon^2), \varphi)\}.$$

Consider the Fourier expansion of ρ:

$$\rho(\varphi) = \rho_0' + \sum_{k>0} \rho_k \sin(k\varphi) + \rho_{-k} \cos(k\varphi).$$

Theorem 1.8 (Ramírez-Ros [89]) *If Ω_ε has an integrable rational caustic $\Gamma_{1/q}$ of rotation number $1/q$, for any ε sufficiently small, then we have $\rho_{kq} = \rho_{-kq} = 0$ for any integer k.*

Let us now assume that the domains Ω_ε are 2-rationally integrable for all sufficiently small ε and ignore for a moment the dependence on the parametrisation: then the above theorem implies that $\rho'_k = \rho''_k = 0$ for $k > 2$, i.e.,

$$\rho(\varphi) = \rho'_0 + \rho'_1 \cos \varphi + \rho''_1 \sin \varphi + \rho'_2 \cos 2\varphi + \rho''_2 \sin 2\varphi$$
$$= \rho'_0 + \rho^*_1 \cos(\varphi - \varphi_1) + \rho^*_2 \cos 2(\varphi - \varphi_2)$$

where φ_1 and φ_2 are appropriately chosen phases.

Remark 1.21 Observe that

- ρ_0 corresponds to an homothety;
- ρ^*_1 corresponds to a translation in the direction forming an angle φ_1 with the polar axis $\{\varphi = 0\}$;
- ρ^*_2 corresponds to a deformation of the circle into an ellipse of small eccentricity, whose major axis forms an angle φ_2 with the polar axis.

This implies that, infinitesimally (as $\varepsilon \to 0$), rationally integrable deformations of a circle are tangent to the 5-parameter family of ellipses.

In order to extend these ideas to the case of an integrable perturbation (not necessarily a deformation) of an ellipse, a more elaborate strategy is needed, involving more quantitative estimates and approximation procedure (we refer to [8, 60] for more technical details). In particular, Fourier modes are replaced by new functions determined by the dynamics inside the approximating ellipse, that we call *dynamical modes* $\{c_q, s_q\}_{q \geq 3}$, which are given by:

$$c_q(\varphi) := \frac{\cos\left(\frac{2\pi q}{4K(k_q)} F(\varphi; k_q)\right)}{\sqrt{1 - k_q^2 \sin^2 \varphi}}$$

$$s_q(\varphi) := \frac{\sin\left(\frac{2\pi q}{4K(k_q)} F(\varphi; k_q)\right)}{\sqrt{1 - k_q^2 \sin^2 \varphi}}$$

where k_q denotes the eccentricity of the confocal ellipse corresponding to the caustic of rotation number $1/q$, while

$$F(\varphi; k) := \int_0^\varphi \frac{d\theta}{\sqrt{1 - k^2 \sin^2 \tau}} \quad \text{and} \quad K(k) := F\left(\frac{\pi}{2}; k\right).$$

are the elliptic integrals of first kind (see, for example, [1] for more details on these functions and their properties).

The core of the proof consists in showing that these dynamical modes together with the infinitesimal generators of homotheties, translations, rotations and hyperbolic rotations (i.e., those transformations preserving the set of ellipses), form a basis of $L^2(\mathbb{R}/2\pi\mathbb{Z})$. This is one of the main difficulties (maybe the hardest one) involved in the extension of the perturbative result in [8] to the case of perturbations of any ellipse, as studied in [60]. While in the former case, one can take advantage of the fact that these functions can be considered small perturbations of the Fourier modes, in the latter new strategies need to be exploited.

In [60], we consider analytic extensions of the action-angle coordinates of the elliptic billiard, more specifically, of the boundary parametrizations induced by each integrable caustic (these functions can be explicitly expressed in terms of elliptic integrals and Jacobi elliptic functions.). A detailed study of their complex singularities and the size of their maximal strips of analiticity, allowed us to deduce their linear independence (both for finite and infinite combinations) and, by a suitable codimension argument, to show that they form a complete set of generators, thus completing the proof that they are a basis of $L^2(\mathbb{R}/2\pi\mathbb{Z})$.

1.3.7.2 Local Birkhoff Conjecture for Nearly Circular Domains (Theorem 1.7)

The main difficulty in this case—in comparison with the one discussed in Theorem 1.6 and Sect. 1.3.7.1—is that we cannot use the preservation of integrable rational caustics for all rotation number $1/q$, with $q \geq 3$; hence, we need to recover the missing conditions on the corresponding Fourier coefficients of the perturbation.

Our key idea is the following: for ellipses of small eccentricity $e > 0$, we study the Taylor expansion, with respect to e, of the corresponding action-angle coordinates. Using this expansion, we derive the necessary condition for the preservation of integrable rational caustics, in terms of the Fourier coefficients of the perturbation, up to the precision of order e^{2N}, for some positive integer $N = N(q_0)$.

Let us outline our strategy, starting from some special cases.

- CASE $q_0 = 3$: We lose a pair of conditions corresponding to Fourier coefficients of order 3. We exploit the conditions obtained from the existence of integrable rational caustics of rotation numbers $1/5, 1/7, 2/7$: we use the corresponding expansions, with respect to e, up to the precision $O(e^6)$, to derive a system of linear equations for the 3rd, 5th, 7th Fourier coefficients. Solving this linear system will provide us with the needed estimates for Fourier coefficients of order 3.
- CASE $q_0 = 4$: In this case we lose two pairs of conditions corresponding to Fourier coefficients of order $q = 3, 4$. These will be recovered in two steps:
  – To recover the one corresponding to Fourier coefficients of order 3, we study the necessary conditions for the existence of integrable rational caustics

of rotation numbers 1/5, 1/7, 1/9, 2/9, written in terms of the Fourier coefficients of the perturbation, and consider their expansions, with respect to e, up to order $O(e^8)$. We then derive a linear system for the 3rd, 5th, 7th, 9th Fourier coefficients, whose solution will provide us with the needed estimates for the Fourier coefficients of order 3.
- To recover the one corresponding to Fourier coefficients of order 4, we study the necessary conditions for the existence of integrable rational caustics of rotation numbers 1/6, 1/8, 1/10, 1/12, 1/14, 3/14, which give rise to a system of linear equation for the 4th, 6th, 8th, 10th, 12th, 14th Fourier coefficients; as before, the solution of this linear system will give us the needed estimates for the Fourier coefficients of order 4.

- THE GENERAL CASE: Along the same lines described in the previous two items, we outlined in [56] a general (conditional) procedure to deal with this problem for any $q_0 \geq 3$; the implementation of this scheme is based on the assumption that certain explicit non-degeneracy conditions for the corresponding linear systems hold. We remark however that all of these conditions are very explicit and the algorithm is explicitly described, so to be implemented on a computer.

1.3.8 More Recent Advances on Birkhoff Conjecture

We recall here some more recent breakthroughs on Birkhoff conjecture that appeared after the CIME summer school.

1. In [14] Bialy and Mironov proved the Birkhoff conjecture for *centrally-symmetric* C^2-smooth convex planar billiards. More specifically, they assume that the domain between the invariant curve the invariant curve foliated by 4-periodic orbits and the boundary of the phase cylinder is foliated by C^0-invariant curves and prove that the billiard table must be elliptic. In [62] the authors proved that this condition is equivalent to having integrable rational caustics of rotation number $\frac{1}{q}$, for every $q \geq 4$. The main ingredients of the proof are the use of a non-standard generating function for convex billiards and the observation that invariant curve consisting of 4 -periodic orbits enjoys some special properties; combining these ingredients with the integral-geometry approach for rigidity results that was introduced by Bialy in [9], they establish a Hopf-type rigidity for billiards in ellipses. See also [15] for an effective version of this result.
2. Combining the method of Bialy and Mironov [14] and Kaloshin and Sorrentino [60], recently Kaloshin et al. [62] proved a perturbative version of Birkhoff conjecture nearby centrally symmetric strictly convex domain, under the assumption that the billiard admits integrable rational caustics for rotation numbers $\frac{1}{3}$ and for all $0 < \frac{p}{q} \leq \frac{1}{4}$.
3. Recently, Koval [67] proved a Local Birkhoff conjecture for nearly elliptic domains. More precisely, for any positive integer q_0 and any eccentricity e

outside of locally finite set in [0, 1), a small q_0-integrable perturbation of an ellipse of eccentricity e is an ellipse.

1.3.9 Local Period-Two Birkhoff Conjecture

Notice that when we discuss integrability in the context of Birkhoff conjecture, usually we refer to integrability near the boundary, namely, integrability for nearly glancing orbits (i.e., caustics with small rotation numbers).

There is an alternative notion of local integrability defined as follows. Fix an ellipse $\mathcal{E}$ of positive eccentricity and let AB denote its minor axis. This corresponds to an elliptic period-two orbit of the billiard map in $\mathcal{E}$. We say that it is locally integrable in the sense that there is a neighborhood foliated by local invariant curves for the square of the billiard map. It turns out that for the square of the billiard map there is a twist. Passing to local polar coordinates $(r, \varphi) \in (\mathbb{R}_+, \mathbb{T})$ with $r = 0$ corresponding to say A. Then, one can define *the polar rotation number* of each of such invariant curves. The billiard trajectories for these invariant curves are tangent to confocal hyperbolae, for this reason we call them *hyperbolic caustics*. Notice that a hyperbolic caustic consists of two connected components, both on the billiard table $(s, \theta) \in \mathbb{A}$ and on the plane $\mathcal{E} \subset \mathbb{R}^2$. Call the projection of a hyperbolic caustic onto the boundary (via the map $\pi(s, \theta) = s$) its support.

A natural extension of the notion of rational integrability is the following one.

Let Ω be convex domain of non-constant width has a period two orbit, whose length is strictly less than the diameter of Ω. We say that Ω *period-two locally integrable* if there exists $\delta > 0$ such that in a neighborhood of period-two periodic orbit there exists invariant curves of polar rotation number $p/2q$ for each $0 < p/2q < \delta$. It is natural to ask:

Question 1. *Let $(\Omega_\tau)_\tau$ be an analytic deformation of an ellipse $\Omega_0 = \mathcal{E}$ that is period-two locally integrable. Is it true that Ω_τ is an ellipse for every τ?*

Question 2. *Let Ω be an analytic convex period-two locally integrable domain. Is it true that Ω is an ellipse?*

In the case of smooth deformations one needs to impose an additional requirement. Let Ω_0 be an ellipse, and $\omega_0^+ = \omega^+(\Omega_0)$ be the rotation number of the period two orbit. One can check there is a twist. Then in the polar coordinates all rotation numbers in $(0, \omega_0^+)$ can be realized near the period-two orbit. For a perturbation Ω of Ω_0, there exists an interval $(\omega^-(\Omega), \omega^+(\Omega))$ of rotation numbers that are admissible.

Fix an interval of polar rotation numbers $[\rho_1, \rho_2]$, $\rho_1 < \rho_2 \subset [0, \omega_0^+]$. We say that the billiard Ω is rationally integrable on $[\rho_1, \rho_2]$ if for all rational rotation numbers in $[\rho_1, \rho_2] \cap (\omega^-(\Omega), \omega^+(\Omega))$, there is a hyperbolic caustic with this rotation number.

Associate to this interval the union of the supports of the associated hyperbolic caustics. Similarly, we can associate to a smooth deformation of an ellipse $(\Omega_\tau)_\tau$, $\Omega_0 = \mathcal{E}$ its support, namely, the part of the boundary, where $\Omega_\tau \setminus \mathcal{E} \neq \emptyset$.

Question 3. *Let $(\Omega_\tau)_\tau$ be a smooth deformation of an ellipse $\Omega_0 = \mathcal{E}$ which is $[\rho_1, \rho_2]$-integrable for an interval of admissible polar rotation numbers $\subset [0, \omega_\tau]$. Is it true that Ω_τ restricted to the support of the hyperbolic caustics coincide with an ellipse?*

1.3.10 Integrable Riemannian Geodesic Flows on the Torus

We want to conclude this section by drawing some connections between Birkhoff conjecture and a problem in Riemannian geometry. The Birkhoff conjecture can be also thought as an analogue, in the case of billiards, of the following task: classifying integrable (Riemannian) geodesic flows on $\mathbb{T}^2$. The complexity of this question, of course, depends on the notion of integrability that one considers. If one assumes that the whole phase space is foliated by invariant Lagrangian graphs (i.e., the system is C^0-*integrable*, see [3, Définition 4.19], in particular, the integral of motion is only assumed to be continuous), then it follows from Hopf's result [54] (see also [23] for the proof in dimension greater than 2) that the associated metric must be flat. Bialy and Wojtkowski's results in the billiard setting, can be considered as the analogs of this result.

However, the question becomes more challenging—and it is still open—if one considers integrability only on an open and dense set (global integrability), or assumes the existence of an open set foliated by invariant Lagrangian graphs (local integrability). Example of globally integrable (non-flat) geodesic flows on $\mathbb{T}^2$ are those associated to *Liouville-type metrics*, namely metrics of the form

$$ds^2 = (f_1(x_1) + f_2(x_2))(dx_1^2 + dx_2^2).$$

A folklore conjecture states that these metrics are the only globally (resp. locally) integrable metrics on $\mathbb{T}^2$, which, in some sense, can be interpreted as the analogue of Birkhoff conjecture, in the realm of integrable geodesic flows on $\mathbb{T}^2$.

A partial answer to this conjecture (global case) is provided in [22], where the authors prove it under the assumption that the system admits an integral of motion which is quadratic in the momenta. Observe that while the case of quadratic integral of motion reduces to a system of linear PDEs, the case of higher degree integrals of motions is very challenging and it turns out to be equivalent to delicate questions on non-linear PDEs of hydrodynamic type (see, for example, [11, 12]).

Recently, some advances about deformational rigidity of some Liouville metrics on the torus have been provided by Henheik in [51].

This notion of integrability is related to the so-called *algebraic integrability*, namely the existence of integrals of motion that are polynomial in the velocity. The relation between this notion of integrability and the Birkhoff conjecture (*algebraic Birkhoff conjecture*) has been studied and has lead to interesting results [13, 21]. Recently, using previous results of [13], Glutsyuk [40] proved the algebraic Birkhoff conjecture.

Finally, we point out that the topological structure of the torus plays a fundamentel role in the above-mentioned conjectures and results. For example, on the two dimensional sphere there are plenty of non-trivial integrable metrics: the so-called *Zoll surfaces*. A Zoll surface is a surface homeomorphic to the 2-sphere, equipped with a Riemannian metric all of whose geodesics are closed and of equal length (the first non-trivial example was discovered by Zoll in [111]). While the usual unit-sphere metric on $\mathbb{S}^2$ obviously has this property, there also exists an infinite-dimensional family of geometrically distinct deformations that are still Zoll surfaces. In particular, most Zoll surfaces do not have constant curvature. See [70] for more details.

1.4 Lecture IV: Aubry-Mather Theory and Billiard Dynamics

In this section we would like to discuss how the study of *action-minimizing properties* of billiards can be used to shed some light on their dynamical properties. In particular, we shall see how many of the questions discussed in the previous sections can be rephrased in these terms. Let us start by briefly recalling the main ideas at the heart of this approach.

1.4.1 Aubry-Mather Theory for Twist Maps of the Annulus

At the beginning of the eighties Serge Aubry and John Mather developed, independently, what nowadays is commonly called *Aubry–Mather theory*. This novel approach to the study of the dynamics of twist diffeomorphisms of the annulus, pointed out the existence of many *action-minimizing orbits* for any given rotation number. For a more detailed introduction, see for example [36, 94, 96]).

More precisely, let $a, b \in \mathbb{R} \cup \{\pm\infty\}$, with $a < b$, and let

$$f : \mathbb{R}/\mathbb{Z} \times (a, b) \longrightarrow \mathbb{R}/\mathbb{Z} \times (a, b)$$

be a monotone twist map, i.e., a C^1 diffeomorphism such that its lift to the universal cover $\tilde{f}$ satisfies the following properties (we denote $(x_1, y_1) = \tilde{f}(x_0, y_0)$):

(i) $\tilde{f}(x_0 + 1, y_0) = \tilde{f}(x_0, y_0) + (1, 0)$ and $x_0 \leq x_1 < x_0 + 1$;
(ii) $\tilde{f}$ is orientation preserving and it preserves the boundaries of $\mathbb{R} \times (a, b)$:

$$y_1(x_0, y_0) \to a \quad \text{as} \quad y_0 \to a \quad \text{and} \quad y_1(x_0, y_0) \to b \quad \text{as} \quad y_0 \to b;$$

(iii) If $a > -\infty$, then $\tilde{f}$ extends continuously to $\mathbb{R} \times \{a\}$ by a rotation:
$$\tilde{f}(x, a) = (x + \omega_-, a);$$
similarly, If $b < +\infty$, then $\tilde{f}$ extends continuously to $\mathbb{R} \times \{b\}$ by a rotation:
$$\tilde{f}(x, b) = (x + \omega_+, b);$$

(iv) $\frac{\partial x_1}{\partial y_0} \geq c > 0$ (monotone twist condition),

(v) $\tilde{f}$ admits a (periodic) generating function h (i.e., it is an exact symplectic map):
$$y_1 \, dx_1 - y_0 \, dx_0 = dh(x_0, x_1).$$

We call the interval $(\omega_-, \omega_+) \subset \mathbb{R}$ the twist interval of f (we remark that if $a = -\infty$, then $\omega_- = -\infty$ and if $b = +\infty$, then $\omega_+ = +\infty$).

In particular, it follows from (v) that:
$$\begin{cases} y_1 = \frac{\partial h}{\partial x_1}(x_0, x_1) \\ y_0 = -\frac{\partial h}{\partial x_0}(x_0, x_1). \end{cases} \quad (1.6)$$

Remark 1.22 The billiard map is an example of monotone twist map (to fit with the above definition, one can normalize the boundary length to be equal to 1). In particular, as we have already pointed out, its generating function is given by $h(x_0, x_1) = -\ell(x_0, x_1)$, where $\ell(x_0, x_1)$ denotes the Euclidean distance between the two points on the boundary of the billiard domain corresponding to $\gamma(x_0)$ and $\gamma(x_1)$.

Exercise 1.12 As it follows from Proposition 1.5, a billiard map f associated to a Birkhoff billiard of perimeter 1 and given in $(s, -\cos\varphi)$-coordinates is an exact-symplectic twist map: we already saw the existence of the generating function. Prove that it also satisfies the other conditions, in particular the twist condition.

Exercise 1.13 (Completely Integrable Map and Standard Map)

(i) The map $F : \mathbb{R}^2 \to \mathbb{R}^2$ defined for all (x, y) by
$$F(x, y) = (x + y, y)$$
is the lift to $\mathbb{R} \times \mathbb{R}$ of an exact-symplectic twist map whose generation function is the map $H : (\mathbb{R}/\mathbb{Z})^2 \to \mathbb{R}$ given by
$$H(x, x') := \frac{1}{2}(x' - x)^2.$$

Describe its dynamics and characterize it orbits in terms of their rotation number (compare with the billiard in a disk).

(ii) This example can be generalized as follows. Consider $v : \mathbb{R} \to \mathbb{R}$, a 1-periodic smooth map with zero average. Then $F : \mathbb{R}^2 \to \mathbb{R}^2$ defined for all (x, y) by

$$F(x, y) = (x + y + v(x), y + v(x))$$

is the lift of an exact-symplectic twist map; show that its generating function is given by

$$H(x, x') = \frac{1}{2}(x' - x)^2 + V(x)$$

where $V : \mathbb{R}/\mathbb{Z} \to \mathbb{Z}$ satisfies $V' = v$.

As it follows from (1.6), orbits $(x_i)_{i \in \mathbb{Z}}$ of the monotone twist diffeomorphism f correspond to 'critical points' of the *action functional*

$$\{x_i\}_{i \in \mathbb{Z}} \longmapsto \sum_{i \in \mathbb{Z}} h(x_i, x_{i+1}).$$

Birkhoff's theorem stated in Theorem 1.1 still holds in this more general setting:

Theorem 1.9 *Let $f : \mathbb{A} \to \mathbb{A}$ be an exact-symplectic twist map. Then for any rational $m/n \in (0, 1)$ there exist at least two distinct periodic orbits of rotation number m/n.*

We recall this interesting result about invariant curves of twist maps (hence, it applies to billiard maps): the so-called *graph property*. This is a famous result due to Birkhoff [16, 18].

Theorem 1.10 (Birkhoff Invariant Curve Theorem) *Let $f: \mathbb{R}/\mathbb{Z} \times (a, b) \longrightarrow \mathbb{R}/\mathbb{Z} \times (a, b)$ be an exact-symplectic twist map. Assume that f admits an embedded, homotopically nontrivial, invariant curve $\gamma \subset \mathbb{R}/\mathbb{Z} \times (a, b)$. Then, it the graph of a Lipschitz function.*

Numerous proofs of this result have been given, see for instance [36, 63, 65].

These results are the starting point of an important and deep theory called Aubry-Mather theory, developed by S. Aubry and J. Mather in 1980s (see [6, 7, 75]). Aubry-Mather theory is concerned with the study of orbits that minimize this action-functional amongst all configurations with a prescribed rotation number; recall that the rotation number of an orbit $\{x_i\}_{i \in \mathbb{Z}}$ is given by $\omega = \lim_{i \to \pm\infty} \frac{x_i}{i}$, if this limit exists (in the billiard case, this definition leads to the same notion of rotation number introduced in Sect. 1.2). In this context, *minimizing* is meant in the statistical mechanical sense, i.e., every finite segment of the orbit minimizes the action functional with fixed end-points.

Theorem (Aubry [6, 7], Mather [36, 75]) *A monotone twist map possesses minimal orbits for every rotation number in its twist interval* (ω_-, ω_+). *For rational numbers there are always at least two periodic minimal orbits. Moreover, every minimal orbit lies on a Lipschitz graph over the x-axis.*

We refer to [36, 94, 96] for self-contained presentations on Aubry-Mather theory for twist maps and Hamiltonian flows.

Let us now introduce the *minimal average action* (or *Mather's β-function*).

Definition 1.8 Let $x^\omega = \{x_i\}_{i \in \mathbb{Z}}$ be any minimal orbit with rotation number ω. Then, the value of the *minimal average action* at ω is given by (this value is well-defined, since it does not depend on the chosen orbit):

$$\beta(\omega) := \lim_{N \to +\infty} \frac{1}{2N} \sum_{i=-N}^{N-1} h(x_i, x_{i+1}). \tag{1.7}$$

This function $\beta : \mathbb{R} \longrightarrow \mathbb{R}$ enjoys many properties and encodes interesting information on the dynamics. In particular:

(i) β is strictly convex and, hence, continuous (see [36]);
(ii) β is differentiable at all irrationals (see [77]);
(iii) β is differentiable at a rational p/q if and only if there exists an invariant circle consisting of periodic minimal orbits of rotation number p/q (see [77]).

In particular, β being a convex function, one can consider its convex conjugate:

$$\alpha(c) = \sup_{\omega \in \mathbb{R}} [\omega c - \beta(\omega)].$$

This function—which is generally called *Mather's α-function*—also plays an important rôle in the study of minimal orbits and in Mather's theory (particularly in higher dimension, see for example [74, 97]). We refer interested readers to surveys [36, 94, 96].

Observe that for each ω and c one has:

$$\alpha(c) + \beta(\omega) \geq \omega c,$$

where equality is achieved if and only if $c \in \partial \beta(\omega)$ or, equivalently, if and only if $\omega \in \partial \alpha(c)$ (the symbol ∂ denotes in this case the set of 'subderivatives' of the function, which is always non-empty and is a singleton if and only if the function is differentiable).

1.4.2 Action-Minimizing Properties of Billiards

In the billiard case, since the generating function of the billiard map is the euclidean distance $-\ell$, the action of the orbit coincides—up to a sign—to the length of the trajectory that the ball traces on the table Ω. In particular, these two functions encode many dynamical properties of the billiard (see [94] for more details):

- For each $0 < p/q \leq 1/2$, one has: $\beta(p/q) = -\frac{1}{q} M L_\Omega^{\max}(p/q)$.
- β is differentiable at p/q if and only if there exists an invariant circle of rotation number p/q foliated by periodic orbits.
- If Γ_ω is a convex caustic with rotation number $\omega \in (0, 1/2]$, then β is differentiable at ω and $\beta'(\omega) = -\text{length}(\Gamma_\omega) =: -|\Gamma_\omega|$ (see [94, Theorem 3.2.10]). In particular, β is always differentiable at 0 and $\beta'(0) = -|\partial\Omega|$, where $|\partial\Omega|$ denotes the length of the boundary of Ω.
- If Γ_ω is a convex caustic with rotation number $\omega \in (0, 1/2]$, then its Lazutkin invariant $Q(\Gamma_\omega)$ (see Sect. 1.3.4) can be related to the value of the α-function. In fact, one can show that (see [94, Theorem 3.2.10]):

$$Q(\Gamma_\omega) = \alpha(\beta'(\omega)) = \alpha(-|\Gamma_\omega|).$$

In [94, 95] properties of Mather's β and α functions have been studied more in depth. In particular, explicit expressions for their (formal) Taylor expansions at, respectively, $\omega = 0$ and $c = -|\partial\Omega|$ have been obtained. The coefficients in these expressions will be obtained in terms of the curvature of the boundary and its derivatives.

Theorem 1.11 *Let Ω be a strictly convex planar domain with smooth boundary. Denote by $k(s) > 0$ the curvature of $\partial\Omega$ with arc-length parametrization s. Let $\ell_0 := |\partial\Omega|$ be the length of the boundary and denote:*

$$I_1 := \int_0^{\ell_0} ds = \ell_0$$

$$I_3 := \int_0^{\ell_0} k^{2/3} ds$$

$$I_5 := \int_0^{\ell_0} \left(9 k^{4/3} + \frac{8 \dot{k}^2}{k^{8/3}}\right) ds$$

$$I_7 := \int_0^{\ell_0} \left(9 k^2 + \frac{24 \dot{k}^2}{k^2} + \frac{24 \ddot{k}^2}{k^4} - \frac{144 \dot{k}^2 \ddot{k}}{k^5} + \frac{176 \dot{k}^4}{k^6}\right) ds$$

$$I_9 := \int_0^{\ell_0} \left[\frac{281}{44800} k^{8/3} + \frac{281\,\dot{k}^2}{8400\,k^{4/3}} + \frac{167\,\ddot{k}^2}{4200\,k^{10/3}} - \frac{167\,\dot{k}^2\,\ddot{k}}{700\,k^{13/3}} + \frac{\dddot{k}^2}{42\,k^{16/3}} \right.$$

$$+ \frac{559\,\dot{k}^4}{2100\,k^{16/3}} - \frac{473\,\ddot{k}^3}{4725\,k^{19/3}} - \frac{10\,\dddot{k}\,\ddot{k}\,\dot{k}}{21\,k^{19/3}} + \frac{5\,\ddot{k}\,\dot{k}^3}{7\,k^{22/3}} + \frac{13142\,\dot{k}^2\,\ddot{k}^2}{4725\,k^{22/3}}$$

$$\left. - \frac{10777\,\dot{k}^4\,\ddot{k}}{1575\,k^{25/3}} + \frac{521897\,\dot{k}^6}{127575\,k^{28/3}} \right] ds.$$

Then:

- *the formal Taylor expansion of β at $\omega = 0$, $\beta(\omega) \sim \sum_{k=0}^{\infty} \beta_k \frac{\omega^k}{k!}$, has coefficients:*

$$\beta_{2k} = 0 \quad \text{for all } k$$

$$\beta_1 = -I_1$$

$$\beta_3 = \frac{1}{4} I_3^3$$

$$\beta_5 = -\frac{1}{144} I_3^4 I_5$$

$$\beta_7 = \frac{1}{320} I_3^5 \left(\frac{14}{81} I_5^2 - I_3 I_7 \right) = \frac{I_3^5 \left(14\,I_5^2 - 81\,I_3 I_7 \right)}{25920}$$

$$\beta_9 = -7 I_3^6 \left(I_3^2 I_9 - \frac{1}{5600} I_3 I_5 I_7 + \frac{7}{583200} I_5^3 \right);$$

- *the (formal) Taylor expansion of $(c + \ell_0)^{-3/2}\alpha(c)$ at $c = -\ell_0$ (note that α has in fact a square-root type singularity at the boundary), $(c + \ell_0)^{-3/2}\alpha(c) \sim \sum_{k=0}^{\infty} \alpha_k \frac{(c+\ell_0)^k}{k!}$, has coefficients:*

$$\alpha_0 = \frac{4\sqrt{2}}{3} I_3^{-3/2}$$

$$\alpha_1 = \frac{\sqrt{2}}{135} I_3^{-7/2} I_5$$

$$\alpha_2 = \frac{1}{56700\sqrt{2}} \left(\frac{72\,I_3 I_7 + 7\,I_5^2}{I_3^{11/2}} \right)$$

$$\alpha_3 = \frac{1}{826686000\sqrt{2}} \left(\frac{261273600\,I_3^2 I_9 + 21384\,I_3 I_5 I_7 + 1001\,I_5^3}{I_3^{15/2}} \right).$$

Remark 1.23

(i) The techniques used in the proof of the Theorem 1.11, allow one to obtain explicit expressions up to any arbitary high order (we restrict to order 11 just for the sake of this presentation).
(ii) The coefficients β_k are algebraically related to the set of spectral invariants introduced by Marvizi and Melrose [73] for strictly convex planar regions in order to investigate and give some partial answers to Kac's question on the isospectrality of planar domains. These computations provide explicit expressions for those invariants as well (see the expressions for $\mathcal{I}_k$'s).

An easy consequence of these formulae is the following corollary, which is a direct consequence of the isoperimetric inequality (see [95, Corollary 1] and [94]).

Corollary 1.1 *Let Ω be a strictly convex planar domain with smooth boundary. Then:*

$$\beta_3 + \pi^2 \beta_1 \leq 0$$

and equality holds if and only if Ω is a disc.

Proof The proof easily follows from the expressions of β_1 and β_3, found in Theorem 1.11. In fact, observe that:

$$\beta_3 + \pi^2 \beta_1 \leq 0 \quad \Longleftrightarrow \quad \mathcal{I}_3^3 - 4\pi^2 \mathcal{I}_1 \leq 0.$$

Now, using Hölder inequality (with $p = \frac{3}{2}$ and $q = 3$):

$$\mathcal{I}_3 = \int_0^{\ell_0} k^{2/3} ds \leq \left(\int_0^{\ell_0} (k^{2/3})^{3/2} ds \right)^{2/3} \left(\int_0^{\ell_0} 1^3 ds \right)^{1/3}$$
$$= (2\pi)^{2/3} \ell_0^{1/3} = (4\pi^2 \mathcal{I}_1)^{1/3}.$$

Moreover, equality holds if and only if it holds in Hölder inequality. This means that k must be constant (and strictly positive) and therefore, the curve must be a circle.
□

Remark 1.24 In particular, the above corollary says that if the first two coefficients β_1 and β_3 coincide to those of the β-function of a disc, then the domain must be a disc. Therefore, the β-function univocally determines discs amongst all possible Birkhoff billiards. It would be interesting to find a similar characterization for elliptic billiards. We can prove the following result: the β-function determines univocally a given ellipse in the family of all ellipses.

Proposition 1.10 *If $\mathcal{E}_1$ and $\mathcal{E}_2$ are two ellipses such that $\beta_{\mathcal{E}_1} \equiv \beta_{\mathcal{E}_2}$, then $\mathcal{E}_1$ and $\mathcal{E}_2$ are the same ellipse. More generally: if the Taylor coefficients $\beta_{\mathcal{E}_1,1} = \beta_{\mathcal{E}_2,1}$ and $\beta_{\mathcal{E}_1,3} = \beta_{\mathcal{E}_2,3}$, then the same conclusion remains true.*

The proof easily follows from expressing these coefficients by means of elliptic integrals (see [95, Proposition 1])

1.4.3 Birkhoff Conjecture and Spectral Rigidity Questions (Revisited)

We can now rephrase the Spectral Rigidity Question for the maximal Length spectrum (see Sect. 1.2.2) and Birkhoff Conjecture (see Sect. 1.3.6) in terms of these new objects.

Spectral Rigidity Question (Revisited) *Let Ω_1 and Ω_2 be two strictly convex planar domains with smooth boundaries and assume that $\beta_{\Omega_1} \equiv \beta_{\Omega_2}$. Is it true that Ω_1 and Ω_2 are isometric?*
More generally: if $\beta_{\Omega_1}(\omega) = \beta_{\Omega_2}(\omega)$ for all $\omega \in (0, \varepsilon)$ for some small $\varepsilon > 0$, is it true that Ω_1 and Ω_2 are isometric?

Similarly, keeping into account the relation between the differentiability properties of Mather's β-function at rational rotation numbers and the existence of invariant circles foliated by periodic points (see Sect. 1.4.2), we can also rephrase Birkhoff conjecture in this context.

Birkhoff Conjecture (Revisited) *Let Ω be a strictly convex planar domain with smooth boundary and assume that β_Ω is differentiable in $[0, 1/2)$. Is it true that Ω is an ellipse? More generally: if β_Ω is differentiable in $[0, \varepsilon)$ for some small $0 < \varepsilon < 1/2$, is it true that Ω is an ellipse?*

In fact, if β_Ω is differentiable in an open interval, then the billiard map is locally integrable in an open set. In fact, β_Ω will be differentiable at all rationals in that interval and therefore there will be caustics corresponding to these rotation numbers. By semi-continuity arguments, one obtains caustics corresponding to irrational rotation numbers and hence a family of caustics that foliate an open set. Observe that if β is differentiable in the whole domain of definition $(0, 1/2]$, then it must be a circle by the aforementioned result by Bialy.

The relation between the integrability of the billiard map and the differentiability of the corresponding Mather's β function, implies that a solution to Birkhoff conjecture would lead to a solution to the question whether ellipses are uniquely spectrally determined among convex domain.

Exercise 1.14 Rephrase the results in [8, 60, 67] in terms of Mather's β-function and spectral rigidity of ellipses.

Remark 1.25 Compare this result with the previously mentioned result by Hezari and Zelditch [53] (see Sect. 1.2.3), where it is proved that ellipses of sufficiently small eccentricities are Laplace spectrally unique (up to isometry) among all smooth domains (withouth any assumption on symmetry, convexity, or closeness to other ellipses).

Acknowledgments The authors are sincerely grateful to the organizers of the summer school *"Modern Aspects of Dynamical Systems"* (Cetraro, August 2–6, 2021) and to the Fondazione C.I.M.E. "Roberto Conti" for supporting this event. AS wishes to acknowledge the support of the PRIN Projects *"Regular and stochastic behavior in dynamical systems"* and *"Stability in Hamiltonian dynamics and beyond"*, as well as the MIUR Department of Excellence grant MatMod@TOV. VD acknowledges the support of the ERC Grant # 885707.

References

1. Akhiezer, N.I.: Elements of the Theory of Elliptic Functions. Translations of Mathematical Monographs, vol. 79. viii+237pp. American Mathematical Society, Providence (1990)
2. Andersson, K.G., Melrose, R.B.: The propagation of singularities along gliding rays. Invent. Math. **41**(3), 197–232 (1977)
3. Arnaud, M.C.: Fibrés de Green et régularité des graphes C^0-Lagrangiens invariants par un flot de Tonelli. Ann. Henri Poincaré **9**(5), 881–926 (2008)
4. Arnaud, M.C., Massetti, J.E., Sorrentino, A.: On the fragility of periodic tori for families of symplectic twist maps. Adv. Math. **429**, 109175, 39pp. (2023)
5. Arnold, M., Bialy, M.: Nonsmooth convex caustics for Birkhoff billiards. Pac. J. Math. **295**(2), 257–269 (2018)
6. Aubry, S.: The twist map, the extended Frenkel-Kontorova model and the devil's staircase. Phys. D **7**(1–3), 240–258 (1983)
7. Aubry, S.: The discrete Frenkel-Kontorova model and its extensions. I. Exact results for the ground-states. Phys. D **8**(3), 381–422 (1983)
8. Avila, A., De Simoi, J., Kaloshin, V.: An integrable deformation of an ellipse of small eccentricity is an ellipse. Ann. Math. **184**, 527–558 (2016)
9. Bialy, M.: Convex billiards and a theorem by E. Hopf. Math. Z. **124**(1), 147–154 (1993)
10. Bialy, M.: Hopf rigidity for convex billiards on the hemisphere and hyperbolic plane. Discrete Contin. Dyn. Syst. **33**(9), 3903–3913 (2013)
11. Bialy, M., Mironov, A.: Cubic and quartic integrals for geodesic flow on 2-torus via a system of the hydrodynamic type. Nonlinearity **24**(12), 3541–3554 (2011)
12. Bialy, M., Mironov, A.: Rich quasi-linear system for integrable geodesic flows on 2-torus. Discrete Contin. Dyn. Syst. **29**(1), 81–90 (2011)
13. Bialy, M., Mironov, A.: Angular billiard and algebraic birkhoff conjecture. Adv. Math. **313**, 102–126 (2017)
14. Bialy, M., Mironov, A.: The Birkhoff-Poritsky conjecture for centrally-symmetric billiard tables. Ann. Math. **196**(1), 389–413 (2022)
15. Bialy, M., Tsodikovich, D.: Locally maximising orbits for the non-standard generating function of convex billiards and applications. Nonlinearity **36**(3), 2001–2019 (2023)
16. Birkhoff, G.D.: Surface transformations and their dynamical applications. Acta Math. **43**, 1–119 (1922)
17. Birkhoff, G.D.: On the periodic motions of dynamical systems. Acta Math. **50**(1), 359–379 (1927)
18. Birkhoff, G.D.: Sur quelques combes fermées remarquables. Bull. Soc. Math. de France **60**, 1–26 (1932)
19. Birkhoff, G.D.: Collected Mathematical Papers, vol. II. American Mathematical Society, Providence (1950)

20. Blumen, V., Kim, K., Nance, Zharnitsky, V.: Three-period orbits in billiards on the surfaces of constant curvature. Int. Math. Res. Not. **2012**, 5014–5024 (2012). https://doi.org/10.1093/imrn/rnr228
21. Bolotin, S.: Integrable Birkhoff billiards. Mosc. Univ. Mech. Bull. **45**(2), 10–13 (1990)
22. Bolsinov, A.V., Fomenko, A.T., Matveev, V.S.: Two-dimensional Riemannian metrics with an integrable geodesic flow. Local and global geometries. Mat. Sb. **189**(10), 5–32 (1998); translation in Sb. Math. **189**(9–10), 1441–1466 (1998)
23. Burago, D., Ivanov, S.: Riemannian tori without conjugate points are flat. Geom. Funct. Anal. **4**(3), 259–269 (1994)
24. Callis, K.G.: Absolutely periodic billiard orbits of arbitrarily high order (2022). arXiv: 2209.11721
25. Chen, J., Kaloshin, V., Zhang, H.K.: Length spectrum rigidity for piecewise analytic bunimovich billiards. Commun. Math. Phys. **404**, 1–50 (2023)
26. Croke, C.B.: Rigidity for surfaces of nonpositive curvature. Comment. Math. Helv. **65**(1), 150–169 (1990)
27. Croke, C.B., Sharafutdinov, V.A.: Spectral rigidity of a compact negatively curved manifold. Topology **37**(6), 1265–1273 (1998)
28. de la Llave, R., Marco, J.M., Moriyón, R.: Canonical perturbation theory of Anosov systems and regularity results for the Livšic cohomology equation. Ann. Math. **123**(3), 537–611 (1986)
29. De Simoi, J., Kaloshin, V., Wei, Q., (Appendix B coauthored with H. Hezari): Deformational spectral rigidity among Z_2-symmetric domains close to the circle. Ann. Math. **186**, 277–314 (2017)
30. De Simoi, J., Kaloshin, V., Leguil, M.: Marked length spectral determination of analytic chaotic billiards with axial symmetries. Invent. Math. **233**, 829–901 (2019)
31. de Verdière, C.: Sur les longueurs des trajectoires périodiques d'un billard. In: Dazord, P., Desolneux-Moulis, N. (eds.) Geometrie Symplectique et de Contact: Autour du Theorème de Poincaré-Birkhoff, Hermann, Paris, pp. 122–139 (1984)
32. Delshams, A., Ramírez-Ros, R.: Poincaré-Melnikov-Arnold method for analytic planar maps. Nonlinearity **9**(1), 1–26 (1996)
33. Fierobe, C.: Complex caustics of the elliptic billiard. Arnold Math. J. **7**, 1–30 (2021)
34. Fierobe, C.: On projective billiards with open subsets of triangular orbits. To appear in Isr. J. Math.
35. Fierobe, C.: Examples of reflective projective billiards. https://arxiv.org/pdf/2002.09845.pdf
36. Forni, G., Mather, J.N.: Action minimizing orbits in Hamiltonian systems. In: Transition to Chaos in Classical and Quantum Mechanics (Montecatini Terme, 1991). Lecture Notes in Mathematics, vol. 1589, pp. 92–186. Springer, Berlin (1994)
37. Glutsyuk, A.: On quadrilateral orbits in complex algebraic planar billiards. Moscow Math. J. **14**, 239–289 (2014)
38. Glutsyuk, A.: On Odd-periodic orbits in complex planar billiards. J. Dyn. Control Syst. **20**(3), 293–306 (2014)
39. Glutsyuk, A.: On 4-reflective complex analytic billiards. J. Geom. Anal. **27**, 83–238 (2017)
40. Glutsyuk, A.: On polynomially integrable Birkhoff billiards on surfaces of constant curvature. J. Eur. Math. Soc. **23**, 995–1049 (2020)
41. Glutsyuk, A.A., Kudryashov, Y.G.: On quadrilateral orbits in planar billiards. Dokl. Math. **83**(3), 371–373 (2011)
42. Glutsyuk, A.A., Kudryashov, Y.G.: No planar billiard possesses an open set of quadrilateral trajectories. J. Mod. Dyn. **6**(3), 287–326 (2012)
43. Gole, C.: Symplectic Twist Maps, Global Variational Techniques. Advanced Series in Nonlinear Dynamics, vol. 18. World Scientific, Singapore (2001)
44. Gordon, C., Webb, D.L., Wolpert, S.: One cannot hear the shape of a drum. Bull. Am. Math. Soc. **27**(1), 134–138 (1992)
45. Guillarmou, C., Lefeuvre, T.: The marked length spectrum of Anosov manifolds. Ann. Math. **190**(1), 321–344 (2019)

46. Guillemin, V., Kazhdan, D.: Some inverse spectral results for negatively curved 2-manifolds. Topology **19**(3), 301–312 (1980)
47. Guillemin, V., Melrose, R.: The Poisson summation formula for manifolds with boundary. Adv. Math. **32**(3), 204–232 (1979)
48. Gutkin, E.: Billiard dynamics: a survey with the emphasis on open problems. Regul. Chaotic Dyn. **8**(1), 1–13 (2003)
49. Gutkin, E., Katok, A.: Caustics for inner and outer billiards. Commun. Math. Phys. **173**, 101–133 (1995)
50. Halpern, B.: Strange billiard tables. Trans. Am. Math. Soc. **232**, 297–305 (1977)
51. Henheik, J.: Deformational rigidity of integrable metrics on the torus (2022). arXiv: 2210.02961
52. Hezari, H., Zelditch, S.: C^∞ spectral rigidity of the ellipse. Anal. PDE **5**(5), 1105–1132 (2012)
53. Hezari, H., Zelditch, S.: One can hear the shape of ellipses of small eccentricity. Ann. Math. **196**(3), 1083–1134 (2022)
54. Hopf, E.: Closed surfaces without conjugate points. Proc. Nat. Acad. Sci. U. S. A. **34**, 47–51 (1948)
55. Huang, G., Kaloshin, V., Sorrentino, A.: Nearly circular domains which are integrable close to the boundary are ellipses. Geom. Funct. Anal. **28**(2), 334–392 (2018)
56. Huang, G., Kaloshin, V., Sorrentino, A.: On marked length spetrums of generic strictly convex billiard tables. Duke Math. J. **167**(1), 175–209 (2018)
57. Innami, N.: Convex curves whose points are vertices of billiard triangles. Kodai Math. J. **11**, 17–24 (1988)
58. Innami, N.: Geometry of geodesics for convex billiards and circular billiards. Nihonkai Math. J. **13**, 73–120 (2002)
59. Ivrii, V.Y.: The second term of the spectral asymptotics for a Laplace–Beltrami operator on manifolds with boundary. Funktsional. Anal. i Prilozhen **14**(2), 25–34 (1980)
60. Kaloshin, V., Sorrentino, A.: On the local Birkhoff conjecture for convex billiards. Ann. Math. **188**(1), 315–380 (2018)
61. Kaloshin, V., Koval, I., Vig, A.: Wave trace cancellations for hyperbolic orbits in convex nearly elliptic billiard tables (preprint, 2023)
62. Kaloshin, V., Koudjinan, C.E., Zhang, K.: Birkhoff Conjecture for nearly centrally symmetric domains (preprint, 2023)
63. Katok, A., Hasselblatt, B.: Introduction to the Modern Theory of Dynamical Systems (Encyclopedia of Mathematics and its Applications). Cambridge University Press, Cambridge (1995)
64. Katok, A., Strelcyn, J.M., Ledrappier, F., Przytycki, F.: Invariant Manifolds, Entropy and Billiards; Smooth Maps with Singularities. Lecture Notes in Mathematics, vol. 1222, viii+283pp. Springer, Berlin (1986)
65. Katznelson, Y., Ornstein, D.S.: Twist Maps and Aubry-Mather Sets. Lipa's legacy, Contemporary Mathematics, vol. 211. American Mathematical Society, New York (1995)
66. Knill, O.: On nonconvex caustics of convex billiards. Elem. Math. **53**, 89–106 (1998)
67. Koval, I.: Local strong Birkhoff conjecture and local spectral rigidity of almost every ellipse (2021). arXiv:2111.12171
68. Kozlov, V.V., Treshchëv, D.V.: Billiards, a Genetic Introduction to the Dynamics of Systems with Impacts. Translations of Mathematical Monographs, vol. 89. American Mathematical Society, Providence (1991)
69. Lazutkin, V.F.: Existence of caustics for the billiard problem in a convex domain (in Russian). Izv. Akad. Nauk SSSR Ser. Mat. **37**, 186–216 (1973)
70. LeBrun, C., Mason, L.J.: Zoll manifolds and complex surfaces. J. Differ. Geom. **61**(3), 453–535 (2002)

71. Levallois, P.: Non-intégrabilité des billiards définis par certaines perturbations algébriques d'une ellipse et du flot géodésique de certaines perturbations algébriques d'un ellipsoıde. Ph.D. Thesis, Univ. Paris VII (1993)
72. Levallois, P., Tabanov, M.: Spáration des séparatrices du billard elliptique pour une perturbation algbrique et symétrique de l'ellipse. C. R. Acad. Sci. Paris Sér. I Math. **316**(6), 589–592 (1993)
73. Marvizi, S., Melrose, R.: Spectral invariants of convex planar regions. J. Differ. Geom. **17**, 475–502 (1982)
74. Massart, D., Sorrentino, A.: Differentiability of Mather's average action and integrability on closed surfaces. Nonlinearity **24**, 777–1793 (2011)
75. Mather, J.N.: Existence of quasiperiodic orbits for twist homeomorphisms of the annulus. Topology **21**(4), 457–467 (1982)
76. Mather, J.N.: Glancing billiards. Erg. Theory Dynam. Syst. **2**(3–4), 397–403 (1982)
77. Mather, J.N.: Differentiability of the minimal average action as a function of the rotation number. Bol. Soc. Brasil. Mat. **21**(1), 59–70 (1990)
78. Moser, J.: Selected Chapters of the Calculus of Variations. Lectures in Mathematics. ETH, Zurich (2003)
79. Osgood, B., Phillips, R., Sarnak, P.: Compact isospectral sets of surfaces. J. Funct. Anal. **80**(1), 212–234 (1988)
80. Osgood, B., Phillips, R., Sarnak, P.: Extremals of determinants of Laplacians. J. Funct. Anal. **80**(1), 148–211 (1988)
81. Osgood, B., Phillips, R., Sarnak, P.: Moduli space, heights and isospectral sets of plane domains. Ann. Math. **129**(2), 293–362 (1989)
82. Otal, J.P.: Le spectre marqué des longueurs des surfaces à courbure négative. Ann. Math. **131**(1), 151–162 (1990)
83. Petkov, V., Stojanov, L.: On the number of periodic reflecting rays in generic domains. Erg. Theory Dynam. Syst. **8**, 81–91 (1988)
84. Petkov, V.M., Stoyanov, L.N.: Geometry of Reflecting Rays and Inverse Spectral Problems. Pure and Applied Mathematics (New York). Wiley, Chichester (1992)
85. Popov, G.: Invariants of the length spectrum and spectral invariants of planar convex domains. Commun. Math. Phys. **161**, 335–364 (1994)
86. Popov, G., Topalov, P.: Invariants of isospectral deformations and spectral rigidity. Commun. Partial Differ. Equ. **37**(3), 369–446 (2012)
87. Popov, G., Topalov, P.: From K.A.M. Tori to isospectral invariants and spectral rigidity of billiard tables. arXiv e-prints (2016)
88. Poritsky, H.: The billiard ball problem on a table with a convex boundary — an illustrative dynamical problem. Ann. Math. **51**, 446–470 (1950)
89. Ramírez-Ros, R.: Break-up of resonant invariant curves in billiards and dual billiards associated to perturbed circular tables. Phys. D **214**, 78–87 (2006)
90. Rychlik, M.R.: Periodic points of the billiard ball map in a convex domain. J. Differ. Geom. **30**, 191–205 (1989)
91. Safarov, Y., Vassilev, D.: The Asymptotic Distribution of Eigenvalues of Partial Differential Operators. Translations of Mathematical Monographs, vol. 155. American Mathematical Society, Providence (1996)
92. Safarov, Y., Vassiliev, D.: The Asymptotic Distribution of Eigenvalues of Partial Differential Operators, pp. 98–106. American Mathematical Society, Providence (1996)
93. Sarnak, P.: Determinants of Laplacians; Heights and Finiteness. Analysis, et cetera, pp. 601–622. Academic Press, Boston (1990)
94. Siburg, K.F.: The Principle of Least Action in Geometry and Dynamics. Lecture Notes in Mathematics, vol.1844, xiii+ 128pp. Springer, Berlin (2004)
95. Sorrentino, A.: Computing Mather's beta-function for Birkhoff billiards. Discrete Contin. Dyn. Syst. A **35**(10), 5055–5082 (2015)

96. Sorrentino, A.: Action-minimizing methods in hamiltonian dynamics. In: An Introduction to Aubry-Mather Theory. Mathematical Notes Series, vol. 50. Princeton University Press, Princeton (2015)
97. Sorrentino, A., Viterbo, C.: Action minimizing properties and distances on the group of Hamiltonian diffeomorphisms. Geom. Topol. **14**, 2383–2403 (2010)
98. Stojanov, L.: Note on the periodic points of the billiard. J. Differ. Geom. **34**, 835–837 (1991)
99. Sunada, T.: Riemannian coverings and isospectral manifolds. Ann. Math. **121**(1), 169–186 (1985)
100. Tabachnikov, S.: Billiards. Panorama Synthesizer, vol. 1, vi+ 142pp. Socété mathématique de France, Marseille (1995). https://books.google.it/books/about/Billiards.html?id=cCBwQgAACAAJ&redir_esc=y
    Panorama Synthesizer, vol. 1, vi+ 142pp. (1995)
101. Tabachnikov, S.: Geometry and Billiards. Student Mathematical Library, vol.30, xii+ 176pp. American Mathematical Society, Providence (2005)
102. Treschev, D.: Billiard map and rigid rotation. Phys. D **255**, 31–34 (2013)
103. Vasiliev, D.: Two-term asymptotics of the spectrum of a boundary value problem in interior reflection of general form. Funct. Anal. Appl. **18**, 267–277 (1984)
104. Vignéras, M.F.: Variétés riemanniennes isospectrales et non isométriques. Ann. Math. **112**(1), 21–32 (1980)
105. Vorobets, Y.B.: On the measure of the set of periodic points of a billiard. Math. Not. **55**, 455–460 (1994)
106. Weyl, H.: Das asymptotische Verteilungsgesetz der Eigenwerte linearer partieller Differentialgleichungen (mit einer Anwendung auf die Theorie der Hohlraumstrahlung). Math. Ann. **71**, 441–479 (1912)
107. Wojtkowski, M.P.: Two applications of Jacobi fields to the billiard ball problem. J. Differ. Geom **40**, 155–164 (1994)
108. Zelditch, S.: Survey of the inverse spectral problem. arXiv Mathematics e-prints (2004)
109. Zelditch, S.: Inverse spectral problem for analytic domains. II. Z^2-symmetric domains. Ann. Math. **170**(1), 205–269 (2009)
110. Zelditch, S.: Survey on the inverse spectral problem. ICCM Not. **2**(2), 1–20 (2014)
111. Zoll, O.: Über Flächen mit Scharen geschlossener geodätischer Linien (in German). Math. Ann. **57**, 108–133 (1903)

Chapter 2
Introduction to Fatou Components in Holomorphic Dynamics

Xavier Buff and Jasmin Raissy

Abstract This chapter is an introduction to the classification of Fatou components in holomorphic dynamics. We start with the description of the Fatou and Julia sets for rational maps of the Riemann sphere, and finish with an updated account of the recent results on Fatou components for polynomial skew-products in complex dimension two, where we focus on the key steps in the construction giving the existence of a wandering domain for a polynomial endomorphism of $\mathbb{C}^2$.

2.1 Introduction

Discrete dynamics concerns maps $f : X \to X$ from a space to itself. One of the main goals is to understand sequences of the form

$$x_0 \in X, \quad x_1 := f(x_0), \quad x_2 := f(x_1) = f^{\circ 2}(x_0), \quad x_3 := f(x_2) = f^{\circ 3}(x_0), \ldots$$

obtained by *iterating* f. The *forward orbit* of x_0 under f is

$$O^+(x_0) := \bigcup_{n \geq 0} \{f^{\circ n}(x_0)\}.$$

The main problem is to understand the long term behaviour of the sequence $(x_n)_{n \geq 0}$. The answer will depend on the assumptions we make on the space X and the map $f : X \to X$.

X. Buff
Institut de Mathématiques de Toulouse, UMR5219 Université de Toulouse, CNRS, UPS, Toulouse, France
e-mail: xavier.buff@math.univ-toulouse.fr

J. Raissy (✉)
University of Bordeaux, CNRS, Bordeaux INP, IMB, UMR 5251, Talence, France
Institut Universitaire de France (IUF), Paris, France
e-mail: jasmin.raissy@math.u-bordeaux.fr

If X is a topological space, we may wonder whether the sequence converges. If this is the case, and in addition $f : X \to X$ is a continuous map, then the limit ℓ is a fixed point of f, i.e., $f(\ell) = \ell$.

If X is compact, then thanks to the Bolzano-Weierstrass Theorem, we may extract converging subsequences. We shall denote by $\omega(x_0)$ the *ω-limit set of* x_0, that is the set of possible limit values:

$$\omega(x_0) := \bigcap_{n \geq 0} \overline{O^+(x_n)}.$$

If $f : X \to X$ is continuous, then for any $x \in X$, the ω-limit set $\omega(x)$ is f-invariant.

Exercise 2.1 Prove it.

In the rest of these notes, we will study the case where X is a complex manifold and $f : X \to X$ is a holomorphic map. We shall first consider the case where $X = \widehat{\mathbb{C}} = \mathbb{C} \cup \{\infty\}$ is the Riemann sphere and $f : \widehat{\mathbb{C}} \to \widehat{\mathbb{C}}$ is a polynomial or a rational map.

This subject has a fairly long history, with contributions by Koenigs [43], Schröder [56], Böttcher [14] in the late nineteenth century, and the great memoirs of Fatou [30–32] and Julia [41] around 1920. For the history of this early period we refer the reader to the book of Alexander [1]. Followed a dormant period, with notable contributions by Cremer [22, 23] (1936) and Siegel [58] (1942), and a rebirth in the 1960s (Brolin, Guckenheimer, Jakobson). Since the early 1980s, partly under the impetus of computer graphics, the subject has grown vigorously, with major contributions by Douady, Hubbard, Sullivan, Thurston, and more recently Lyubich, McMullen, Milnor, Shishikura, Yoccoz ... Although the subject still presents many open problems, there is now a substantial body of knowledge and several books have appeared: Beardon [5], Carleson and Gamelin [21], Milnor [50] and Steinmetz [59] and the more specialized books by McMullen [48, 49]. Surveys by Blanchard [12], Devaney [25], Douady [26], Keen [42] and Lyubich [46] are also highly recommended.

There is a small collection of polynomials, for instance

$$f(z) = az + b, \quad f(z) = z^d \quad \text{and} \quad f(z) = z^2 - 2,$$

whose dynamics can be fairly easily understood.

Exercise 2.2 Assume $f(z) = az + b$ with $a \in \mathbb{C}$ and $b \in \mathbb{C}$.

1. Show that when $|a| \neq 1$ or when $a = 1$ and $b \neq 0$, the map

$$\mathbb{C} \ni z \mapsto \omega(z) \in \widehat{\mathbb{C}}$$

   is constant.
2. Show that in all other cases, $\omega(z)$ depends on z and is either a finite set, or a euclidean circle.

Exercise 2.3 Assume $f(z) = z^d$ with $d \geq 2$.

1. Show that $\omega(z) = \{0\}$ if z belongs to the unit disk $\mathbb{D}$ and $\omega(z) = \{\infty\}$ if $z \in \widehat{\mathbb{C}} \setminus \overline{\mathbb{D}}$.
2. Show that
    (a) $\omega(z) \subset S^1$ if z belongs to the unit cirle S^1,
    (b) $\omega(z)$ can be equal to S^1 and
    (c) $\omega(z)$ can be a finite set of arbitrary cardinality.

Exercise 2.4 Assume $f(z) = z^2 - 2$.

1. Show that every $z \in \mathbb{C} \setminus \{0\}$ may be written $z = w + 1/w$ for some $w \in \mathbb{C}$ with $|w| = 1$ if $z \in [-2, 2]$ and $|w| > 1$ if $z \in \mathbb{C} \setminus [-2, 2]$.
2. Show that
    (a) $\omega(z) = \{\infty\}$ if $z \in \widehat{\mathbb{C}} \setminus [-2, 2]$ and
    (b) $\omega(z) \subset [-2, 2]$ if $z \in [-2, 2]$.

Exercise 2.5 Assume $f : \widehat{\mathbb{C}} \to \widehat{\mathbb{C}}$ is a Tchebychev polynomial of degree d, i.e., such that $\cos(d\theta) = f(\cos(\theta))$. Show that $\omega(z) = \{\infty\}$ if $z \in \widehat{\mathbb{C}} \setminus [-1, 1]$ and $\omega(z) \subset [-1, 1]$ if $z \in [-1, 1]$.

Those examples are really atypical examples, since most cases exhibit *fractal* and *chaotic* behaviour, the analysis of which requires tools from complex analysis, dynamical systems, topology, combinatorics, ...

2.2 Fatou Sets and Julia Sets

Both Fatou's and Julia's work have as a major theme that the Riemann sphere $\widehat{\mathbb{C}}$ breaks up sharply into two complementary subsets: the open Fatou set $\mathcal{F}_f$ with orderly dynamics, and the Julia set $\mathcal{J}_f$ on which the dynamical behaviour of f is chaotic. In practice, understanding the dynamics of a rational map has come to mean understanding the topology of the Julia set, its geometry if possible, and classifying the components of the Fatou set.

2.2.1 Definition

There are several possible definitions of the Fatou and Julia sets. We will adopt the one based on normal families, which seems very natural to us: a normal family of analytic maps is "well-behaved". But the very fact that it has become standard reflects the origins of holomorphic dynamics in complex analysis, and may explain why the subject has never quite entered the main stream of dynamical systems.

Definition 2.1 (Normal Families) A family $(f_\alpha : U \to \widehat{\mathbb{C}})_{\alpha \in \mathcal{A}}$ of holomorphic maps on an open subset $U \subset \widehat{\mathbb{C}}$ is *normal* if from any sequence $(f_{\alpha_n})_{n \geq 0}$ one can extract a subsequence converging uniformly (for the spherical metric in the range) on compact subsets of U.

A criterion that will enable us to deal with normality is due to Montel.

Theorem 2.1 (Montel) *A family* $(f_\alpha : U \to \widehat{\mathbb{C}})_{\alpha \in \mathcal{A}}$ *of holomorphic maps on an open subset* $U \subset \widehat{\mathbb{C}}$ *which omits at least three values in* $\widehat{\mathbb{C}}$ *is normal.*

In particular, a bounded sequence of holomorphic maps $(f_n : U \to \mathbb{C})_{n \in \mathbb{N}}$ is normal.

Definition 2.2 (Fatou Sets and Julia Sets) The *Fatou set* $\mathcal{F}_f$ of a rational map $f : \widehat{\mathbb{C}} \to \widehat{\mathbb{C}}$ is the largest open subset of $\widehat{\mathbb{C}}$ on which the family of iterates $(f^{\circ n})_{n \in \mathbb{N}}$ is normal. The *Julia set* $\mathcal{J}_f$ is the complement of the Fatou set.

Example For $f(z) = z^d$, with $d \geq 2$ or $d \leq -2$, the Julia set $\mathcal{J}_f$ is the unit circle

$$S^1 := \{z \in \mathbb{C} \,;\, |z| = 1\}.$$

Indeed, the family of iterates $(f^{\circ n} : \widehat{\mathbb{C}} \setminus S^1 \to \widehat{\mathbb{C}})_{n \in \mathbb{N}}$ takes its values in $\widehat{\mathbb{C}} \setminus S^1$, hence it omits more than three values in $\widehat{\mathbb{C}}$. By Montel's theorem, it is normal and so, $\mathcal{J}_f \subseteq S^1$. To prove that $S^1 \subseteq \mathcal{J}_f$, one may argue that the sequence $\left(f^{\circ(2n)}\right)_{n \in \mathbb{N}}$ converges uniformly on any compact subset of $\mathbb{D}$ to the constant map 0 and on any compact subset of $\widehat{\mathbb{C}} \setminus \overline{\mathbb{D}}$ to the constant map ∞. As any neighborhood of a point $z \in S^1$ contains points in $\mathbb{D}$ and points in $\widehat{\mathbb{C}} \setminus \overline{\mathbb{D}}$, there is no neighborhood of z on which the sequence $\left(f^{\circ n}\right)_{n \in \mathbb{N}}$ is normal.

Figure 2.1 shows the Julia set of the quadratic polynolimal $f(z) = z^2 + 1$ drown in the complex plane.

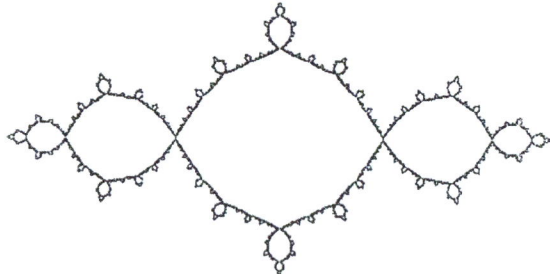

Fig. 2.1 The Julia set of the quadratic polynomial $f : \widehat{\mathbb{C}} \to \widehat{\mathbb{C}}$ defined by $f(z) = z^2 - 1$

2.2.2 The Polynomial Case

In the case of a polynomial, there is a slightly different way of understanding things. In this case, ∞ is a fixed point and this fixed point has no preimage other than itself. It is an exceptional point.

Definition 2.3 Let $f : \widehat{\mathbb{C}} \to \widehat{\mathbb{C}}$ be a polynomial of degree $d \geq 2$. The *filled-in Julia set* of f is the set $\mathcal{K}_f$ of points with bounded orbits:

$$\mathcal{K}_f = \left\{ z \in \mathbb{C} \,\middle|\, \left(f^{on}(z)\right)_{n \in \mathbb{N}} \text{ is bounded} \right\}.$$

This set is not only very natural for a dynamical system, but fits in very nicely with the general theory.

Proposition 2.1 *The filled-in Julia set $\mathcal{K}_f$ is a non-empty compact set. The Julia set is the topological boundary of $\mathcal{K}_f$.*

Proof Since f is of degree $d \geq 2$, $f(z)/z \to \infty$ as $z \to \infty$, and so, there exists a constant $R > 0$ such that $|f(z)| > 2|z|$ whenever $|z| > R$. Set $D_R := \{z \in \mathbb{C} \mid |z| < R\}$. Then, the nested intersection of compact sets

$$\mathcal{K}_f = \bigcap_{n \in \mathbb{N}} f^{-n}\left(\overline{D}_R\right)$$

is a non-empty compact set.

Furthermore, define U_R to be the open set $U_R := \widehat{\mathbb{C}} \setminus \overline{D}_R$. Then, the sequence $(f^{on})_{n \geq 0}$ converges uniformly to infinity on U_R. Clearly, the same property holds on the open set

$$\Omega_\infty = \bigcup_{n \in \mathbb{N}} f^{-n}(U_R)$$

Observe that $\Omega_\infty = \widehat{\mathbb{C}} \setminus \mathcal{K}_f$ and $\partial \Omega_\infty = \partial \mathcal{K}_f$.

We will first show that $\partial \mathcal{K}_f \subseteq \mathcal{J}_f$. Choose a point $z_0 \in \partial \Omega_\infty = \partial \mathcal{K}_f$ and assume that there is a neighborhood U of z_0 on which the family of iterates of f is normal. Then, U intersects Ω_∞ and the sequence $(f^{on})_{n \geq 0}$ must converge to infinity on the whole set U. This contradicts the fact that the sequence $(f^{on}(z_0))_{n \geq 0}$ is bounded.

We will now show that $\mathcal{J}_f \subseteq \partial \mathcal{K}_f$. As mentioned previously, the sequence (f^{on}) converges locally uniformly to a map which is constant and equal to ∞ outside $\mathcal{K}_f$. Thus, the complement of $\mathcal{K}_f$ is contained in the Fatou set $\mathcal{F}_f$ and $\mathcal{J}_f \subseteq \mathcal{K}_f$. Similarly, on the interior of $\mathcal{K}_f$, the sequence (f^{on}) is bounded, thus normal by Montel's theorem. As a consequence, the interior of $\mathcal{K}_f$ is contained in $\mathcal{F}_f$ and $\mathcal{J}_f \subseteq \partial \mathcal{K}_f$. □

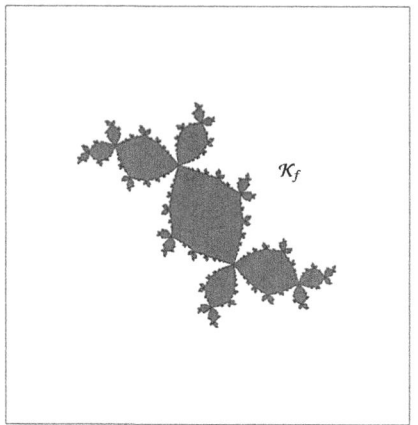 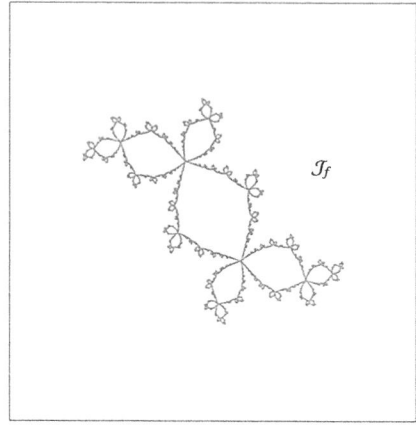

Fig. 2.2 Left: the filled-in Julia set of a quadratic polynomial. It is known as the Douady Rabbit. Right: the Julia set for the same quadratic polynomial. It is the topological boundary of the filled-in Julia set

Figure 2.2 show the filled-in Julia set and the Julia set of a quadratic polynomial. The reason why $\mathcal{K}_f$ is called a filled-in Julia set is the following.

Definition 2.4 A compact subset $K \subset \mathbb{C}$ is called *full* if it is connected and its complement is connected.

Proposition 2.2 *Let* $f : \widehat{\mathbb{C}} \to \widehat{\mathbb{C}}$ *be a polynomial. Each connected component of* $\mathcal{K}_f$ *is full.*

Proof Of course, any component K_0 is compact and connected. Any component U of the complement must satisfy $\partial U \subset K_0$, and if the complement is not connected at least one such component must be bounded. Then, by the maximum principle

$$\sup_U |f^{\circ n}(z)| = \sup_{\partial U} |f^{\circ n}(z)| \leq \sup_{K_0} |f^{\circ n}(z)|.$$

Hence, it is uniformly bounded, so $U \subset K_0$. □

2.2.3 Invariance of Fatou Sets and Julia Sets

Proposition 2.3 *Both the Julia set and the Fatou set are completely invariant:*

$$f(\mathcal{J}_f) = \mathcal{J}_f, \quad f^{-1}(\mathcal{J}_f) = \mathcal{J}_f, \quad f(\mathcal{F}_f) = \mathcal{F}_f, \quad f^{-1}(\mathcal{F}_f) = \mathcal{F}_f.$$

Proof As the Julia set is the complement of the Fatou set, it is enough to prove the statements about the Fatou set. This is an immediate consequence of the following lemma.

Lemma 2.1 *Let U be an open subset of $\widehat{\mathbb{C}}$, and let $\phi : U \to \widehat{\mathbb{C}}$ be holomorphic and non-constant. Then the family $(g_\alpha : \phi(U) \to \widehat{\mathbb{C}})_{\alpha \in \mathcal{A}}$ is normal if and only if $(f_\alpha := g_\alpha \circ \phi : U \to \widehat{\mathbb{C}})_{\alpha \in \mathcal{A}}$ is normal.* □

Remark 2.1 Since ϕ is a non-constant analytic map, it is an open map. Thus $\phi(U)$ is open and the family g_α is defined on an open set. □

Proof ($\Rightarrow$) This is obvious. If the sequence g_{α_n} converges to g, then $f_{\alpha_n} := g_{\alpha_n} \circ \phi$ converges to $g \circ \phi$.

($\Leftarrow$) Assume that the sequence $f_{\alpha_k} := g_{\alpha_k} \circ \phi$ is uniformly convergent on every compact subset of U. Let $K \subset \phi(U)$ be compact, and choose a cover of U by open subsets U_i with compact closure. Since ϕ is open, $\{\phi(U_i)\}$ is an open cover of K, and there exists a finite subcover $\{\phi(U_{i_j})\}$. The set

$$K' := \phi^{-1}(K) \cap \bigcup_j \overline{U}_{i_j}$$

is a compact subset of U such that $\phi(K') = K$. Then,

$$\sup_{z \in K} \operatorname{dist}_{\widehat{\mathbb{C}}}(g_{\alpha_n}(z), g_{\alpha_m}(z)) = \sup_{w \in K'} \operatorname{dist}_{\widehat{\mathbb{C}}}(f_{\alpha_n}(w), f_{\alpha_m}(w)) \to 0$$

as $n, m \to \infty$. The lemma follows. □

To prove $\mathcal{F}_f = f^{-1}(\mathcal{F}_f)$, apply this lemma to the families

$$\left\{ f^{\circ(m+1)} : f^{-1}(\mathcal{F}_f) \to \widehat{\mathbb{C}} \right\} \quad \text{and} \quad \left\{ f^{\circ m} : \mathcal{F}_f \to \widehat{\mathbb{C}} \right\}$$

and to the holomorphic map $f : f^{-1}(\mathcal{F}_f) \to \mathcal{F}_f$. Applying f to both sides of the equality $\mathcal{F}_f = f^{-1}(\mathcal{F}_f)$, we deduce that $f(\mathcal{F}_f) = \mathcal{F}_f$. □

Proposition 2.4 *The Julia and Fatou sets of the k-th iterate $f^{\circ k}$ are equal to the Julia and Fatou sets of f:*

$$\mathcal{J}_{f^{\circ k}} = \mathcal{J}_f \quad \text{and} \quad \mathcal{F}_{f^{\circ k}} = \mathcal{F}_f$$

Proof It is enough to prove this proposition for the Fatou sets. If the family of iterates $f^{\circ n}$ is normal on an open set U, then the subfamily $f^{\circ(kn)}$ is also normal. Hence, $\mathcal{F}_f \subseteq \mathcal{F}_{f^{\circ k}}$.

Now, assume that $z \in \mathcal{F}_{f^{\circ k}}$ and let U be a neighborhood of z on which the family of iterates $f^{\circ(kn)}$ is normal. For any sequence $(n_i)_{i \in \mathbb{N}}$, there exists an integer N such that $n_i = N \mod (k)$ for infinitely many i's. Let us extract this subsequence which, with an abuse of notations, we still denote by n_i. By assumption, the sequence $f^{\circ(n_i-N)} : U \to \widehat{\mathbb{C}}$ is normal and we can extract a subsequence converging to a map $g : U \to \widehat{\mathbb{C}}$. As $f^{\circ n_i} = f^{\circ N} \circ f^{\circ(n_i-N)}$, we can extract a subsequence of the sequence $f^{\circ n_i} : U \to \widehat{\mathbb{C}}$ converging to $f^{\circ N} \circ g : U \to \widehat{\mathbb{C}}$. This shows that the family $\left(f^{\circ n} : U \to \widehat{\mathbb{C}}\right)_{n \in \mathbb{N}}$ is normal and that $\mathcal{F}_{f^{\circ k}} \subseteq \mathcal{F}_f$. □

2.3 Conjugacy

In every part of mathematics, the *morphisms* are the maps that preserve the structure. Thus the morphisms of a dynamical system $f : X \to X$ to a dynamical system $g : Y \to Y$ are the maps sending sequences of iterates of f to sequences of iterates of g. This is just what maps $\phi : X \to Y$ such that $\phi \circ f = g \circ \phi$ do; such maps are called *semi-conjugacies* between f and g. Indeed, we see by induction that then $\phi \circ f^{\circ n} = g^{\circ n} \circ \phi$ for all $n \geq 1$: by hypothesis it is true for $n = 1$, and if it is true for $n - 1$, then

$$\phi \circ f^{\circ n} = \phi \circ f \circ f^{\circ(n-1)} = g \circ \phi \circ f^{\circ(n-1)} = g \circ g^{\circ(n-1)} \circ \phi = g^{\circ n} \circ \phi.$$

When ϕ is invertible, $\phi \circ f = g \circ \phi$ can be rewritten $f = \phi^{-1} \circ g \circ \phi$, so f and g are *conjugate*: conjugacies are the isomorphisms of dynamical systems. When studying dynamical systems, we constantly try to construct conjugacies or semi-conjugacies to simpler dynamical systems, linear ones in particular.

Definition 2.5 (Conjugate Rational Maps) Two rational maps f and g are *topologically* (respectively *analytically*) *conjugate* if there exists a homeomorphism $\phi : \widehat{\mathbb{C}} \to \widehat{\mathbb{C}}$ (respectively an analytic isomorphism) such that $\phi \circ f = g \circ \phi$.

One should think of ϕ as a change of coordinates on $\widehat{\mathbb{C}}$. The only analytic isomorphisms of $\widehat{\mathbb{C}}$ are the Möbius transformations $z \mapsto (az+b)/(cz+d)$ and the only isomorphisms of $\mathbb{C}$ are the affine maps $z \mapsto az + b$. Being a zero or a pole of a rational map depends on the coordinates, whereas being a fixed point or a critical point of a rational map does not depend on the choice of coordinates.

Proposition 2.5 *If f and g are conjugate (topologically or analytically) by $\phi : \widehat{\mathbb{C}} \to \widehat{\mathbb{C}}$, then $\phi(\mathcal{F}_f) = \mathcal{F}_g$ and $\phi(\mathcal{J}_f) = \mathcal{J}_g$.*

Proof The family $(f^{\circ n} : U \to \widehat{\mathbb{C}})_{n \in \mathbb{N}}$ is normal if and only if the family $(\phi \circ f^{\circ n} \circ \phi^{-1} : \phi(U) \to \widehat{\mathbb{C}})_{n \in \mathbb{N}}$ is normal. □

Example We will see below that the Newton's methods

$$N_P : z \mapsto z - \frac{P(z)}{P'(z)} \quad \text{and} \quad N_Q : z \mapsto z - \frac{Q(z)}{Q'(z)}$$

are analytically conjugate as soon as $Q(z) = \lambda P(\alpha z + \beta)$.

Exercise 2.6 Let $a \neq b$ be complex numbers and set $P(z) := (z-a)(z-b)$.

1. Show that the Newton's method N_P is conjugate to $f : w \mapsto w^2$ via the isomorphism $z \mapsto w := (z-a)/(z-b)$.
2. Determine the Julia set of N_P and the set $\omega(z)$ for $z \in \widehat{\mathbb{C}} \setminus \mathcal{J}_{N_P}$.

2.4 Periodic Point and Critical Points

Periodic points and critical points are important objects in holomorphic dynamics as we shall see in this section.

2.4.1 Periodic Points

Definition 2.6 (Multiplier at a Fixed Point) A *fixed point* of f is a point α such that $f(\alpha) = \alpha$. The derivative of f at α is an endomorphism of the tangent line $D_\alpha f : T_\alpha \widehat{\mathbb{C}} \to T_\alpha \widehat{\mathbb{C}}$, thus a multiplication by a number $\lambda \in \mathbb{C}$. The eigenvalue λ of the endomorphism $D_\alpha f : T_\alpha \widehat{\mathbb{C}} \to T_\alpha \widehat{\mathbb{C}}$ is called the *multiplier* of f at α. The fixed point is called

- *superattracting* if $\lambda = 0$;
- *attracting* if $0 < |\lambda| < 1$;
- *repelling* if $|\lambda| > 1$;
- *indifferent* if $|\lambda| = 1$, and more precisely
  - *parabolic* if λ is a root of 1,
  - *elliptic* otherwise, that is if $\lambda = e^{2\pi i \theta}$ with $\theta \in \mathbb{R} \setminus \mathbb{Q}$.

Exercise 2.7 Let α be a fixed point of a holomorphic map f.

1. Show that if $\alpha \neq \infty$, then the multiplier of f at α is $\lambda = f'(\alpha)$.
2. Show that if $\alpha = \infty$ and if $f(z) \sim az$ as $z \to \infty$, then the multiplier is $\lambda = 1/a$, whereas if $f(z) \sim az^k$ with $k \geq 2$, the multiplier is $\lambda = 0$.
3. Show that if f is a polynomial of degree $d \geq 2$, then ∞ is a superattracting fixed point of f.

The multiplier at a fixed point is invariant under analytic conjugacy. In other words, if $\phi : (\widehat{\mathbb{C}}, \alpha) \to (\widehat{\mathbb{C}}, \beta)$ is a local isomorphism conjugating $f : (\widehat{\mathbb{C}}, \alpha) \to (\widehat{\mathbb{C}}, \alpha)$ to $g : (\widehat{\mathbb{C}}, \beta) \to (\widehat{\mathbb{C}}, \beta)$, then the multiplier of f at α is equal to the multiplier of g at β. Indeed, the derivative of $\phi \circ f = g \circ \phi$ at α is

$$D_\alpha \phi \circ D_\alpha f = D_\beta g \circ D_\alpha \phi.$$

In other words, $D_\alpha \phi$ conjugates $D_\alpha f : T_\alpha \widehat{\mathbb{C}} \to T_\alpha \widehat{\mathbb{C}}$ to $D_\beta g : T_\beta \widehat{\mathbb{C}} \to T_\beta \widehat{\mathbb{C}}$ and the two endomorphisms have the same eigenvalue.

Definition 2.7 (Periodic Points) A *periodic point* of f is a fixed point of $f^{\circ p}$ for some $p \geq 1$. The smallest such integer $p \geq 1$ is called the period of α. In this case, we say that $\langle \alpha, f(\alpha), \ldots, f^{\circ (p-1)}(\alpha) \rangle$ is a *cycle of f*.

Proposition 2.6 *If $\langle \alpha_1, \alpha_2, \ldots, \alpha_p \rangle$ is a cycle, then the multiplier of $f^{\circ p}$ at all points of the cycle is the same.*

Proof According to the Chain Rule, the derivative of $f^{\circ p}$ at any point of the cycle is the composition of the derivatives of f, as linear transformations, along the cycle. If the derivative of f vanishes at a point of the cycle, then the multiplier is 0 at all points of the cycle. If the derivative never vanishes, then f is locally invertible at each point of the cycle and f conjugates $f^{\circ p}$ at α_i to $f^{\circ p}$ at α_{i+1}. Thus, the multiplier of $f^{\circ p}$ at α_i is equal to the multiplier of $f^{\circ p}$ at α_{i+1}. □

Definition 2.8 (Multiplier Along a Cycle) The *multiplier* of f along a cycle of period p is the multiplier of $f^{\circ p}$ at any point of the cycle. The cycle is *superattracting (respectively attracting, indifferent or repelling)* for f if the points of the cycle are superattracting (respectively attracting, indifferent or repelling) fixed points of $f^{\circ p}$.

If $\langle \alpha_1, \alpha_2, \ldots, \alpha_p \rangle$ is a cycle of period p avoiding ∞, then its multiplier is given by the product

$$\lambda = (f^{\circ p})'(\alpha_1) = \prod_{i=1}^{p} f'(\alpha_i).$$

Proposition 2.7 *Let $f : \widehat{\mathbb{C}} \to \widehat{\mathbb{C}}$ be a rational map, and let α be a periodic point of f. If α is superattracting or attracting, then $\alpha \in \mathcal{F}_f$. If α is repelling, then $\alpha \in \mathcal{J}_f$.*

Proof Replacing f by $f^{\circ p}$ if necessary, we may assume that α is a fixed point of f. Conjugating by a Möbius transformation if necessary, we may assume that $\alpha = 0$. If α is (super)attracting, i.e., $|f'(\alpha)| < 1$, then there exists a bounded neighborhood U of α which is mapped into itself. In particular, for all $n \in \mathbb{N}$, $f^{\circ n}(U) \subset U$.

Hence, the family $(f^{\circ n}: U \to \widehat{\mathbb{C}})_{n \in \mathbb{N}}$ is normal, which implies that $\alpha \in \mathcal{F}_f$. If α is repelling, then

$$(f^{\circ n})'(\alpha) = [f'(\alpha)]^n \xrightarrow[n \to +\infty]{} \infty.$$

So, the family $(f^{\circ n})_{n \in \mathbb{N}}$ cannot be normal in a neighborhood of α. □

A cycle is superattracting if and only if it contains a point at which the derivative of f vanishes, i.e., a *critical point* of f.

2.4.2 Critical Points and Critical Values

We will see that the dynamical properties of a rational map $f : \widehat{\mathbb{C}} \to \widehat{\mathbb{C}}$ are strongly related to the dynamical behaviour of the critical points of f. For example, the Julia set of a polynomial f is connected if and only if the orbit of every critical point of f is bounded.

Definition 2.9 (Critical Points and Critical Values) Let U and V be two Riemann surfaces and $f : (U, x) \to (V, y)$ be an analytic map which is not locally constant at x. The point x is a *critical point* if the derivative $D_x f : T_x U \to T_y V$ is zero. In that case, y is a *critical value*.

More precisely, in local coordinates z vanishing at x and w vanishing at $y = f(x)$, the expression of f is of the form

$$z \mapsto aw^k + O(w^{k+1}) \quad \text{for some integer } k \geq 1 \text{ and some complex number } a \neq 0.$$

The integer k does not depend on the choice of coordinates and is called the *local degree* of f at x. The point x is a critical point if and only if $k \geq 2$. In that case, x is called a *critical point of f of multiplicity $k - 1$*.

> **Example** The map $z \mapsto z^3$ has a critical point at 0. The local degree of the map at 0 is 3, and the multiplicity of this critical point is 2.

We shall use the following result which is a consequence of the Riemann-Hurwitz Formula (which will not be proved in those notes).

Definition 2.10 (Euler Characteristic) If U is an open subset of $\widehat{\mathbb{C}}$ with m boundary components, the *Euler characteristic of U* is

$$\chi(U) := 2 - m.$$

Theorem 2.2 *Let* $f : \widehat{\mathbb{C}} \to \widehat{\mathbb{C}}$ *be a rational map of degree d, $V \subseteq \widehat{\mathbb{C}}$ be an open set and $U := f^{-1}(V)$. Then,*

$$\chi(U) = d \cdot \chi(V) - N$$

where N is the number of critical points of f in U, counting multiplicities.

Exercise 2.8 Let $f : \widehat{\mathbb{C}} \to \widehat{\mathbb{C}}$ be a rational map of degree d.

1. Prove that f has $2d - 2$ critical points counting multiplicities.
2. Prove that f has at least 2 distinct critical values.
3. Prove that if f is a polynomial, then ∞ is a critical point of multiplicity $d - 1$ and that there are $d - 1$ critical points in $\mathbb{C}$ counting multiplicities.

2.5 Description of the Julia Set

2.5.1 Topology of the Julia Set

Proposition 2.8 *The Julia set $\mathcal{J}_f$ is non empty and compact.*

Proof If the Julia set were empty, there would be a subsequence of $f^{\circ m}$ converging uniformly on $\widehat{\mathbb{C}}$; let f_0 be the limit. The rational map f_0 must have some finite degree. Since the degree is an invariant of homotopy, the approximating iterates $f^{\circ m_i}$ must have the same degree. But the degree of $f^{\circ m_i}$ is d^{m_i}, which leads to a contradiction.

Evidently the Fatou set $\mathcal{F}_f$ is open, hence the Julia set $\mathcal{J}_f$ is closed. Since $\widehat{\mathbb{C}}$ is compact, the Julia set $\mathcal{J}_f$ is also compact. □

Proposition 2.9 *Either the Julia set $\mathcal{J}_f$ has empty interior, or it is the entire Riemann sphere.*

Proof Suppose $U \subset \mathcal{J}_f$ is an open subset of $\widehat{\mathbb{C}}$. The family $\{f^{\circ m} : U \to \widehat{\mathbb{C}}\}$ must avoid $\mathcal{F}_f$, which is open and will contain more than two points if it is not empty. But this would make $\{f^{\circ m} : U \to \widehat{\mathbb{C}}\}$ normal, a contradiction. Thus if the interior of $\mathcal{J}_f$ is non empty, $\mathcal{J}_f$ is the entire Riemann sphere. □

We have seen that the Julia set $\mathcal{J}_f$ is completely invariant. We will now determine whether there are any other completely invariant closed sets than $\mathcal{J}_f$. For this purpose, we will use the Riemann Hurwitz formula.

Proposition 2.10 *Let E be a completely invariant closed set. If E is finite, then E contains at most two points.*

Proof If E is finite, then its complement U is connected and completely invariant. Its Euler characteristic $\chi(U) = 2 - \text{card}(E)$ is finite. Let us denote by N the number of critical points of f in U counted with multiplicity. The Riemann-Hurwitz formula applied to $f : U \to U$ gives

$$\chi(U) = d \cdot \chi(U) - N.$$

Thus

$$\chi(U) = \frac{N}{d-1} \geq 0,$$

which leads to $\text{card}(E) \leq 2$. □

A union of completely invariant sets is still completely invariant, therefore the following definition makes sense.

Definition 2.11 The exceptional set E_f is the largest finite completely invariant set.

Remark 2.2 The grand-orbit of a point is a totally invariant set. If it is a finite set, then, by definition of the exceptional set E_f, we have $z \in E_f$.

Observe that when two rational maps $f : \widehat{\mathbb{C}} \to \widehat{\mathbb{C}}$ and $g : \widehat{\mathbb{C}} \to \widehat{\mathbb{C}}$ are conjugate by a homeomorphism ϕ, then the homeomorphism ϕ sends the forward (resp. backward) orbit of a point $z \in \widehat{\mathbb{C}}$ under the action of f to the forward (resp. backward) orbit of $\phi(z)$ under the action of g. Hence, f-invariant sets are mapped by ϕ to g-invariant sets. Thus, the largest finite completely f-invariant set is mapped by ϕ to the largest finite completely g-invariant set. In other words, when it is not empty, the exceptional set E_f is mapped by ϕ to the exceptional set E_g.

The exceptional set E_f is a minor irritant, which will come back to annoy us on several occasions. The following description should remove any doubts about its lack of significance.

Proposition 2.11 *Up to conjugacy, the exceptional set E_f is not empty in precisely two cases:*

- *when $f(z) = z^d$ with $|d| \geq 2$, $E_f = \{0, \infty\}$ and*
- *when f is a polynomial but is not conjugate to $z \mapsto z^d$, $E_f = \{\infty\}$.*

Proof It is clear from the construction that E_f is completely invariant; so if it has only one point, this point must be fixed, and if it consists of two points, these must be fixed or exchanged by f. We will examine these three cases one at a time.

If E_f has just one point a, move it to ∞, for example by making the change of coordinates $w = 1/(z - a)$. Then $f = P/Q$ is a rational map such that infinity is its only inverse image. Hence, the polynomial Q has no roots, it is constant, and f is a polynomial.

If E_f has two points a and b, both fixed, then we can put one at ∞ and the other at 0 (make the change of coordinates $w = (z - b)/(z - a)$). We see that f is a polynomial and that 0 is the only root of the polynomial equation $f(w) = 0$, so $f(w) = \lambda w^d$ for some λ. Choose a number ν such that $\nu^{d-1} = \lambda$, and make the change of coordinates $z = \nu w$. Then, it is easy to see that the expression of f becomes $z \mapsto z^d$.

If E_f has two points exchanged by f, put one at ∞ and the other at 0, and write $f = P/Q$. Since ∞ is the only inverse image of 0, we see that P has no roots, so it is a constant, which we may take to be 1. Since 0 is the only inverse image of ∞, we see that $Q(w) = \lambda w^d$ for some $\lambda \neq 0$, so that $f(w) = 1/(\lambda w^d)$. We can do a change of coordinate $z = \nu w$, where $\nu^{d+1} = \lambda$. Again it is easy to check that the expression of f becomes $z \mapsto z^{-d}$. □

Proposition 2.12 *The exceptional set E_f is always contained in the Fatou set $\mathcal{F}_f$.*

Proof In all three cases, E_f is a union of superattracting cycles. Proposition 2.7 shows that such cycles are always contained in the Fatou set $\mathcal{F}_f$. □

We can now give a new characterization of the Julia set.

Proposition 2.13 *The Julia set of a rational map is the smallest completely invariant closed set containing at least three points.*

Proof The Julia set is a completely invariant closed set. It contains at least three points, since otherwise it would be contained in the exceptional set $E_f \subset \mathcal{F}_f$. Furthermore, if E is a completely invariant closed set containing at least 3 points, the complement $\Omega = \widehat{\mathbb{C}} \setminus E$ of E is a completely invariant open set omitting more than three points. Hence, by Montel's theorem, the family of iterates $f^{\circ n}$ is normal on Ω. Thus $\Omega \subset \mathcal{F}_f$ and $\mathcal{J}_f \subset E$. □

The following proposition asserts that if z does not belong to the exceptional set E_f, then the closure of the backward orbit of z,

$$\mathcal{O}^-(z) := \bigcup_{m \geq 0} f^{-m}\{z\},$$

contains the Julia set $\mathcal{J}_f$.

Proposition 2.14 *For any $z \notin E_f$, we have*

$$\mathcal{J}_f \subseteq \overline{\mathcal{O}^-(z)}.$$

Proof Take a point $z_0 \in \mathcal{J}_f$ and let U be an arbitrary neighborhood of z_0. We must show that the set $\Omega = \bigcup_{n \in \mathbb{N}} f^{\circ n}(U)$ contains z. We will show that its complement $E = \widehat{\mathbb{C}} \setminus \Omega$ is contained in the exceptional set E_f. Indeed, Ω is a forward invariant open set which omits at most two points, since otherwise the family $f^{\circ n}$ would be normal on U. Thus, $E = \widehat{\mathbb{C}} \setminus \Omega$ is a backward invariant set. Since E is finite, it is permuted by f, so it is forward invariant. Hence, by definition of the exceptional set, it is contained in E_f. □

Proposition 2.15 *The Julia set $\mathcal{J}_f$ is perfect.*

Proof Since the Julias set is closed, it suffices to prove that it has no isolated points. The Julia set $\mathcal{J}_f$ is not finite. Thus, it contains a non-isolated point z_0. Since the

2 Introduction to Fatou Components in Holomorphic Dynamics

backward orbit of z_0 is dense in $\mathcal{J}_f$, any point in $\mathcal{J}_f$ is non-isolated. This proves that $\mathcal{J}_f$ is perfect, hence uncountable. □

2.5.2 Julia Sets and Periodic Points

Fatou and Julia both proved that the Julia set is the closure of the set of repelling periodic points. This "dynamical" definition of the Julia set is probably more natural to specialists of dynamical systems, and relates holomorphic dynamics to "Axiom A attractors" and hyperbolic dynamics. Fatou's and Julia's proofs are different but both involve results which will be proved later. Fatou's proof is based on the fact that there are only finitely many non-repelling cycles. Julia's proof is based on the existence of a repelling cycle.

In these notes, we will only prove the following weaker result which follows directly from Montel's Theorem. This result is the first step in Fatou's proof. The density of repelling cycles in the Julia set then follows from the finiteness of non-repelling cycles.

Proposition 2.16 *The closure of the set of periodic points contains the Julia set.*

Proof Let $z_0 \in \mathcal{J}_f$ be arbitrary. Since $\mathcal{J}_f$ is perfect, we may assume that z_0 is not a critical value of $f^{\circ 2}$, so there exist a neighborhood U of z_0 and three branches g_1, g_2, g_3 of f^{-2} with mutually disjoint images. We are considering the second iterate because if $d = 2$, there would only be two branches of f^{-1}.

Now consider the family of maps

$$h_m(z) := \frac{f^{\circ m}(z) - g_1(z)}{f^{\circ m}(z) - g_2(z)} \bigg/ \frac{g_3(z) - g_1(z)}{g_3(z) - g_2(z)}.$$

If $f^{\circ m}(z) \neq g_i(z), i = 1, 2, 3$ for all z in some neighborhood U of z, then the maps $h_m(z)$ are never $0, 1$ or ∞ in U, hence they form a normal family. But then the family $f^{\circ m}$ is also normal, since we can express $f^{\circ m}$ in terms of h_m:

$$f^{\circ m} = \frac{g_1(g_3 - g_2) - g_2(g_3 - g_1)h_m}{(g_3 - g_2) - (g_3 - g_1)h_m}.$$

This is a contradiction, so in every neighborhood of z there are roots of $f^{\circ m}(z) = g_i(z)$ for some i and some m, i.e., roots of $f^{\circ(m+2)}(z) = z$. □

2.5.3 Connectivity of Julia Sets of Polynomials

Here is a first relation between the behaviour of critical orbits and the topology of the Julia set.

Theorem 2.3 (Fatou) *The filled-in Julia set $\mathcal{K}_f$ of a polynomial f is connected if and only if all critical points of f belong to $\mathcal{K}_f$.*

Remark 2.3 Note that since the Julia set $\mathcal{J}_f$ is the boundary of $\mathcal{K}_f$, $\mathcal{J}_f$ is connected if and only if $\mathcal{K}_f$ is connected.

Proof Since f is a polynomial of degree $d \geq 2$, we have that $f(z)/z$ tends to infinity as z tends to infinity, hence there exists $R > 0$ so that $|f(z)| \geq 2|z|$ for $|z| > R$. Set $U_0 := \widehat{\mathbb{C}} \setminus \overline{D}_R$ and define inductively $U_{n+1} := f^{-1}(U_n)$. By the Riemann-Hurwitz formula, we have

$$\chi(U_{n+1}) = d\chi(U_n) - (d-1),$$

where the number $d - 1$ is the multiplicity of the critical point at ∞. Since U_0 is simply connected, we have $\chi(U_0) = 1$, and by induction, we see that when all the critical points of f belong to $\mathcal{K}_f$, then $\chi(U_n) = 1$ for all $n \geq 0$. It follows that

$$\Omega_\infty := \bigcup_{n \geq 0} U_n$$

is simply connected and its complement $\mathcal{K}_f$ is connected.

Conversely, let $\Omega_\infty \subset \widehat{\mathbb{C}}$ be the connected component of the Fatou set $\mathcal{F}_f$ which contains ∞. Since the only inverse image of ∞ is ∞ itself, $f^{-1}(\Omega_\infty) = \Omega_\infty$. Applying the Riemann-Hurwitz formula to $f : \Omega_\infty \to \Omega_\infty$, we get

$$\chi(\Omega_\infty) = d\chi(\Omega_\infty) - (d-1) - k,$$

where the number $d-1$ is contributed by the critical point at ∞, and the number k is the number of other critical points in Ω_∞ (i.e., not in $\mathcal{K}_f$) counted with multiplicity. If $\chi(\Omega_\infty)$ is finite, then $\chi(\Omega_\infty) = 1 + k/(d-1)$. But the Euler characteristic of a connected plane domain is at most 1. So $\chi(\Omega_\infty) = 1$ if no critical point is attracted to ∞; otherwise $\chi(\Omega_\infty)$ is infinite (Fig. 2.3). □

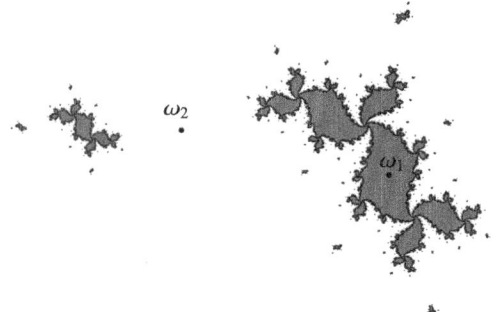

Fig. 2.3 The filled-in Julia set of a cubic polynomial. The critical point ω_1 is in $\mathcal{K}_f$. The other critical point ω_2 is in the basin of infinity. The Julia set $\mathcal{J}_f$ is not connected

2.6 Description of the Fatou Set

We now return to the case where f is a rational map, not necessarily a polynomial. Our goal is to give a classification of periodic Fatou components.

Definition 2.12 A map $f : U \to V$ is a *proper map* if for any compact set $K \subseteq V$, the preimage $f^{-1}(K)$ is compact.

Proposition 2.17 *If U is a connected component of $\mathcal{F}_f$, then $f(U)$ is also a connected component of $\mathcal{F}_f$ and $f : U \to f(U)$ is a proper map.*

Proof The rational map $f : \widehat{\mathbb{C}} \to \widehat{\mathbb{C}}$ is a proper map. Hence, for any open set $V \subset \widehat{\mathbb{C}}$, $f : f^{-1}(V) \to V$ is a proper map. In particular, this holds for $U = \mathcal{F}_f$. But since $f^{-1}(\mathcal{F}_f) = \mathcal{F}_f$ by Proposition 2.3, we get that $f : \mathcal{F}_f \to \mathcal{F}_f$ is a proper map. Then the proposition follows, since the restriction of a proper map to a connected component of the domain is still proper. □

Definition 2.13 A connected component U of $\mathcal{F}_f$ is called a *Fatou component of f*.

Definition 2.14 A Fatou component is *periodic* if there exists an integer $k \geq 1$ such that

$$f^{\circ k}(U) = U.$$

Note that if U is periodic of period k for f, then it is invariant for $f^{\circ k}$.

Definition 2.15 An invariant Fatou component U of f is called

- a *(super)attracting domain* if there is a (super)attracting fixed point $\alpha \in U$ ($f(\alpha) = \alpha$ and $0 \leq |f'(\alpha)| < 1$) and the sequence $f^{\circ n}$ converges uniformly to α on every compact subset of U;
- a *parabolic domain* if there is a fixed point $\alpha \in \partial U$ with $f'(\alpha) = 1$, and the sequence $f^{\circ n}$ converges uniformly to α on every compact subset of U;
- a *Siegel disk* if it is simply connected, and if there exists an isomorphism $h : U \to \mathbb{D}$ such that

$$h \circ f \circ h^{-1}(z) = e^{2i\pi\theta} z$$

with $\theta \in \mathbb{R} \setminus \mathbb{Q}$;
- a *Herman ring* if it is doubly connected, and if there exists a radius r and an isomorphism

$$h : U \to \mathcal{A}_r := \{z \in \mathbb{C} \,;\, r < |z| < 1\},$$

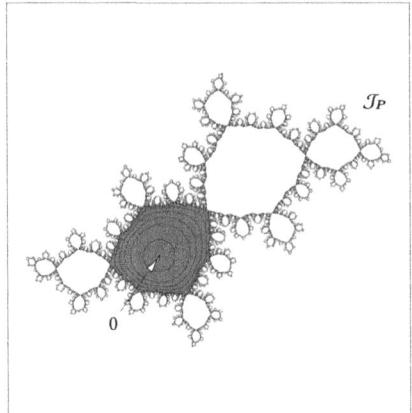 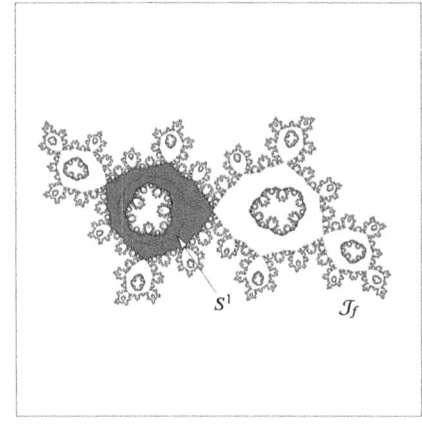

Fig. 2.4 Left: the polynomial $P: z \mapsto e^{i\pi(\sqrt{5}-1)}z + z^2$ has a Siegel disk (colored grey). We have drawn the orbits of some points in the Siegel disk. Each orbit accumulates on a $\mathbb{R}$-analytic circle. Right: if $t \in \mathbb{R}/\mathbb{Z}$ is chosen carefully enough, the rational map $f(z) = e^{2i\pi t}z^2(z-4)/(1-4z)$ has a Herman ring. It leaves the circle S^1 invariant and is conjugate to an irrational rotation on S^1. The picture is drawn for $t = 0.61517321588\ldots$

such that

$$h \circ f \circ h^{-1}(z) = e^{2i\pi\theta}z$$

with $\theta \in \mathbb{R}\setminus\mathbb{Q}$.

Remark 2.4 Clearly, the possibilities are mutually exclusive. Theorem 2.4 below asserts that these are the only possibilities and it is known that they all occur. The existence of Siegel disks was proved in 1942 by Siegel, and the existence of Herman rings was proved by Herman in 1981.

Figure 2.4 shows the Julia set of a polynomial having a Siegel disk and the Julia set of a rational map having a Herman ring.

Proposition 2.18 *If U is an invariant Fatou component of a rational map f of degree $d \geq 2$, then it is either simply connected, or doubly connected or its complement has infinitely many connected components. It is doubly connected if and only if it is a Herman ring.*

Proof We can apply the Riemann-Hurwitz formula to $f : U \to U$. It gives

$$\chi(U) = d\chi(U) - n$$

where d is the degree of $f : U \to U$ and n is the number of critical points of f in U, counting multiplicities. If $\chi(U)$ is finite, it follows that $\chi(U) = n/(d-1)$ is non negative, and since U is not the entire Riemann sphere, we have $\chi(U) = 0$ or 1.

Moreover, if $\chi(U) = 0$, i.e., if U is doubly connected, then $n = 0$ and $f : U \to U$ is a covering map. In addition, the degree of $f : U \to U$ is 1 since it preserves the modulus of U. Thus, $f : U \to U$ is an isomorphism. This isomorphism cannot be of finite order, since otherwise an iterate of f would have infinitely many fixed points. Therefore U is a Herman ring. □

Remark 2.5 If f is a polynomial and U is a bounded Fatou component, then thanks to the maximum principle U is simply connected. In particular, there are no Herman rings.

The following result is due to Fatou. We will not give the detailed proof.

Theorem 2.4 (Classification of Invariant Fatou Components) *Let $f : \widehat{\mathbb{C}} \to \widehat{\mathbb{C}}$ be a rational function of degree $d \geq 2$. An invariant Fatou component U of f is a (super)attracting domain, a parabolic domain, a Siegel disk or a Herman ring.*

Proof The classification starts by studying what are the possible limit values of the sequence of iterates $(f^{\circ n} : U \to U)$.

Case 1. There is a non constant limit value.
Case 1.1. There is a sequence (n_k) such that $f^{\circ n_k} \to \text{Id}$. This is a first place where we do not provide the details. One proves that

- either f has finite order (which is not the case since if $f^{\circ p} = \text{Id}$ on U for some $p \geq 1$, then $f^{\circ p} = \text{Id}$ on $\widehat{\mathbb{C}}$ by analytic continuation, which is not possible since $f^{\circ p}$ is a rational map of degree $d^p > 1$);
- or U is simply connected, in which case it is a Siegel disk;
- or U is doubly connected, in which case it is a Herman ring.

Case 1.2. There is a sequence (n_k) such that $f^{\circ n_k} \to \phi$ with ϕ non constant. By continuity, ϕ takes its values in $\overline{U}$. According to the Hurwitz Theorem, ϕ takes its values in U. Extracting a subsequence, we may assume that $n_{k+1} - n_k \to +\infty$ and that $f^{\circ(n_{k+1}-n_k)} \to \psi$. Then, $\psi \circ \phi = \phi$, so that ψ is the identity on the image of ϕ, thus on the whole component U by analytic continuation. In other words, we are in the previous situation : U is a Siegel disk or a Herman ring.

Case 2. Every limit value of the sequence $(f^{\circ n})$ is a constant. Note that if the sequence $(f^{\circ n_k})$ converges to α, then α is a fixed point of f. Indeed, since $f(z) \in U$ for all $z \in U$, we have that

$$\alpha \xleftarrow[k \to +\infty]{} f^{\circ n_k}(f(z)) = f(f^{\circ n_k}(z)) \xrightarrow[k \to +\infty]{} f(\alpha).$$

Note also that any point $z \in U$ can be joined to its image $f(z) \in U$ by a path compactly contained in U. It follows that the set of limit values of the sequence $(f^{\circ n}(z))$ is a continuum. Since f has finitely many fixed points, the whole sequence $(f^{\circ n})$ converges to a fixed point α of f.

Case 2.1. If $\alpha \in U$, the derivative of f^{on} at α tends to 0 as $n \to \infty$. As a consequence, α is a (super)attracting fixed point and U is a (super)attracting domain.

Case 2.2. If $\alpha \in \partial U$, then α cannot be (super)attracting since it belongs to the Julia set $\mathcal{J}_f$. It cannot be repelling since the sequence (f^{on}) converges to α. So, it is indifferent. We claim that $D_\alpha f = 1$ and so, U is a parabolic domain.

Fix $z_0 \in U$ and set $z_n = f^{on}(z_0)$. Let $\tilde{\rho} : [0, 1] \to U$ be a continuous path with $\tilde{\rho}(0) = z_0$ and $\tilde{\rho}(1) = z_1$. Then we can extend $\tilde{\rho}$ to a continuous path $\rho : [0, +\infty) \to U$ by setting $\rho(n + t) = f^{on} \circ \tilde{\rho}(t)$ for all $n \in \mathbb{N}$ and all $t \in [0, 1)$. Then $\lim_{t \to +\infty} \rho(t) = \alpha$ and ρ satisfies $\rho(t + 1) = f \circ \rho(t)$ for all $t \in [0, +\infty)$ and we conclude using the following lemma.

Lemma 2.2 (Snail Lemma) *Let V be a neighborhood of the origin in $\mathbb{C}$ and let $f : V \to f(V)$ be a holomorphic function such that $f(\alpha) = \alpha$ and $f'(\alpha) \neq 0$. If there is a continuous path $\rho : [0, +\infty) \to V \setminus \{\alpha\}$ such that $\lim_{t \to +\infty} \rho(t) = \alpha$ and $f \circ \rho(t) = \rho(t + 1)$ for all $t \in [0, +\infty)$ then either $|f'(\alpha)| < 1$ or $f'(\alpha) = 1$.* □

This is a second place where we do not provide the details. For a proof of this result see for example [50, Lemma 16.2]. □

The following theorem, due to Fatou (1905), is probably the result which started the entire field of holomorphic dynamics.

Theorem 2.5 *A (super)attracting domain always contains at least one critical point.*

Proof If U is a superattracting domain, then it contains a fixed critical point, and the result is obvious. If U is an attracting domain, it contains an attracting fixed point α with multiplier λ. We can show that there is a map $\phi : (\mathbb{C}, 0) \to (\widehat{\mathbb{C}}, \alpha)$ which conjugates the multiplication by λ to f. If U does not contain any critical point, then ϕ extends to an entire map $\phi : \mathbb{C} \to U$ via the relation $\phi(\lambda z) = f \circ \phi(z)$. This contradicts the Liouville Theorem. □

Remark 2.6 The same is true for parabolic domains, but the proof is more difficult and requires a detailed study of the local theory near a fixed point with multiplier 1. A proof can be found for example in [50, Theorem 10.15].

Fatou's classification of invariant Fatou components clearly provides also a classification of *periodic* Fatou components since they are invariant for an iterate of the rational map, and even *pre-periodic* Fatou components, that is components U so that there exist $m \geq 0$ and $p \geq 1$ such that $f^{o(m+p)}(U) = f^{om}(U)$. The following fundamental result due to Sullivan completes the description for rational maps of degree $d \geq 2$.

Theorem 2.6 (Sullivan [60]) *Let $f : \widehat{\mathbb{C}} \to \widehat{\mathbb{C}}$ be a rational map of degree $d \geq 2$. Then every Fatou component of f is pre-periodic.*

The proof of Sullivan's non-wandering Theorem 2.6 strongly relies on the Ahlfors-Bers mesurable mapping Theorem for quasi-conformal functions and we refer to the original paper of Sullivan [60] for it. The recent notes [19] present a proof due to Adam Epstein and based on a density result of Bers for quadratic differentials. Such results are strongly one-dimensional and do not have an analogue in higher dimension, making impossible to mimic Sullivan's proof there. Besides this observation, little was known about this problem until recently. In Sect. 2.8 we will present some recent results on Fatou components in dimension 2.

2.7 Parabolic Implosion in Dimension 1

In this section, we recall the main ingredients in the proof that the Julia set $\mathcal{J}_f$ does not depend continuously on f for the Hausdorff topology on the space of compact subsets of $\mathbb{C}$. Lavaurs proved the following result (see also [27]).

Theorem 2.7 (Lavaurs [44]) *Assume* $f : \mathbb{C} \to \mathbb{C}$ *is a polynomial fixing* 0 *with* $f(z) = z + z^2 + O(z^3)$. *Then,*

$$\mathcal{J}_f \subsetneq \limsup_{\delta \to 0} \mathcal{J}_{f+\delta}.$$

The proof is based on the description of limits of iterates of $f + \delta$ in terms of maps called *Fatou coordinates* that will be introduced in Sect. 2.7.2. Those limits are called *Lavaurs maps*.

2.7.1 Parabolic Basin

In the rest of this section, we assume that $f : \mathbb{C} \to \mathbb{C}$ is a polynomial whose expansion near 0 is of the form

$$f(z) = z + z^2 + az^3 + O(z^4) \quad \text{with} \quad a \in \mathbb{C}.$$

To understand the local dynamics of f near 0, it is convenient to consider the change of coordinates $Z := -1/z$. In the Z-coordinate, the expression of f becomes

$$F(Z) = Z + 1 + \frac{b}{Z} + O\left(\frac{1}{Z^2}\right) \quad \text{with} \quad b := 1 - a.$$

In particular, if $R > 0$ is large enough, F maps the right half-plane $\{\Re(Z) > R\}$ into itself, so that if $r > 0$ is close enough to 0, f maps the disk $D(-r, r)$ into itself. In addition, the orbit under f of any point $z \in D(-r, r)$ converges to 0 tangentially to the real axis. Similarly, if $r > 0$ is sufficiently close to 0, there is a branch of f^{-1} which maps the disk $D(r, r)$ into itself and the orbit under that branch of f^{-1} of any point $z \in D(r, r)$ converges to 0 tangentially to the real axis.

Fig. 2.5 For the cubic polynomial $f(z) = z + z^2 + 0.95z^3$, the basin $\mathcal{B}_f$ (grey) is the interior of the filled-in Julia set $\mathcal{K}_f$

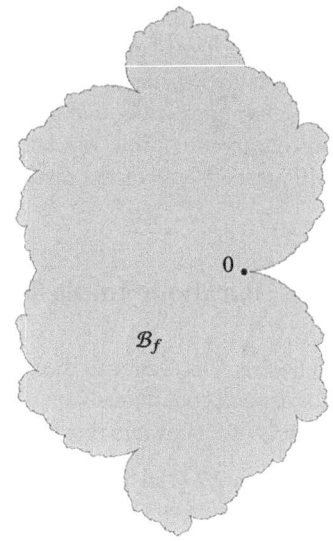

Definition 2.16 The *basin* $\mathcal{B}_f$ is the open set of points whose orbit under iteration of f intersects the disks $D(-r, r)$ for all $r > 0$.

Figure 2.5 shows the basin of attraction of the cubic polynomial $f(z) = z + z^2 + 0.95z^3$.

2.7.2 Fatou Coordinates

In order to understand further the local dynamics of f near 0, it is customary to use local *attracting and repelling Fatou coordinates*. In the case of a polynomial, those Fatou coordinates have global properties.

Proposition 2.19 *There exists a (unique) attracting Fatou coordinate* $\Phi_f : \mathcal{B}_f \to \mathbb{C}$ *which semi-conjugates* $f : \mathcal{B}_f \to \mathcal{B}_f$ *to the translation* $T_1 : \mathbb{C} \ni Z \mapsto Z+1 \in \mathbb{C}$:

$$\Phi_f \circ f = T_1 \circ \Phi_f.$$

and satisfies the normalization:

$$\Phi_f(z) = -\frac{1}{z} - b \log\left(-\frac{1}{z}\right) + o(1) \quad as \quad \Re\left(-\frac{1}{z}\right) \to +\infty.$$

Proof For $z \in \mathcal{B}_f$, set

2 Introduction to Fatou Components in Holomorphic Dynamics

$$Z := -\frac{1}{z}, \quad Z_n := -\frac{1}{f^{\circ n}(z)} \quad \text{and} \quad \Phi_n(z) := Z_n - n - b \log Z_n.$$

We have that

$$\Re(Z_n) \to +\infty \quad \text{and} \quad Z_{n+1} = Z_n + 1 + \frac{b}{Z_n} + O\left(\frac{1}{Z_n^2}\right),$$

so that

$$\frac{1}{Z_n} = O\left(\frac{1}{Z+n}\right) \quad \text{as} \quad n \to +\infty.$$

As a consequence,

$$\Phi_{n+1}(z) - \Phi_n(z) = Z_{n+1} - Z_n - 1 - b \log \frac{Z_{n+1}}{Z_n} = O\left(\frac{1}{Z_n^2}\right) = O\left(\frac{1}{(Z+n)^2}\right).$$

So, the sequence (Φ_n) converges to a limit $\Phi_f : \mathcal{B}_f \to \mathbb{C}$ which satisfies

$$\Phi_f(z) = \Phi_0(z) + O\left(\frac{1}{Z}\right) = -\frac{1}{z} - b \log\left(-\frac{1}{z}\right) + o(1) \quad \text{as} \quad \Re\left(-\frac{1}{z}\right) \to +\infty.$$

Passing to the limit on the relation $\Phi_n \circ f = T_1 \circ \Phi_{n+1}$ yields the required result. □

Figure 2.6 gives a rough idea of the behaviour of the Fatou coordinate Φ_f for the cubic polynomial $f(z) := z + z^2 + 0.95z^3$. The basin $\mathcal{B}_f$ contains the two critical points of f. Those points and their iterated preimages form the critical points of Φ_f. Denote by c^+ the critical point with positive imaginary part and by c^- its complex conjugate. Set $v^\pm := \Phi_f(c^\pm)$. Points $z \in \mathcal{B}_f$ are colored according to the location $\Phi_f(z)$: dark grey when $\Im(\Phi_f(z)) < \Im(v^-)$, light grey when $\Im(v^-) < \Im(\Phi_f(z)) < \Im(v^+)$ and medium grey when $\Im(v^+) < \Im(\Phi_f(z))$.

Proposition 2.20 *There exists a (unique) repelling Fatou parametrization $\Psi_f : \mathbb{C} \to \mathbb{C}$ which semi-conjugates $T_1 : \mathbb{C} \to \mathbb{C}$ to $f : \mathbb{C} \to \mathbb{C}$:*

$$\Psi_f \circ T_1 = f \circ \Psi_f.$$

and satisfies the normalization:

$$\Psi_f(Z) = -\frac{1}{Z + b \log(-Z) + o(1)} \quad \text{as} \quad \Re(Z) \to -\infty.$$

Proof Choose $r > 0$ sufficiently close to 0 so that there is a branch g of f^{-1} which maps the disk $D(r, r)$ into itself. For $z \in D(r, r)$, set

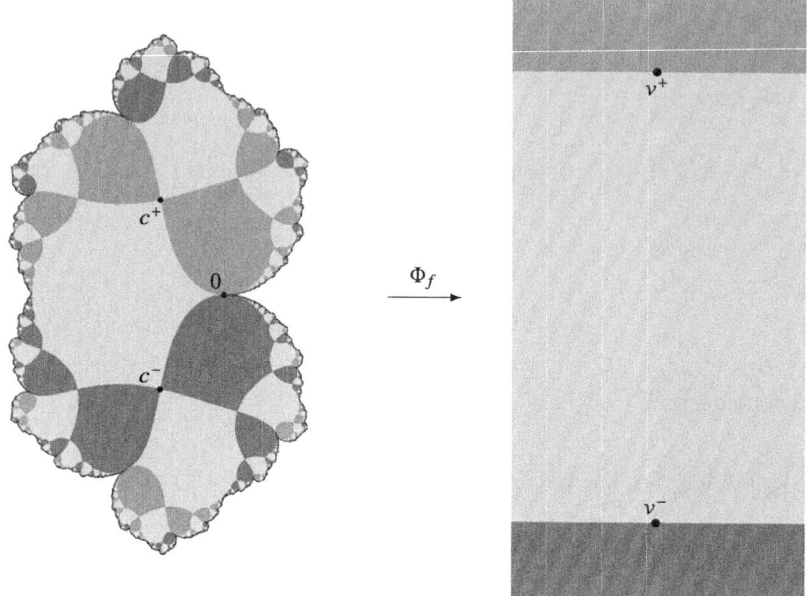

Fig. 2.6 Behavior of Φ_f for $f(z) := z + z^2 + 0.95z^3$. Right: three regions delimited by the horizontal lines passing through the critical values of Φ_f. Left: the basin $\mathcal{B}_f$ is tiled by the preimages of those three regions by Φ_f

$$Z := -\frac{1}{z}, \quad Z_n := -\frac{1}{g^{\circ n}(z)} \quad \text{and} \quad \Phi_n(z) := Z_n + n - b\log(-Z_n).$$

Note that

$$Z_n = Z_{n+1} + 1 + \frac{b}{Z_{n+1}} + O\left(\frac{1}{Z_{n+1}^2}\right).$$

So, as in the previous proof

$$\Phi_{n+1}(z) - \Phi_n(z) = Z_{n+1} - Z_n + 1 + b\log\frac{Z_{n+1}}{Z_n} = O\left(\frac{1}{Z_{n+1}^2}\right) = O\left(\frac{1}{(Z-n)^2}\right).$$

The sequence Φ_n converges in the left half-plane $\{\Re(Z) < -1/(2r)\}$ to a limit Φ_g which satisfies

$$\Phi_g(z) = \Phi_0(z) + O\left(\frac{1}{Z}\right) = Z - b\log(-Z) + o(1) \quad \text{as} \quad \Re\left(-\frac{1}{z}\right) \to -\infty.$$

Passing to the limit on the equation $\Phi_{n+1} \circ f = T_1 \circ \Phi_n$ shows that Φ_g conjugates f to T_1. The inverse Ψ_f of Φ_g conjugates T_1 to f and

$$Z = \Phi_g \circ \Psi_f(Z) = -\frac{1}{\Psi_f(Z)} - b\log\left(-\frac{1}{\Psi_f(Z)}\right) + o(1)$$

$$= -\frac{1}{\Psi_f(Z)} - b\log(-Z) + o(1)$$

as $\Re(Z) \to -\infty$. □

2.7.3 Lavaurs Maps

Definition 2.17 The *Lavaurs map* $\mathcal{L}_f : \mathcal{B}_f \to \mathbb{C}$ is the map

$$\mathcal{L}_f := \Psi_f \circ \Phi_f : \mathcal{B}_f \to \mathbb{C}.$$

Note that the Lavaurs map $\mathcal{L}_f$ commutes with f. Indeed, $\Phi_f \circ f = T_1 \circ \Phi_f$ and $\Psi_f \circ T_1 = f \circ \Psi_f$, so that

$$\mathcal{L}_f \circ f = \Psi_f \circ \Phi_f \circ f = \Psi_f \circ T_1 \circ \Phi_f = f \circ \Psi_f \circ \Phi_f = f \circ \mathcal{L}_f.$$

Figure 2.7 gives a rough idea of the behaviour of the Lavaurs map $\mathcal{L}_f$ for the cubic polynomial $f(z) := z + z^2 + 0.95z^3$. Points in the basin $\mathcal{B}_f$ are colored according to the location of their image by $\mathcal{L}_f$: grey when $\mathcal{L}_f(z) \in \mathcal{B}_f$, white otherwise. The restriction of $\mathcal{L}_f$ to each bounded white domain is a covering of $\mathbb{C}\setminus\overline{\mathcal{B}_f}$.

Fig. 2.7 Behavior of $\mathcal{L}_f$ for $f(z) := z + z^2 + 0.95z^3$

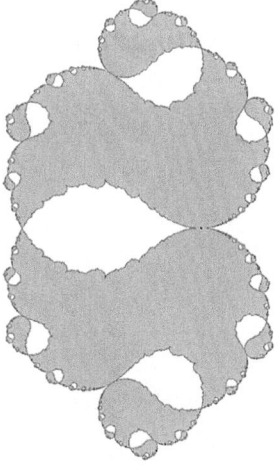

Fig. 2.8 The set $\mathcal{K}(\mathcal{L}_f)$ for $f(z) = z + z^2 + 0.95z^3$

2.7.4 Discontinuity of the Julia Set

Set

$$\mathcal{K}(\mathcal{L}_f) := \bigcap_{n \geq 0} \mathcal{L}_f^{-n}(\mathcal{K}_f) \quad \text{and} \quad \mathcal{J}(\mathcal{L}_f) := \partial \mathcal{K}(\mathcal{L}_f).$$

Figure 2.8 shows $\mathcal{K}(\mathcal{L}_f)$ for the cubic polynomial $f(z) = z + z^2 + 0.95z^3$. The Lavaurs map $\mathcal{L}_f$ has two complex conjugate sets of attracting fixed points. The fixed points of $\mathcal{L}_f$ are indicated and their basins of attraction are colored (dark grey for one of the fixed points, and light grey for the others). Those basins form the interior of $\mathcal{K}(\mathcal{L}_f)$. The black set $\mathcal{J}(\mathcal{L}_f)$ is the topological boundary of $\mathcal{K}(\mathcal{L}_f)$.

The following result may be considered as the main reason why Lavaurs maps are studied in holomorphic dynamics.

Proposition 2.21 (Lavaurs [44]) *Let* $f : \mathbb{C} \to \mathbb{C}$ *be a polynomial whose expansion at 0 is* $f(z) = z + z^2 + O(z^3)$. *Let* (N_n) *be a sequence of integers tending to* $+\infty$ *and* (ε_n) *be a sequence of complex numbers tending to 0, such that*

$$N_n - \frac{\pi}{\varepsilon_n} \to 0.$$

Then,

$$(f + \varepsilon_n^2)^{\circ N_n} \to \mathcal{L}_f \quad \text{locally uniformly on } \mathcal{B}_f.$$

2 Introduction to Fatou Components in Holomorphic Dynamics

In addition,
$$\mathcal{J}_f \subsetneq \mathcal{J}(\mathcal{L}_f) \subseteq \liminf \mathcal{J}_{f+\varepsilon_n^2} \quad \text{and} \quad \limsup \mathcal{K}_{f+\varepsilon_n^2} \subseteq \mathcal{K}(\mathcal{L}_f) \subsetneq \mathcal{K}_f.$$

We do not present the proof of this result which is rather technical, and can be found in [27] or [57] for example. Instead, we will show that the result holds for the Möbius transformation

$$g(z) = \frac{z}{1-z} = z + z^2 + O(z^3).$$

In that case, all maps involved are Möbius transformations and the computations are explicit. If we perform the change of coordinates $Z = -1/z$, the Möbius transformation g gets conjugated to T_1. Thus,

$$\Phi_g(z) = -\frac{1}{z} \quad \text{and} \quad \Psi_g(Z) = -\frac{1}{Z}, \quad \text{so that} \quad \mathcal{L}_g = \text{Id}.$$

Note that $g + \varepsilon^2$ is also a Möbius transformation. It has two fixed points

$$\alpha^\pm = \pm i\varepsilon + O(\varepsilon^2) \quad \text{with multipliers} \quad \lambda^\pm = \exp\bigl(\pm 2i\varepsilon + O(\varepsilon^3)\bigr).$$

So, if $N \to +\infty$ and

$$N - \frac{\pi}{\varepsilon} \to 0, \quad \text{so that} \quad \varepsilon = \frac{\pi}{N + o(1)} = \frac{\pi}{N} + o\left(\frac{1}{N^2}\right),$$

then $(g + \varepsilon^2)^{\circ N}$ is a Möbius transformation fixing $\alpha^\pm$ with multipliers

$$\mu^\pm := (\lambda^\pm)^N = \exp\left(\pm 2\pi i + o(1/N)\right) = 1 + o\left(\frac{1}{N}\right).$$

This Möbius transformation is

$$z \mapsto \alpha^+ + \frac{\mu^+ \cdot (z - \alpha^+)}{1 - \dfrac{\mu^+ - 1}{\alpha^+ - \alpha^-}(z - \alpha^+)}.$$

Since

$$\mu^+ - 1 = o\left(\frac{1}{N}\right) = o(\alpha^+ - \alpha^-),$$

we see that indeed,

$$(g + \varepsilon^2)^{\circ N} \xrightarrow[N \to +\infty]{} \mathcal{L}_g.$$

Recently, in [63], Vivas studied a non-autonomous parabolic implosion in dimension 1 for the map $f(z) = \frac{z}{1-z}$.

In higher dimension, (semi-)parabolic implosion was recently studied for dissipative polynomial automorphisms of $\mathbb{C}^2$ by Bedford, Smillie and Ueda in [9] (see also [29]) and their strategy was recently adapted by Bianchi in [11] for a perturbation of a class of holomorphic endomorphisms tangent to identity, establishing a two-dimensional Lavaurs theorem for such a class.

2.8 Fatou Components for Polynomial Maps in Dimension 2

We end these notes by giving an updated account of the recent results on Fatou components for polynomial skew-products in complex dimension two in a neighbourhood of a periodic fiber. We divide our discussion according to the different possible kinds of periodic fibers.

2.8.1 Preliminaries

Let $F : \mathbb{C}^2 \to \mathbb{C}^2$ be a holomorphic endomorphism of $\mathbb{C}^2$, and consider the discrete holomorphic dynamical system given by the iteration of F. In the investigation of the global behaviour of such a system it is natural to give the generalize the definition of the *Fatou set of F* as the largest open set $\mathcal{F}(F)$ where the family of iterates $(F^{\circ n})_{n \in \mathbb{N}}$ of F is normal. A connected component of the Fatou set is called a *Fatou component*.

We have seen in the previous sections that in complex dimension one, Fatou components of rational maps of degree at least 2 on the Riemann sphere are well understood, thanks to Theorems 2.4 and 2.6.

In complex dimension two, the understanding of Fatou components is far less complete. A considerable progress in the classification of periodic Fatou components has been achieved thanks to Bedford and Smillie [6–8], Fornæss and Sibony [33], Lyubich and Peters [47] and Ueda [61].

The question of the existence of wandering (i.e., not pre-periodic) Fatou components in higher dimension was put forward by several authors since the 1990s (see e.g., [35]). Higher dimensional *transcendental* (i.e., non polynomial) maps with wandering domains can be constructed from one-dimensional examples by taking direct products. An example of a transcendental *biholomorphism* of $\mathbb{C}^2$ with a wandering Fatou component oscillating to infinity was constructed by Fornæss and

Sibony in [34]. Nonetheless, until recently very little was known about the existence of wandering Fatou components for holomorphic endomorphisms of $\mathbb{P}^2(\mathbb{C})$ or for polynomial endomorphisms of $\mathbb{C}^2$.

A first natural class of polynomial endomorphisms of $\mathbb{C}^2$ to consider are *direct product*, that is maps $F : \mathbb{C}^2 \to \mathbb{C}^2$ of the form

$$F(z, w) = (f(z), g(w)),$$

where f and g are complex polynomials in one variable. This allows us to recover the generalizations of one-dimensional dynamical behaviours in dimension two. However, direct products are a very particular class and they do not give us a complete understanding of all possible behaviours of polynomial endomorphisms in $\mathbb{C}^2$.

A more interesting class to consider is given by polynomial *skew-products* in $\mathbb{C}^2$, namely polynomial maps $F : \mathbb{C}^2 \to \mathbb{C}^2$ of the form

$$F(z, w) = (f(z, w), g(w)), \tag{2.1}$$

where g is a complex polynomial in one variable and f is a complex polynomial in two variables. Since they leave invariant the fibration $\{w = \text{const.}\}$, skew-products allow us to *build on one-dimensional dynamics* and to get a first flavour of the richness of the higher dimension setting we are working in. This idea has been used by several authors to construct maps with particular dynamical properties. Dujardin, for example, used in [28] specific skew-products to construct a non-laminar Green current. Boc-Thaler, Fornæss and Peters constructed in [13] a map having a Fatou component with a punctured limit set. Last but not least, as we will explain in Sect. 2.8.2, skew-products are one of the key ingredients in the construction we recently obtained in [3] with Astorg, Dujardin, and Peters of holomorphic endomorphisms of $\mathbb{P}^2(\mathbb{C})$ having a wandering Fatou component.

The investigation of the holomorphic dynamics of polynomial skew-products was started by Heinemann [37] and then continued by Jonsson [40]. The topology of Fatou components of skew-products has been studied by Roeder in [54].

Given a Fatou component Ω of a polynomial skew-product F in $\mathbb{C}^2$, its projection on the second coordinate $\Omega_2 = \pi_2(\Omega)$ is contained in a Fatou component for g and hence thanks to Sullivan's non-wandering Theorem 2.6, up to considering an iterate of F, it has to fall into one of the three cases given by Theorem 2.4, and moreover, since we are considering polynomials, Herman rings cannot occur. Therefore, since (pre-)periodic points for g correspond to (pre-)periodic fibers for F, up to considering an iterate of F, we can restrict ourselves to study what happens in neighbourhoods of invariant fibers of the form $\{w = w_0\}$. One-dimensional theory also describes the dynamics on the invariant fiber, which is given by the action of the one-dimensional polynomial $f(z, w_0) := f_{w_0}(z)$, and hence the Fatou components of f_c will be again all pre-periodic and, up to consider an iterate, we can assume that they are either attracting basins, or parabolic basins or Siegel disks. This structure leads us to two immediate questions.

1. Do all Fatou components of f_c bulge to two-dimensional Fatou components of F?
2. Is it possible to have wandering Fatou components for F in a neighbourhood of an invariant fiber?

In the following we shall call an invariant fiber $\{w = w_0\}$ *attracting, parabolic* or *elliptic* according to whether w_0 is an attracting, parabolic or elliptic fixed point for g. A *bulging* Fatou component will be a Fatou component Ω of F such that $\Omega \cap \{w = w_0\}$ consists of one-dimensional Fatou components of f_{w_0} on the invariant fiber $\{w = w_0\}$. We shall say that a Fatou component Ω_{w_0} of f_{w_0} on the invariant fiber $\{w = w_0\}$ *is bulging* if there exists a bulging Fatou component Ω of F so that $\Omega_{w_0} \subseteq \Omega \cap \{w = w_0\}$.

2.8.2 Attracting Invariant Fiber

Let us consider a polynomial skew-product $F : \mathbb{C}^2 \to \mathbb{C}^2$ of degree $d \geq 2$

$$F(z, w) = (f(z, w), g(w)),$$

with an *attracting invariant fiber*. We can assume without loss of generality that the invariant fiber is $\{w = 0\}$. Therefore we have $g(0) = 0$ and $|g'(0)| < 1$. In this case we know that there exists an *attracting basin*, containing the origin, of points whose iterates converge to the origin. The rate of convergence to the fixed point depends on whether the fixed point is *superattracting*, i.e. $g'(0) = 0$, or *attracting* or *geometrically attracting*, i.e., $g'(0) \neq 0$.

2.8.2.1 Superattracting Case

This setting was studied by Lilov in [45] who was able to answer both questions stated in the introduction. He first proved the following result giving a positive answer to Question 1.

Theorem 2.8 (Lilov [45]) *Let $F : \mathbb{C}^2 \to \mathbb{C}^2$ be a polynomial skew-product of the form (2.1) of degree $d \geq 2$. Let $\{w = w_0\}$ be a superattracting invariant fiber for F. Then all one-dimensional Fatou components of f_{w_0} bulge to Fatou components of F.*

Idea of the Proof We can assume $w_0 = 0$ without loss of generality. Thanks to Theorems 2.4 and 2.6, all Fatou components of the restriction $f_0(z) = f(z, 0)$ of $f(z, w)$ to the invariant fiber are (pre-)periodic and are either attracting basins, or parabolic basins or Siegel domains. The strategy of the proof is to prove separately for each of these cases that the corresponding component is contained in a two-dimensional Fatou component of F. The bulging of one-dimensional Fatou components of attracting periodic points of $f_0(z)$ is well-known and follows for instance from the results of Rosay and Rudin [55]. For the remaining cases, by [45, Theorem 3.17] there exists a strong stable manifold through all point in the one-dimensional Fatou components of parabolic or elliptic periodic points of $f_0(z)$, and so the corresponding bulging Fatou components simply consist of the union of such manifolds. □

Then Lilov proved the following result implying the non-existence of wandering Fatou components in a neighbourhood of a superattracting invariant fiber.

Theorem 2.9 (Lilov [45]) *Let $F : \mathbb{C}^2 \to \mathbb{C}^2$ be a polynomial skew-product of the form (2.1) of degree $d \geq 2$. Let $\{w = w_0\}$ be a superattracting invariant fiber for F and let $\mathcal{B}$ be the immediate basin of the superattracting fixed point w_0. Take $w \in \mathcal{B}$ and let D_w be a one-dimensional open disk lying in the fiber over w. Then the forward orbit of D_w must intersect one of the bulging Fatou components of f_{w_0}.*

The proof relies on the repeated use of [45, Lemma 3.2.4] applied to the orbit of a disk lying in a fiber over a point in the attracting basin, in order to obtain estimates from below for the radii of the images. Thanks to [45, Proposition 3.2.8], by studying the geometry of the bulging Fatou components, it is also possible to obtain an upper bound on the largest possible disk lying in a fiber over a point in the attracting basin that can lie in the complement of a bulging Fatou component, depending on the distance to the invariant fiber. The conclusion then follows combining these two estimates.

All bulging Fatou components are (pre-)periodic, therefore all Fatou components for F in a neighbourhood of a superattracting invariant fiber are (pre-)periodic, and then the non-existence of wandering Fatou components in a neighbourhood of a superattracting invariant fiber follows immediately.

Corollary 2.1 (Lilov [45]) *Let $F : \mathbb{C}^2 \to \mathbb{C}^2$ be a polynomial skew-product of the form (2.1) of degree $d \geq 2$. Let $\{w = w_0\}$ be a superattracting invariant fiber for F and let $\mathcal{B}$ be the immediate basin of the superattracting fixed point w_0. Then there are no wandering Fatou components in $\mathcal{B} \times \mathbb{C}$.*

2.8.2.2 Geometrically Attracting Case

The geometrically attracting case was first partially addressed by Lilov in [45] even if not stated explicitly. In fact, the proof of Theorem 2.8 can be readily adjusted to this case obtaining the following statement answering Question 1.

Theorem 2.10 (Lilov [45]) *Let $F : \mathbb{C}^2 \to \mathbb{C}^2$ be a polynomial skew-product of the form (2.1) of degree $d \geq 2$. Let $\{w = w_0\}$ be an attracting invariant fiber for F. Then all one-dimensional Fatou components of f_{w_0} bulge to Fatou components of F.*

On the other hand, the proof of Theorem 2.9 cannot be generalized to this setting, which is indeed more complicated than the superattracting case. In fact,

Theorem 2.9 does not hold in general, as showed by Peters and Vivas with the following result.

Theorem 2.11 (Peters-Vivas [53]) *Let* $F : \mathbb{C}^2 \to \mathbb{C}^2$ *be a polynomial skew-product of the form*

$$F(z, w) = (p(z) + q(w), \lambda w), \tag{2.2}$$

with $0 < |\lambda| < 1$ *and* p *and* q *complex polynomials. Then there exists a triple* (λ, p, q) *and a holomorphic disk* $D \subset \{w = w_0\}$ *whose forward orbit accumulates at a point* $(z_0, 0)$, *where* z_0 *is a repelling fixed point in the Julia set of* f_0.

As a consequence, the forward orbits of D cannot intersect the bulging Fatou components of f_0.

The family $(F|_D^{\circ n})_{n \in \mathbb{N}}$ is normal on the disks D, and so these are Fatou disks. However such disks are completely contained in the Julia set of F, which is the complement in $\mathbb{C}^2$ of the Fatou set (see [53, Theorem 6.1]).

The geometrically attracting case has been further investigated by Peters and Smit in [52]. They focused their investigation on polynomial skew-products such that the action on the invariant attracting fiber is *subhyperbolic*, that is the polynomial does not have parabolic periodic points and all critical points lying on the Julia set are pre-periodic. They proved the following result.

Proposition 2.22 (Peters and Smit [52]) *Let* $F : \mathbb{C}^2 \to \mathbb{C}^2$ *be a polynomial skew-product of the form* (2.1). *Assume that the origin is an attracting, not superattracting, fixed point for* g *with corresponding basin* B_g, *and the polynomial* $f_0(z) := f(z, 0)$ *is subhyperbolic. Then there exists a set* $E \subset \mathbb{C}$ *of full measure, such that for every* $\tilde{w}_0 \in E$ *the forward orbit of every disk in the fiber* $\{w = \tilde{w}_0\}$ *must intersect a bulging Fatou component of* f_0.

Idea of the Proof Notice that it suffices to prove the proposition in a neighbourhood of the attracting fiber $\{w = 0\}$. Therefore, up to considering a smaller neighbourhood, we can assume without loss of generality that $g(w) = \lambda w$, and

$$f(z, w) = a_0(w) + a_1(w)z + \cdots + a_d(w)z^d$$

where $a_0(w), \ldots, a_d(w)$ are holomorphic functions in w. The subhyperbolicity of the polynomial f_0 implies that its Fatou set is the union of finitely many attracting basins, and the orbits of the critical points contained in the Fatou set converge to one of these attracting cycles. The proof can be divided into 5 main steps.

Step 1. Fix $R > 0$ large enough so that for all z such that $|z| > R$ we have $|f_0(z)| > 2|z|$ and set

$$W_0 = \{|z| > R\} \cup \bigcup_{y \in \text{Att}(f_0)} W_y$$

where Att(f_0) is the set of all attracting periodic points of f_0, and for each $y \in$ Att(f_0) the set W_y is an open neighbourhood of the orbit of y such that $\overline{f_0(W_y)} \subset W_y$. Fix a neighbourhood U of the post-critical set of f_0. Then by Peters and Smit [52, Proposition 15], there exists a set $E \subset \mathbb{C}$ of full measure in a neighbourhood of the origin such that for all $\widetilde{w}_0 \in E$ there exists a constant $C = C(\widetilde{w}_0, U)$ such that for all $n \in \mathbb{N}$ we have

$$\operatorname{Card}\left\{z : \frac{\partial F_1^{\circ n}}{\partial z}(z, \widetilde{w}_0) = 0 \text{ and } F_1^{\circ n}(z, \widetilde{w}_0) \notin W_0 \times U\right\} \leq C\sqrt{n}, \quad (2.3)$$

where $F_1^{\circ n}$ is the first component of the n-th iterate of F.

Step 2. Assume by contradiction that a fiber $\{w = \widetilde{w}_0\}$, with $\widetilde{w}_0 \in E$, contains a disk D whose forward orbit avoids the bulging Fatou components of f_0. Then the restriction of $F^{\circ n}$ to D is bounded and hence a normal family. Therefore, up to shrinking D there exists a subsequence $F^{\circ n_j}$ such that $F^{\circ n_j}|_D$ converges, uniformly on compact subsets of D, to a point ζ in the Julia set of f_0. Moreover, there exists $\varepsilon > 0$ so that $F^{\circ n}(D) \cap (W_0 \times D(0, \varepsilon))$ is empty for all $n \in \mathbb{N}$.

Step 3. Each critical point x contained in the Julia set is eventually mapped into a repelling periodic point, and up to considering an iterate of F we may assume that it is eventually mapped into a repelling fixed point with multiplier μ, with $|\mu| > 1$. The main tool to control the orbits of the critical points of F is obtained using a linearization map of the unstable manifold of the repelling fixed point, given by a map $\Phi : \mathbb{C} \to \mathbb{C}$ satisfying $\Phi(\mu t) = f_0^{\circ k} \circ \Phi(t)$ for some $k \in \mathbb{N}$. Thanks to [52, Proposition 10], there exist $\widetilde{C} > 1$ and $0 < \gamma < 1$ so that

$$\operatorname{Area}(F^{\circ n}(D)) \leq \widetilde{C}\gamma^n. \quad (2.4)$$

Step 4. We may assume that ζ does not lie in the post-critical set, and we may choose U and $r > 0$ such that $D(\zeta, r) \cap (U \cup W_0) = \emptyset$. Let $j_1 \in \mathbb{N}$ be such that $F_1^{\circ n_j}(D) \subseteq D(\zeta, \frac{r}{2})$ for all $j \geq j_1$, and consider O_j the connected component of $(F^{\circ n_j})^{-1}(D(\zeta, r) \times \{\lambda^{n_j}\widetilde{w}_0\})$ containing D. Then $D \subseteq O_j \subseteq D(0, R) \times \{\widetilde{w}_0\}$, and we can study the proper holomorphic function $F_1^{\circ n_j} : O_j \to D(\zeta, r)$. Thanks to (2.3), such a map has at most $d_j = C\sqrt{n_j}$ critical points.

Step 5. It is possible (see [52, Proposition 28]) to find a uniform constant $C_1 > 0$ so that if $f : \mathbb{D} \to \mathbb{D}$ is a proper holomorphic function of degree d, the set $R \subset \mathbb{D}$ has Poincaré area equal to A, and $d \cdot A^{1/2d} < 8$, then the Poincaré area of $f^{-1}(R)$ is at most $C_1 d^3 A^{1/d}$. Then, setting $R_j = F_1^{\circ n_j}(D)$ and denoting by A_j its Poincaré area $\operatorname{Area}_{D(\zeta, r)}(R_j)$ with respect to $D(\zeta, r)$, for $j \geq j_1$, we have $R_j \subseteq D(\zeta, r)$, and we can estimate A_j applying (2.4). Therefore there exists $j_2 \geq j_1$ such that $d_j A_j^{1/2d_j} < 1/8$ for all $j \geq j_2$. This implies

$$\operatorname{Area}_{D(0,R)}(D) \leq \operatorname{Area}_{O_j}(D) \leq C_2 d_j^3 A_j^{1/d_j} \leq M n_j^{3/2} \gamma^{n_j^{3/2}}$$

where $M > 0$. The contradiction follows from the fact that the last expression will converge to zero as j increases towards infinity.

□

Thanks to the fact that in particular E is dense, Peters and Smit are able to give a negative answer to Question 2 when the action on the invariant fiber is subhyperbolic. They also obtain as a corollary that the only Fatou components of F are the bulging ones, since the topological degree of F equals the one of f_0, implying that the only Fatou components that can be mapped onto the bulging Fatou components of f_0 are exactly those bulging Fatou components.

Theorem 2.12 (Peters-Smit [52]) *Let $F : \mathbb{C}^2 \to \mathbb{C}^2$ be a polynomial skew-product of the form (2.1). Assume that the origin is an attracting fixed point for g with corresponding basin $\mathcal{B}_g$, and the polynomial $f_0(z) := f(z, 0)$ is subhyperbolic. Then F has no wandering Fatou component over $\mathcal{B}_g$.*

Very recently, Ji was able in [38] to generalize Lilov's Theorem 2.9 to polynomial skew-products with an invariant geometrically attracting fiber under the hypothesis that the multiplier of the invariant fiber is sufficiently small. More precisely he proved the following result.

Theorem 2.13 (Ji [38]) *Let $F : \mathbb{C}^2 \to \mathbb{C}^2$ be a polynomial skew-product of the form (2.1) of degree $d \geq 2$. Let $\{w = w_0\}$ be an attracting invariant fiber for F and let $\mathcal{B}_{w_0}$ be the basin of the attracting fixed point c. Then there exists $\lambda_0(w_0, f) > 0$ depending only on f and c such that if $|g'(w_0)| < \lambda_0$, then there are no wandering Fatou components in $\mathcal{B}_{w_0} \times \mathbb{C}$.*

The proof of this result follows Lilov's strategy. The main difficulty is due to the breaking down of Lilov's argument in the geometrically attracting case as we pointed out at the beginning of this section. Ji is able to overcome such difficulty by adapting a one-dimensional result due to Denker, Przytycki and Urbanski in [24] in this case. Such result is used to obtain estimates of the size of bulging Fatou components and of the size of forward images of wandering Fatou disks.

Ji also proved in [39] that if $F : \mathbb{C}^2 \to \mathbb{C}^2$ is a polynomial skew-product with an attracting invariant fiber L such that the restriction of F to L is non-uniformly hyperbolic, then the Fatou set in the basin of L coincides with the union of the basins of attracting cycles, and the Julia set in the basin of L has Lebesgue measure zero. Therefore there are no wandering Fatou components in the basin of the invariant attracting fiber.

2.8.3 Parabolic Invariant Fiber and Wandering Domains

A first contribution to the investigation of this case is due to Vivas, who proved a parametrization result [62, Theorem 3.1] for the unstable manifolds for *special parabolic skew-product* of $\mathbb{C}^2$. Vivas used this parametrization as the main tool to prove the analogue of Theorem 2.11 for special parabolic skew-product. However, this construction does not allow to construct a wandering Fatou component in a neighbourhood of the parabolic invariant fiber.

In [3], together with Astorg, Dujardin and Peters, we proved the existence of polynomial skew-products of $\mathbb{C}^2$, extending to holomorphic endomorphisms of $\mathbb{P}^2(\mathbb{C})$, having a wandering Fatou component. The key tool consists in using parabolic implosion techniques on polynomial skew-products, and this idea was initially suggested by Lyubich. The main strategy is to combine slow convergence to an invariant parabolic fiber and parabolic transition in the fiber direction, to produce orbits shadowing those of the so-called Lavaurs map, that we defined in the previous section.

Theorem 2.14 ([3]) *There exists a holomorphic endomorphism* $F : \mathbb{P}^2(\mathbb{C}) \to \mathbb{P}^2(\mathbb{C})$, *induced by a polynomial skew-product mapping* $F : \mathbb{C}^2 \to \mathbb{C}^2$, *having a wandering Fatou component. More precisely, let* $f : \mathbb{C} \to \mathbb{C}$ *and* $g : \mathbb{C} \to \mathbb{C}$ *be polynomials of the form*

$$f(z) = z + z^2 + \mathrm{O}(z^3) \quad \text{and} \quad g(w) = w - w^2 + \mathrm{O}(w^3). \tag{2.5}$$

If the Lavaurs map $\mathcal{L}_f : \mathcal{B}_f \to \mathbb{C}$ *has an attracting fixed point, then the skew-product* $F : \mathbb{C}^2 \to \mathbb{C}^2$ *defined by*

$$F(z, w) := \left(f(z) + \frac{\pi^2}{4} w, g(w) \right) \tag{2.6}$$

has a wandering Fatou component.

The orbits in these wandering Fatou components are bounded and the approach used in the proof is essentially local. Notice that if f and g have the same degree, F extends to a holomorphic endomorphism of $\mathbb{P}^2(\mathbb{C})$. Moreover we can obtain examples in arbitrary dimension $k \geq 2$ by simply considering products mappings of the form (F, Q), where Q has a fixed Fatou component.

A first step in the proof of Theorem 2.14 is to find a parabolic polynomial f whose Lavaurs map $\mathcal{L}_f$ admits an attracting fixed point.

Proposition 2.23 ([3, Proposition B]) *Let* $f : \mathbb{C} \to \mathbb{C}$ *be the cubic polynomial defined by*

$$f(z) = z + z^2 + az^3 \quad \text{with} \quad a \in \mathbb{C}.$$

If $r > 0$ *is sufficiently close to* 0 *and* a *belongs to the disk* $D(1-r, r)$, *then the Lavaurs map* $\mathcal{L}_f : \mathcal{B}_f \to \mathbb{C}$ *admits an attracting fixed point.*

Idea of the Proof We consider

$$\mathcal{U}_f := \psi_f^{-1}(\mathcal{B}_f) \quad \text{and} \quad \mathcal{E}_f := \varphi_f \circ \psi_f : \mathcal{U}_f \to \mathbb{C}.$$

The open set $\mathcal{U}_f$ contains an upper half-plane and a lower half-plane, and it is invariant under T_1. Like the Lavaurs map, the map $\mathcal{E}_f$ commutes with T_1, therefore $\mathcal{E}_f - \text{Id}$ is periodic of period 1 and admits a Fourier expansion in a upper half-plane:

$$\mathcal{E}_f(Z) = Z + \sum_{k \geq 0} c_k e^{2\pi i k Z}.$$

Using the expansion of φ_f and ψ_f near infinity, we obtain with an elementary computation:

$$\mathcal{E}_f(Z) = \varphi_f \circ \psi_f(Z) = Z + (1-a)\log(-Z) + o(1)$$
$$- (1-a)\log\bigl(Z + (1-a)\log(-Z) + o(1)\bigr) + o(1)$$
$$= Z + (1-a)\log(Z) - \pi i(1-a) - (1-a)\log(Z) + o(1)$$
$$= Z - \pi i(1-a) + o(1),$$

and so $c_0 = -\pi i(1-a)$.

Thanks to a more elaborate argument, based on the notion of finite type analytic map introduced by Adam Epstein, it is possible to prove that:

$$\mathcal{E}_f(Z) = Z - \pi i(1-a) + c_1 e^{2\pi i Z} + o(e^{2\pi i Z}) \quad \text{with} \quad c_1 \neq 0.$$

It then follows that for $a \neq 1$ close to 1, $\mathcal{E}_f$ has a fixed point Z_f with multiplier ρ_f satisfying

$$c_1 e^{2\pi i Z_f} \sim \pi i(1-a) \quad \text{and} \quad \rho_f - 1 \sim 2\pi i c_1 e^{2\pi i Z_f} \sim -2\pi^2(1-a) \quad \text{as} \quad a \to 1.$$

Therefore, for $r > 0$ sufficiently close to 0 and $a \in D(1-r, 1)$, the multiplier ρ_f belongs to the unit disk and Z_f is an attracting fixed point of $\mathcal{E}_f$. This concludes the proof since $\psi_f : \mathcal{U}_f \to \mathcal{B}_f$ semi-conjugates $\mathcal{E}_f$ to $\mathcal{L}_f$, and so, the fact that Z_f is an attracting fixed point of $\mathcal{E}_f$ implies that the point $\psi_f(Z_f)$ is an attracting fixed point of $\mathcal{L}_f$. □

The key result in the proof of Theorem 2.14 relies on a non-autonomous analogue of Lavaurs estimates in the setting of skew-products.

Let $\mathcal{B}_f$ and $\mathcal{B}_g$ be the parabolic basins of 0 under iteration of respectively f and g. One of the key points is to choose $(\widetilde{z}_0, \widetilde{w}_0) \in \mathcal{B}_f \times \mathcal{B}_g$ so that the first coordinate of $F^{\circ m}(\widetilde{z}_0, \widetilde{w}_0)$ returns infinitely many times close to the attracting fixed point of $\mathcal{L}_f$. The proof is designed so that the return times are the integers n^2 for $n \geq n_0$. Therefore, we need to analyze the orbit segment between n^2 and $(n+1)^2$, which is of length $2n+1$.

Proposition 2.24 ([3]) *As $n \to +\infty$, the sequence of maps*

$$\mathbb{C}^2 \ni (z, w) \mapsto F^{2n+1}\bigl(z, g^{n^2}(w)\bigr) \in \mathbb{C}^2$$

converges locally uniformly in $\mathcal{B}_f \times \mathcal{B}_g$ to the map

$$\mathcal{B}_f \times \mathcal{B}_g \ni (z, w) \mapsto \big(\mathcal{L}_f(z), 0\big) \in \mathbb{C} \times \{0\}.$$

Idea of the Proof Let $\mathcal{B}_g$ the parabolic basin of 0 under iteration of g. For all $w \in \mathcal{B}_g$, the orbit $g^{\circ m}(w)$ converges to 0 like $1/m$. We want to analyze the behaviour of F starting at $\big(z, g^{\circ n^2}(w)\big)$ during $2n + 1$ iterates. For large n, the first coordinate of F along this orbit segment is approximately

$$f(z) + \varepsilon^2 \quad \text{with} \quad \frac{\pi}{\varepsilon} \simeq 2n.$$

As we saw in the previous section, a rough statement of Lavaurs Theorem from parabolic implosion gives us that if $\frac{\pi}{\varepsilon} = 2n$, then for large n, the $(2n)^{\text{th}}$ iterate of $f(z) + \varepsilon^2$ is approximately equal to $\mathcal{L}_f(z)$ on $\mathcal{B}_f$.

Our setting is different since in our case ε keeps decreasing along the orbit. Indeed on the first coordinate we are taking the composition of $2n + 1$ transformations of the form

$$f(z) + \varepsilon_k^2 \quad \text{with} \quad \frac{\pi}{\varepsilon_k} \simeq 2n + \frac{k}{n} \quad \text{and} \quad 1 \leq k \leq 2n + 1.$$

The key step in the proof of the statement consists in a detailed analysis of this non-autonomous situation, proving that the decay of ε_k is counterbalanced by taking *exactly* one additional iterate of F. □

With this proposition in hand, the proof of Theorem 2.14 is easily completed.

Proof of Theorem 2.14 Let ξ be an attracting fixed point of $\mathcal{L}_f$ and let $V \subset \mathcal{B}_f$ be a disk centered at ξ and such that $\mathcal{L}_f(V)$ is compactly contained in V. Therefore $\mathcal{L}_f^{\circ k}(V)$ converges to ξ as $k \to +\infty$. Let $W \Subset \mathcal{B}_g$ be an arbitrary disk.

Thanks to Proposition 2.24, there exists $n_0 \in \mathbb{N}$ such that for every $n \geq n_0$,

$$\pi_1 \circ F^{\circ(2n+1)}(V \times g^{\circ n^2}(W)) \Subset V,$$

where $\pi_1 : \mathbb{C}^2 \to \mathbb{C}$ denotes the projection on the first coordinate, $\pi_1(z, w) := z$.

Let U be a connected component of the open set $F^{-n_0^2}\big(V \times g^{\circ n_0^2}(W)\big)$. Then for every integer $n \geq n_0$, we have

$$F^{\circ n^2}(U) \subseteq V \times g^{\circ n^2}(W). \tag{2.7}$$

In fact, this holds by assumption for $n = n_0$. Now if the inclusion is true for some $n \geq n_0$, then

$$\pi_1 \circ F^{\circ(n+1)^2}(U) = \pi_1 \circ F^{\circ(2n+1)}\left(F^{\circ n^2}(U)\right)$$

$$\subset \pi_1 \circ F^{\circ(2n+1)}\left(V \times g^{\circ n^2}(W)\right) \subset V,$$

from which (2.7) follows. This yields that the sequence $(F^{\circ n^2})_{n \geq 0}$ is uniformly bounded, and hence normal, on U. Moreover, any cluster value of this sequence of maps is constant and of the form $(z, 0)$ for some $z \in V$, and $(z, 0)$ is a limit value (associated to a subsequence $(n_k)_{k \in \mathbb{N}}$) if and only if $(\mathcal{L}_f(z), 0)$ is a limit value (associated to the subsequence $(1 + n_k)_{k \in \mathbb{N}}$). Therefore the set of cluster limits is totally invariant under $\mathcal{L}_f : V \to V$, and so it must coincide with the attracting fixed point ξ of $\mathcal{L}_f$. Therefore the sequence $(F^{\circ n^2})_{n \geq 0}$ converges locally uniformly to $(\xi, 0)$ on U.

The sequence $(F^{\circ m})_{m \in \mathbb{N}}$ is locally bounded on U if and only if there exists a subsequence $(m_k)_{k \in \mathbb{N}}$ such that $(F^{\circ m_k}|_U)_{k \in \mathbb{N}}$ has the same property. In fact, since $\overline{W}$ is compact, there exists $R > 0$ such that if $|z| > R$, then for every $w \in W$, (z, w) escapes locally uniformly to infinity under iteration. The domain U is therefore contained in the Fatou set of F.

Let Ω be the component of the Fatou set $\mathcal{F}_F$ containing U. For any integer $j \geq 0$, the sequence of maps $(F^{\circ n^2 + j})_{n \in \mathbb{N}}$ converges locally uniformly to $F^{\circ j}(\xi, 0) = (f^{\circ j}(\xi), 0)$ on U and hence on Ω. Therefore the sequence $(F^{\circ n^2})_{n \in \mathbb{N}}$ converges locally uniformly to $(f^{\circ j}(\xi), 0)$ on $F^{\circ j}(\Omega)$. If i, j are nonnegative integers such that $F^{\circ i}(\Omega) = F^{\circ j}(\Omega)$, then $f^{\circ i}(\xi) = f^{\circ j}(\xi)$, and so $i = j$ because ξ cannot be pre-periodic under iteration of f, since it belongs to the parabolic basin $\mathcal{B}_f$. This proves that Ω is not (pre-)periodic under iteration of F, and so it is a wandering Fatou component for F. □

We end this subsection with some explicit examples satisfying the assumption of Theorem 2.14.

Example As a consequence of Proposition 2.23 we obtain that if $f : \mathbb{C} \to \mathbb{C}$ is the cubic polynomial $f(z) = z + z^2 + az^3$, and g is as in (2.5), then the polynomial skew-product F defined in (2.6) admits a wandering Fatou component for $r > 0$ sufficiently small and $a \in D(1 - r, r)$.

It is also interesting to search for real polynomial mappings with wandering Fatou domains intersecting $\mathbb{R}^2$. We also have such examples.

Example (3, Proposition C) Let $f : \mathbb{C} \to \mathbb{C}$ be the degree 4 polynomial defined by

$$f(z) := z + z^2 + bz^4 \text{ with } b \in \mathbb{R}.$$

There exist parameters $b \in (-8/27, 0)$ such that for g as in (2.5), the polynomial skew-product F defined in (2.6) has a wandering Fatou component intersecting $\mathbb{R}^2$.

Astorg, Boc-Thaler and Peters constructed in [4] a second example of polynomial skew-products, arising from similar techniques, but with distinctly different dynamical behaviour. Instead of wandering domains arising from a Lavaurs map with an attracting fixed point, they constructed a domain arising from a Lavaurs map with a fixed point of Siegel type.

In [36], Hahn and Peters constructed polynomial automorphisms of $\mathbb{C}^4$ with wandering Fatou components. Their four-dimensional automorphisms lie in a one-parameter family, depending on a parameter $\delta \in \mathbb{C}\setminus\{0\}$, and degenerating to the two-dimensional polynomial map in Eq. (2.5) of Theorem 2.14 as δ converges to 0.

Berger and Biebler recently proved in [10] the existence of a locally dense set of polynomial automorphisms of $\mathbb{C}^2$ with real coefficients, having a wandering Fatou component. These Fatou components have nonempty real trace and their statistical behaviour is historic with high emergence. The proof is based on a geometric model for parameter families of surface real mappings.

2.8.4 Elliptic Invariant Fiber

With Peters we investigated in [51] the case of invariant fibers at the center of a Siegel disk. More precisely, we considered a polynomial skew-product of the form (2.1) having an *elliptic invariant fiber*. As before, we can assume without loss of generality that the invariant fiber is $\{w = 0\}$ and so we have $g(0) = 0$ and $g'(0) = e^{2\pi i \theta}$ with $\theta \in \mathbb{R}\setminus\mathbb{Q}$. We assume that the origin belongs to a Siegel disk for g and hence g is locally holomorphically linearizable near $w = 0$. Therefore, up to a local change of coordinates we may assume that F is of the form

$$F(z, w) = (f(z, w), \lambda \cdot w),$$

where $f_w(z) := f(z, w)$ is a polynomial in z with coefficients depending holomorphically on w. We assume that the degree of the polynomial f_w is constant near $\{w = 0\}$, and at least 2.

In this case we only have a partial answer to Question 1. In fact, while we already know that the attracting Fatou components of f_0 always bulge, the general situation appears to be more complicated as there might be resonance phenomena. In [51] we proved the following local result, implying that all parabolic Fatou components of a polynomial skew-product with an elliptic invariant fiber bulge if the rotation number satisfies the Brjuno condition.

Proposition 2.25 ([51, Proposition 2]) *Let F be a holomorphic skew-product of the form*

$$F(z, w) = (f_w(z), g(w)) \tag{2.8}$$

with $g(w) = \lambda w + O(w^2)$, $f_0(0) = 0$, and $f_0'(0) = 1$. Assume λ is a Brjuno number. If $f_0(w) \equiv w$, then F is holomorphically linearizable at the origin. If $f_0(z) = z + f_{0,k+1} z^{k+1} + O(z^{k+2})$ with $f_{0,k+1} \neq 0$ for some $k \geq 1$, then for any $h \geq 0$ there exists a local holomorphic change of coordinates near the origin conjugating F to a map of the form $\widetilde{F}(z, w) = (\widetilde{f}(z, w), \lambda w)$ satisfying

$$\widetilde{f}(z, w) = z + f_{0,k+1} z^{k+1} + \cdots + f_{0,k+h+1} z^{k+h+1} + \sum_{j \geq h} z^{k+j+2} \alpha_{k+j+2}(w), \tag{2.9}$$

where, for $j \geq h$, $\alpha_{k+j+2}(w)$ is a holomorphic function in w such that $\alpha_{k+j+2}(0) = f_{0,k+j+2}$.

We recall that a complex number is called *Brjuno* (see [17] for more details) if

$$\sum_{k=0}^{+\infty} \frac{1}{2^k} \log \frac{1}{\omega(2^{k+1})} < +\infty, \tag{2.10}$$

where $\omega(m) = \min_{2 \leq k \leq m} |\lambda^k - \lambda|$ for any $m \geq 2$, and a quadratic polynomial $\lambda w + w^2$ is linearizable near 0 if and only if λ is Brjuno (see [17, 18, 64]). For higher degree polynomials the Brjuno condition (2.10) is sufficient, but necessity is still an open question.

The proof of the previous proposition makes use of a procedure inspired by the Poincaré-Dulac normalization process [2, Chapter 4] aiming to conjugate the given polynomial skew-product to a skew-product in a simpler form. At each step of the usual Poincaré-Dulac normalization procedure we can use a polynomial change of coordinates to eliminate all non-resonant monomials of a given degree. A similar idea is used here, and thanks to the skew-product structure of the germ and the Brjuno assumption we are able to eliminate all non-resonant terms of a given degree in the powers of z by means of changes of coordinates that are polynomial in z with holomorphic coefficients in w. It turns out that there are some differences in between the first degrees as it is pointed out in [51, Section 2].

The arithmetic Brjuno condition is strongly needed in the proof of Proposition 2.25 and it is natural to ask whether parabolic Fatou components of f_0 always bulge, or it is possible to characterize when they bulge. We do not have a complete answer to such question. However we can prove the following result.

Proposition 2.26 ([51, Proposition 10]) *Let $F(z, w) = (f_z(w), g(w))$ be a holomorphic skew-product with an elliptic linearizable invariant fiber. If F does not admit a holomorphic invariant curve on the invariant fiber, then the parabolic Fatou components of f_0 do not bulge.*

Moreover, we can construct examples showing that the Brjuno condition cannot be completely omitted. In fact by taking *Cremer* numbers, that is $\lambda \in \mathbb{C}$ with $|\lambda| = 1$ and such that

$$\limsup_{m \to \infty} \frac{1}{m} \log \frac{1}{\omega(m)} = +\infty,$$

we can construct examples of polynomial skew-product not admitting a holomorphic invariant curve on the invariant fiber. An elementary example is the following polynomial skew-product

$$F(z, w) = (z + w + zw, \lambda w)$$

with λ a Cremer number (see [51, Example 8] for details). More generally, arguing as in Cremer's example [23] we can show (see [51, Proposition 9]) the existence of holomorphic skew-products without invariant holomorphic curves of the form $\{w = \varphi(z)\}$, having an elliptic invariant fiber that is not point-wise fixed.

Describing the general situation can also be complicated by resonance phenomena. For example, an invariant fiber at the center of a Siegel disk was used in [13] to construct a non-recurrent Fatou component with limit set isomorphic to a punctured disk, and in that construction the invariant fiber also contains a Siegel disk, but with opposite rotation number. Moreover, it might happen that Fatou components on the invariant fiber do not bulge. For example, we can consider the skew-product

$$F(z, w) = (\lambda z(1 + azw), \lambda^{-1} w),$$

with $a \in \mathbb{C}^*$, $\lambda = e^{2\pi i \theta}$ and $\theta \in \mathbb{R} \setminus \mathbb{Q}$. We have $F(z, 0) = (\lambda z, 0)$, but the Siegel disk around the origin in $\{w = 0\}$ is not bulging, and in fact it follows from [15] and [16] that there exists a Fatou component of parabolic type having on its boundary the origin of $\mathbb{C}^2$, which is fixed by F.

We also have an answer to Question 2, under the assumption that the multiplier at the elliptic invariant fiber is Brjuno and all critical points of the polynomial acting on the invariant fiber lie in basins of attracting or parabolic cycles.

Theorem 2.15 ([51, Theorem 1]) *Let F be a polynomial skew-product of the form*

$$F(z, w) = (f_w(z), g(w)), \tag{2.11}$$

and let $\{w = w_0\}$ be an elliptic invariant fiber with multiplier λ. If λ is Brjuno and all critical points of the polynomial f_{w_0} lie in basins of attracting or parabolic cycles, then all Fatou components of f_{w_0} bulge, and there is a neighbourhood of the invariant fiber $\{w = w_0\}$ in which the only Fatou components of F are the bulging Fatou components of f_{w_0}. In particular there are no wandering Fatou components in this neighbourhood.

Idea of the Proof The proof relies on the fact that, in dimension 1, if every critical point of the polynomial either (1) lies in the basin of an attracting periodic cycle, (2) lies in the basin of a parabolic periodic cycle, or (3) lies in the Julia set and after finitely many iterates is mapped to a periodic point, then it is possible to construct a conformal metric μ, defined in a backward invariant neighbourhood of the Julia set minus the parabolic periodic orbits, so that f_c is expansive with respect to this metric. It then follows that there can be no wandering Fatou components.

The key step in the proof is that for fibers $\{w = \tilde{w}_0\}$ sufficiently close to the invariant fiber $\{w = w_0\}$ we can define conformal metrics $\mu_{\tilde{w}_0}$ that depend continuously on $\tilde{w}_0$, and so that F acts expansively with respect to this family of metrics (see [51, Section 3] for details). That is, for a point $(\tilde{z}_0, \tilde{w}_0) \in \mathbb{C}^2$ lying in the region where the metrics are defined, and for a non-zero tangent vector $\xi \in T_{\tilde{z}_0}(\mathbb{C}_{\tilde{w}_0})$ we have that

$$\mu_{\tilde{w}_1}(\tilde{z}_1, df_{\tilde{w}_0}\xi) > \mu_{\tilde{w}_0}(\tilde{z}_0, \xi),$$

where $(\tilde{z}_1, \tilde{w}_1) = F(\tilde{z}_0, \tilde{w}_0)$. Therefore, if $(\tilde{z}_0, \tilde{w}_0)$ does not lie in one of the bulging Fatou components and its orbit $(\tilde{z}_n, \tilde{w}_n)$ remains in the neighbourhood where the expanding metrics are defined, for any tangent vector $\xi \in T_{\tilde{z}_0}(\mathbb{C}_{\tilde{w}_0})$ we have that

$$\mu_{\tilde{w}_n} df_{\tilde{z}_{n-1}} \cdots df_{\tilde{z}_0} \xi \to \infty,$$

from which it follows that the family $(F^{\circ n})_{n \in \mathbb{N}}$ cannot be normal on any neighbourhood of $(\tilde{z}_0, \tilde{w}_0)$. This proves that there are no Fatou components but the bulging components near the invariant fiber $\{w = w_0\}$. □

It is unclear whether the same techniques could be used to deal with pre-periodic critical points lying on the Julia set. The difficulty is that the property that critical points are eventually mapped onto periodic cycles is not preserved in nearby fibers.

Acknowledgments We thank John H. Hubbard for letting us use the book in preparation [20] to write this survey. The first seven sections are adapted from the first chapter of this book. We also thank Marco Abate and Fabrizio Bianchi for useful comments on the first draft of these notes.

References

1. Alexander, D.S.: A history of complex dynamics. In: From Schröder to Fatou and Julia. Aspects of Mathematics, Friedrich Vieweg & Sohn, Braunschweig (1994)
2. Arnold, V.I.: Geometrical Methods in the Theory of Ordinary Differential Equations. Springer, New York (1988)
3. Astorg, M., Buff, X., Dujardin, R., Peters, H., Raissy, J.: A two-dimensional polynomial mapping with a wandering Fatou component. Ann. Math. **184**(1), 263–313 (2016)
4. Astorg, M., Boc-Thaler, L., Peters, H.: Wandering domains arising from Lavaurs maps with Siegel disks (2019). Analysis & PDE **16**(1), 35–88 (2023)
5. Beardon, A.F.: Iteration of rational functions. In: Complex Analytic Dynamical Systems. Graduate Texts in Mathematics, vol. 132. Springer, New York (1991)
6. Bedford, E., Smillie, J.: Polynomial diffeomorphisms of $\mathbb{C}^2$: currents, equilibrium measure and hyperbolicity. Invent. Math. **103**(1), 69–99 (1991)
7. Bedford, E., Smillie, J.: Polynomial diffeomorphisms of $\mathbb{C}^2$: II. Stable manifolds and recurrence. J. Am. Math. Soc. **4**(4), 657–679 (1991)
8. Bedford, E., Smillie, J.: External Rays in the Dynamics of Polynomial Automorphisms of $\mathbb{C}^2$. Contemporary Mathematics, vol. 222, pp. 41–79. AMS, Providence (1999)
9. Bedford, E., Smillie, J., Ueda, T.: Semi-parabolic bifurcations in complex dimension two. Commun. Math. Phys. **350**(1), 1–29 (2017)
10. Berger, P., Biebler, S.: Emergence of wandering stable components. J. Am. Math. Soc. **36**(2), 397–482 (2023)
11. Bianchi, F.: Parabolic implosion for endomorphisms of $\mathbb{C}^2$. J. Eur. Math. Soc. **21**(12), 3709–3737 (2019)
12. Blanchard, P.: Complex analtic dynamics on the Riemann sphere. Bull. Am. Math. Soc. **11**, 85–141 (1984)
13. Boc-Thaler, L., Fornæss, J.E., Peters, H.: Fatou components with punctured limit sets. Ergod. Theory Dyn. Syst. **35**, 1380–1393 (2015)
14. Böttcher, L.E.: The principle laws of convergence of iterates and their applications to analysis (in Russian). Izv. Kazan. Fiz.-Mat. Obshch. **14**, 155–234 (1904)
15. Bracci, F., Zaitsev, D.: Dynamics of one-resonant biholomorphisms. J. Eur. Math. Soc. **15**(1), 179–200 (2013)
16. Bracci, F., Raissy, J., Zaitsev, D.: Dynamics of multi-resonant biholomorphisms. Int. Math. Res. Not. IMRN **20**, 4772–4797 (2013)
17. Brjuno, A.D.: Analytic form of differential equations. I. Trans. Mosc. Math. Soc. **25**, 131–288 (1971)
18. Brjuno, A.D.: Analytic form of differential equations. II. Trans. Mosc. Math. Soc. **26**, 199–239 (1972)
19. Buff, X.: Wandering Fatou component for polynomials. In: Annales de la faculté des sciences de Toulouse Sér. 6, 27 no. 2: Numéro spécial à l'occasion de KAWA Komplex Analysis Winter school And workshop, 2014–2016, 445–475 (2018)
20. Buff, X., Hubbard, J.H.: Dynamics in One Complex Variable (in preparation)
21. Carleson, L., Gamelin, T.W.: Complex Dynamics. Universitext: Tracts in Mathematics. Springer, New York (1993)
22. Cremer, H.: Über die Schrödersche Funktionalgleichung und das Schwartsche Eckenabbildungsproblem. Ber. Verh. Sächs. Akad. Wiss. Leipzig, Math.-Phys. Kl. **84**, 291–324 (1932)
23. Cremer, H.: Über die Häufigkeit der Nichtzentren. Math. Ann. **115**, 573–580 (1938)
24. Denker, M., Przytycki, F., Urbański, M.: On the transfer operator for rational functions on the Riemann sphere. Ergodic Theory Dynam. Syst. **16**(02), 255–266 (1996)
25. Devaney, R.: An introduction to Chaotic Dynamical Systems. Addison-Wesley, Boston (1989)
26. Douady, A.: Disques de Siegel et anneaux de Herman. Sémin. Bourbaki **1986/87**. Astérisque No. 152–153 (1987); 4, 151–172 (1988)

27. Douady, A.: Does a Julia set depend continuously on the polynomial? Complex dynamical systems (Cincinnati, OH, 1994). In: Proceedings of Symposia in Applied Mathematics, vol. 49, pp. 91–138. American Mathematical Society, Providence (1994)
28. Dujardin, R.: A non-laminar dynamical Green current. Math.Ann. **365**(1–2), 77–91 (2016)
29. Dujardin, R., Lyubich, M.: Tability and bifurcations for dissipative polynomial automorphisms of $\mathbb{C}^2$. Invent. Math. **200**(2), 439–511 (2015)
30. Fatou, P.: Sur les solutions uniformes de certaines équations fonctionnelles, C.R. Acad. Sci. Paris **143**, 546–548 (1906)
31. Fatou, P.: Sur les équations fonctionnelles (Deuxième mémoire). Bull. Soc. Math. France **48**, 33–94 (1920)
32. Fatou, P.: Sur les équations fonctionnelles (Troisième mémoire). Bull. Soc. Math. France **48**, 208–314 (1920)
33. Fornæss, J.E., Sibony, N.: Classification of recurrent domains for some holomorphic maps. Math. Ann. **301**(4), 813–820 (1995)
34. Fornæss, J.E., Sibony, N.: Fatou and Julia sets for entire mappings in $\mathbb{C}^k$. Math. Ann. **311**(1), 27–40 (1998)
35. Fornæss, J.E., Sibony, N.: Some open problems in higher dimensional complex analysis and complex dynamics. Publ. Mat. **45**(2), 529–547 (2001)
36. Hahn, D., Peters, H.: A polynomial automorphism with a wandering Fatou component. Adv. Math. **382**, paper no. 107650, 46pp. (2021)
37. Heinemann, S.M.: Julia sets of skew-products in $\mathbb{C}^2$. Kyushu J. Math. **52**(2), 299–329 (1998)
38. Ji, Z.: Non-wandering Fatou components for strongly attracting polynomial skew products. J. Geom. Anal. **30**(1), 124–152 (2020)
39. Ji, Z.: Non-uniform hyperbolicity in polynomial skew products. Int. Math. Res. Not. **2023**, 8755–8799 (2023). arXiv:1909.06084
40. Jonsson, M.: Dynamics of polynomial skew-products on $\mathbb{C}^2$. Math. Ann. **314**, 403–447 (1999)
41. Julia, G.: Mémoire sur l'itération des fonctions rationelles. J. Math. Pures Appl. **8**, 47–245 (1918)
42. Keen, L.: Julia sets, in "Chaos and Fractals, The Mathematics behind the Computer Graphics", edit. Devaney & Keen. In: Proceedings of Symposia in Applied Mathematics, vol. 39, pp. 57–74. American Mathematical Society, Providence (1989)
43. Koenigs, G.: Recherches sur les intégrales de certaines équations fonctionnelles. Ann. Sci. École Norm. Sup. **1**, 3–41 (1884)
44. Lavaurs, P.: Systèemes dynamiques holomorphes, explosion de points péeriodiques paraboliques. Ph.D. Thesis, Université Paris-Sud, Orsay (1989)
45. Lilov, K.: Fatou theory in two dimensions. Ph.D. Thesis, University of Michigan (2004)
46. Lyubich, M.Y.: Dynamics of rational transformations: topological picture. Uspekhi Mat. Nauk **4**(250), 35–95, 239 (1986)
47. Lyubich, M.Y., Peters, H.: Classification of invariant Fatou components for dissipative Hénon maps. Geom. Funct. Anal. **24**, 887–915 (2014)
48. McMullen, C.T.: Complex Dynamics and Renormalization. Annals of Math Studies vol. 135. Princeton University Press, Princeton (1994)
49. McMullen, C.T.: Renormalization and 3-Manifolds Which Fibers Over the Circle. Annals of Math Studies, vol. 142. Princeton University Press, Princeton (1996)
50. Milnor, J.: Dynamics in One Complex Variable. Annals of Mathematics Studies, 3rd edn., vol. 160. Princeton University Press, Princeton (2006)
51. Peters, H., Raissy, J.: Fatou components of polynomial elliptic skew products. Ergodic Theory Dynam. Syst. **39**(8), 2235–2247 (2019)
52. Peters, H., Smit, I.M.: Fatou components of attracting skew products. J. Geom. Anal. **28**(1), 84–110 (2018)
53. Peters, H., Vivas, L.: Polynomial skew-products with wandering Fatou-disks. Math. Z. **283**(1–2), 349–366 (2016)
54. Roeder, R.: A dichotomy for Fatou components of polynomial skew-products. Conform. Geom. Dyn. **15**, 7–19 (2011)

55. Rosay, J.P., Rudin, W.: Holomorphic maps from $\mathbb{C}^n$ to $\mathbb{C}^n$. Trans. Am. Math. Soc. **310**(1), 47–86 (1988)
56. Schröder, E.: Ueber iterirte Functionen. Math. Ann. **3**(2), 296–322 (1870)
57. Shishikura, M.: Bifurcation of parabolic fixed points. In: The Mandelbrot Set, Theme and Variations. London Mathematical Society Lecture Note Series, vol. 274, pp. 325–363. Cambridge University Press, Cambridge (2000)
58. Siegel, C.L.: Iteration of analytic functions. Ann. Math. **43**, 607–612 (1942)
59. Steinmetz, N.: Rational iteration. In: Complex Analytic Dynamical Systems. De Gruyter Studies in Mathematics, vol. 16. Walter de Gruyter, Berlin (1993)
60. Sullivan, D.: Quasiconformal Homeomorphisms and Dynamics I. Solution of the Fatou-Julia Problem on Wandering Domains. Ann. Math. **122**(3), 401–418 (1985)
61. Ueda, T.: Fatou sets in complex dynamics on projective spaces. J. Math. Soc. Jpn. **46**, 545–555 (1994)
62. Vivas, L.: Parametrization of unstable manifolds and Fatou disks for parabolic skew-products. Complex Anal. Synerg. **4**(1), paper no. 1 (2018)
63. Vivas, L.: Non-autonomous parabolic bifurcation. Proc. Am. Math. Soc. **148**(6), 2525–2537 (2020)
64. Yoccoz, J.C.: Théorème de Siegel, nombres de Brjuno et polynômes quadratiques. Astérisque **231**, 3–88 (1995)

Chapter 3
Homogeneous Dynamics and its Connection to Diophantine Approximation

Manfred Einsiedler and Thomas Ward

Abstract These notes introduce how homogeneous dynamics may be used to study certain questions in Diophantine approximation.

3.1 Introduction

In these notes we try to set up the basics of homogeneous dynamics, and connect this to the theory of Diophantine approximation. In the interest of keeping this introduction short, we will sometimes only sketch some of the details. We will also often refer to our first joint book project [5] for notions like ergodicity, mixing, and complete proofs of some of the results sketched here and, later, to one of the ongoing follow-up projects [7] for some more advanced results and background. Indeed, instead of attempting to present all the many theorems in the area (which would fill many books), we try to pick certain crucial ones in order to reach some classical, and some more recent, results in the area of Diophantine approximation that have particularly beautiful proofs using dynamical ideas.

Speaking of beauty, the first named author of these notes would like to thank the Centro Internazionale Matematico Estivo (C.I.M.E.) and the organizers of the C.I.M.E. summer school "Modern aspects of dynamical systems" for organizing the school at the beautiful beach in Cetraro. We would also like to thank Andreas Wieser for his comments on an earlier draft of these notes.

These notes come from the Fondazione CIME 'Modern Aspects of Dynamical Systems' Summer school(20–24 July 2020) given by the first named author.

M. Einsiedler (✉)
Departement Mathematik, ETH Zürich, Zürich, Switzerland
e-mail: manfred.einsiedler@math.ethz.ch

T. Ward
Department of Mathematical Sciences, Durham University, Durham, UK

3.2 Hyperbolic Surfaces

We start our introduction to dynamics on homogeneous spaces with dynamics on hyperbolic surfaces, referring to [5, Ch. 9] for more details. 'Homogeneous dynamics' is, roughly speaking, the study of actions of subgroups $H < G$ of a Lie group G on spaces of the form G/Γ where $\Gamma < G$ is a discrete subgroup. This may initially sound intimidating, but we hope that the following concrete discussion will persuade the reader that this set-up is worth studying.

3.2.1 Hyperbolic Geometry

We define the hyperbolic plane in the form of the upper half-plane model

$$\mathbb{H} = \{z \in \mathbb{C} \mid \Im z > 0\},$$

where $\Im z$ denotes the imaginary part of $z \in \mathbb{C}$, which comes equipped with the hyperbolic Riemannian metric

$$\frac{(\mathrm{d}x)^2 + (\mathrm{d}y)^2}{y^2}.$$

This defines in turn:

- The length of a tangent vector $(z, v) \in \mathrm{T}\mathbb{H} = \mathbb{H} \times \mathbb{C}$ by the formula

$$\|(z, v)\|_{\mathrm{hyp}} = \frac{\|v\|}{\Im z},$$

  where $\|v\|$ denotes the Euclidean length of $v \in \mathbb{C}$.
- The length of smooth paths (by integrating the hyperbolic length of their derivative).
- The hyperbolic distance between two points in $\mathbb{H}$, as the infimum of all smooth paths connecting them.

At this point the reader might wonder what is so special about this? The answer lies in the intrinsic symmetry—and resulting beauty—behind these definitions. In fact—just as for Euclidean geometry—there is a huge group of maps acting on $\mathbb{H}$ that preserve this Riemannian metric, meaning they are *isometries* of $\mathbb{H}$.

Exercise 3.2.1 In fact a direct calculation[1] reveals that for

$$g = \begin{pmatrix} a & b \\ c & d \end{pmatrix} \in SL_2(\mathbb{R})$$

the so-called Möbius transformation

$$\mathbb{H} \ni z \longmapsto g \cdot z = \frac{az+b}{cz+d}$$

maps $\mathbb{H}$ to $\mathbb{H}$, and preserves (via the derivative action on $T\mathbb{H}$) the length of vectors in $T\mathbb{H}$. Hence the length of smooth curves, and so also the distance between two points in $\mathbb{H}$, is preserved by this transformation.

Exercise 3.2.2 Another calculation[2] reveals that this defines an action of $SL_2(\mathbb{R})$ on $\mathbb{H}$, which is transitive since

$$\begin{pmatrix} 1 & x \\ & 1 \end{pmatrix} \begin{pmatrix} \sqrt{y} & \\ & \sqrt{y}^{-1} \end{pmatrix} \cdot i = x + iy$$

for any $z = x + iy \in \mathbb{H}$. Moreover, the derived action is also transitive on the *unit tangent bundle of the hyperbolic plane*

$$T^1 \mathbb{H} = \left\{ (z, v) \in T\mathbb{H} \ \Big| \ \|(z, v)\|_{\text{hyp}} = 1 \right\}.$$

Hence

$$\mathbb{H} \cong SL_2(\mathbb{R}) / SO_2(\mathbb{R})$$

and

$$T^1 \mathbb{H} \cong PSL_2(\mathbb{R}) = SL_2(\mathbb{R})/\{\pm I\}$$

since

$$SO_2(\mathbb{R}) = \text{Stab}_{SL_2(\mathbb{R})}(i)$$

and

$$\{\pm I\} = C(SL_2(\mathbb{R})) = \text{Stab}_{SL_2(\mathbb{R})}((i, i)).$$

---

[1] Doing this exercise will also reveal the importance of the factor $(cz + d)^{-2}$ that (for related reasons) plays an important role in the theory of modular forms.
[2] We will omit 0s from matrices where they do not play an interesting role.

Here we used the general isomorphism

$$G/\operatorname{Stab}_G(x_0) \cong \operatorname{Orbit}(x_0)$$

for the action of a group G on a space X containing x_0, the *stabilizer*

$$\operatorname{Stab}_G(x_0) = \{g \in G \mid g{\cdot}x_0 = x_0\},$$

and the *orbit*,

$$\operatorname{Orbit}(x_0) = G{\cdot}x_0 = \{g{\cdot}x_0 \mid g \in G\}$$

of x_0.

Our goal is to define 'algebraic' or 'homogeneous' dynamical systems relying on this geometry. The hyperbolic plane, like the Euclidean one, is non-compact with infinite area.[3] Hence it cannot be expected to exhibit any kind of recurrence, which would be the beginning of interesting dynamical phenomena. Thus it is important to consider quotients by (actions of) discrete subgroups of the isometry group of $\mathbb{H}$ in order to expect interesting dynamical phenomena.

Definition 3.2.3 (Lattices) A discrete subgroup $\Gamma < \operatorname{SL}_2(\mathbb{R})$ is called a *lattice* if $\Gamma \backslash \mathbb{H}$ (or, equivalently, $\Gamma \backslash \operatorname{SL}_2(\mathbb{R})$) has finite volume with respect to the hyperbolic measure $\frac{dx\,dy}{y^2}$ (or with respect to the Haar measure on $\operatorname{SL}_2(\mathbb{R})$, respectively). A discrete subgroup $\Gamma < \operatorname{SL}_2(\mathbb{R})$ is called a *uniform lattice* if $\Gamma \backslash \mathbb{H}$ (or, equivalently, $\Gamma \backslash \operatorname{SL}_2(\mathbb{R})$) is compact.

We note that here, and more generally in homogeneous dynamics, the Haar measure of a group often plays the same role as the Lebesgue measure does in Euclidean geometry and basic real analysis. In particular, Haar measures should be viewed as readily understood objects that can be calculated using explicit formulas.

The easiest example of a lattice (in this hyperbolic context) is in fact the non-uniform lattice $\Gamma = \operatorname{SL}_2(\mathbb{Z})$. A fundamental domain for its action on $\mathbb{H}$ is illustrated in Fig. 3.1.

3.2.2 Dynamics on Quotients of $\mathbb{H}$

The most straightforward dynamical system related to $\mathbb{H}$, its geometry, and its quotients is the *geodesic flow*. In fact, every point $(z, v) \in \mathrm{T}^1\mathbb{H}$ is the tangent vector of a unique unit speed smooth path that connects any of its points with minimal hyperbolic path length. These paths are called geodesic paths, and their images are

---

[3] Indeed, in a geometric sense the hyperbolic plane is even bigger than the Euclidean one.

3 Homogeneous Dynamics and its Connection to Diophantine Approximation

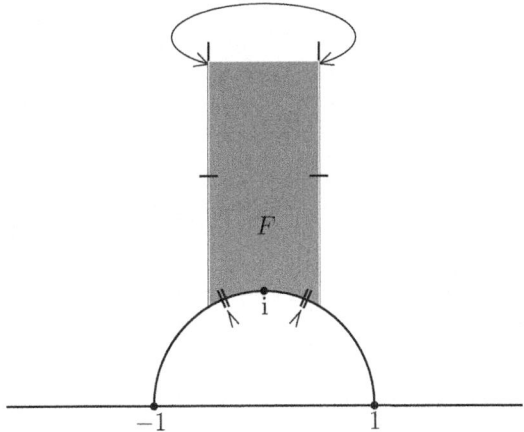

Fig. 3.1 The hyperbolic triangle $F \subseteq \mathbb{H}$ in this picture is (after making clear which points of the boundary ∂F should belong to F) a fundamental domain for the action of $SL_2(\mathbb{Z})$. From this it is easy to show that $SL_2(\mathbb{Z}) \backslash \mathbb{H} \cong F$ indeed has finite hyperbolic volume (defined by $\frac{dx\,dy}{y^2}$). To obtain the quotient $SL_2(\mathbb{Z}) \backslash \mathbb{H}$ as a geometric object from F, you should envision that the right and left boundaries of F are glued together using the isometry $z \mapsto \begin{pmatrix} 1 & 1 \\ & 1 \end{pmatrix} \cdot z = z + 1$, and the bottom boundary is glued to itself using the map $z \mapsto \begin{pmatrix} & -1 \\ 1 & \end{pmatrix} \cdot z = -\frac{1}{z}$ instead

the *geodesics* in $\mathbb{H}$. In the hyperbolic geometry of $\mathbb{H}$, these play the role of straight lines. Some geodesics are illustrated in Fig. 3.2, with a dot denoting the coordinate z and a small arrow denoting the vector v of a point $(z, v) \in T^1 \mathbb{H}$.

We will always use $(i, i) \in T^1 \mathbb{H}$ as our *initial reference vector*. For this initial point $i \in \mathbb{H}$ and direction i the geodesic path, which takes the form

$$\mathbb{R} \ni t \longmapsto \eta(t) = e^t i = \begin{pmatrix} e^{t/2} & \\ & e^{-t/2} \end{pmatrix} \cdot i,$$

is particularly simple.

Exercise 3.2.4 Using the reference vector $(i, i) \in T^1 \mathbb{H}$ and the isomorphism

$$PSL_2(\mathbb{R}) \cong T^1 \mathbb{H}$$

defined by $g \mapsto g \cdot (i, i)$, it is now a brain-twister (but no more than that) to verify that moving for time t along the geodesic tangent to $g \cdot (i, i)$ and starting at $g \cdot (i, i)$ corresponds under the isomorphism to multiplication on the *right* by the matrix

$$\begin{pmatrix} e^{t/2} & \\ & e^{-t/2} \end{pmatrix}.$$

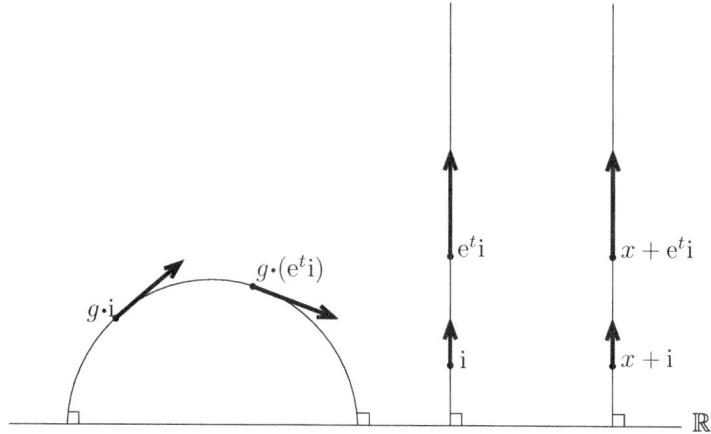

Fig. 3.2 The path defined by $\eta(t) = e^t i$ for $t \in \mathbb{R}$ is a geodesic path. Applying any $g \in SL_2(\mathbb{R})$ to this path gives another geodesic path, which always travels along (vertical) lines or along semi-circles that meet the horizontal axis $\mathbb{R}$ at right angles. In fact every geodesic arises in this way by a suitable choice of $g \in SL_2(\mathbb{R})$.

Also note that the action of some $h \in PSL_2(\mathbb{R})$ by its Möbius transformation on $T^1 \mathbb{H}$ corresponds in the same way to multiplication by h on the *left* in $PSL_2(\mathbb{R})$. In particular, the description of the geodesic flow as multiplication on the right by diagonal matrices therefore descends to any left quotient $\Gamma \backslash PSL_2(\mathbb{R}) \cong \Gamma \backslash T^1 \mathbb{H}$ by a discrete subgroup $\Gamma < PSL_2(\mathbb{R})$.

Note that the canonical action of $g \in PSL_2(\mathbb{R})$ on an element of a quotient $\Gamma \backslash PSL_2(\mathbb{R}) \ni x$ is defined by $g \cdot x = xg^{-1}$. To make the description of the geodesic flow in Exercise 3.2.4 compatible with this, we define

$$a_t = \begin{pmatrix} e^{-t/2} & \\ & e^{t/2} \end{pmatrix}$$

for $t \in \mathbb{R}$. In this sense the natural action of the subgroup

$$A = \{a_t \mid t \in \mathbb{R}\}$$

on $\Gamma \backslash PSL_2(\mathbb{R})$ corresponds to the geodesic flow.

There are a few more distinguished subgroups of $PSL_2(\mathbb{R})$ with geometric significance (which, as in the case of A, will again depend on having made the same choice of reference vector $(i, i) \in T^1 \mathbb{H}$), as follows.

- The action of $PSO_2(\mathbb{R}) = SO_2(\mathbb{R})/\{\pm I\}$ on $\Gamma \backslash PSL_2(\mathbb{R})$ rotates the tangent vector without moving the base point in $\Gamma \backslash \mathbb{H}$, as illustrated on the left in Fig. 3.3.

3 Homogeneous Dynamics and its Connection to Diophantine Approximation 111

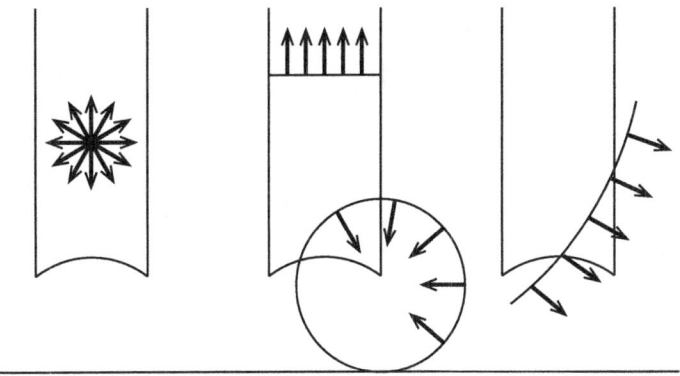

Fig. 3.3 The action of $PSO_2(\mathbb{R})$, U, and V on tangent vectors of the quotient space $SL_2(\mathbb{Z})\backslash\mathbb{H}$

- The action of

$$U = \left\{ \begin{pmatrix} 1 & s \\ & 1 \end{pmatrix} \,\middle|\, s \in \mathbb{R} \right\}$$

is called the *stable horocycle flow* and naturally parameterizes the 'stable manifold' of a point with respect to the geodesic flow, as illustrated in the middle of Fig. 3.3.

- The action of

$$V = \left\{ \begin{pmatrix} 1 & \\ s & 1 \end{pmatrix} \,\middle|\, s \in \mathbb{R} \right\}$$

is, similarly, the *unstable horocycle flow* and naturally parameterizes the 'unstable manifold' of a point with respect to the geodesic flow, as illustrated on the right in Fig. 3.3.

Even though we will for the most part consider higher-dimensional analogues, we hope the discussion above gives the reader some geometric motivation and intuition for our largely algebraic construction.

3.2.3 The Artin–Dani Correspondence

As we wish to emphasise the connection between homogeneous dynamics and Diophantine approximation, we take a moment to point out—informally—a connection found by Artin in 1924 [1] and generalized by Series [12], Dani [4], and others.

For Artin, this was a setting in which to see properties of one of the first ergodic systems using input from the theory of Diophantine approximation. We use

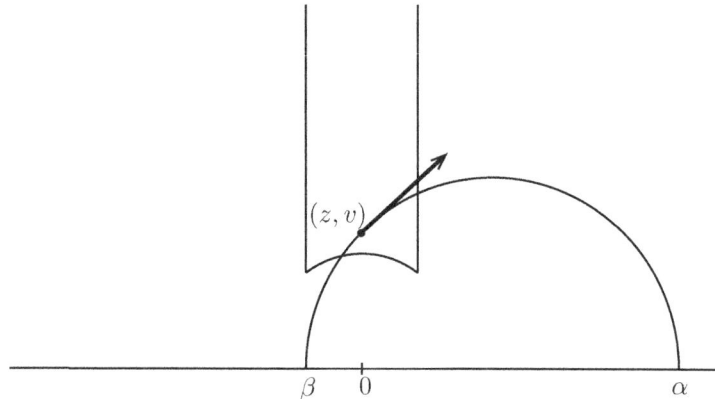

Fig. 3.4 A geodesic with real 'end points' $\alpha > 0$ and $\beta < 0$

this connection in reverse, to allow dynamical proofs of theorems in Diophantine approximation.

We consider a point $(z, v) \in T^1 \mathbb{H}$ with z in the fundamental domain F for the Möbius action of $\mathrm{SL}_2(\mathbb{Z})$. Suppose the geodesic through (z, v) has a 'forward end point' $\alpha > 0$ and a 'backward end point' $\beta < 0$ (that is, the points in $\mathbb{R}$ 'reached' by following the geodesic infinitely far in a forward direction or a backward direction, respectively), as illustrated in Fig. 3.4.

Following the geodesic into the immediate future, the base point z will eventually leave the fundamental domain F. Suppose $\alpha > 1$, so that the geodesic leaves F by going to the right. At this point we can apply the Möbius transformation

$$\begin{pmatrix} 1 & -1 \\ & 1 \end{pmatrix} : z \longmapsto z - 1$$

to move back into the fundamental domain F. Since the corresponding matrix lies in $\mathrm{SL}_2(\mathbb{Z})$, this move in $\mathbb{H}$ is not a move in the space $\mathrm{SL}_2(\mathbb{Z}) \backslash \mathbb{H}$, but an identification of the same point in the fundamental domain. This may of course happen many times in a row, and along the way the end point $\alpha \in \mathbb{R}$ will be replaced by $\alpha - 1, \alpha - 2, \ldots, \alpha - n$ for some $n \in \mathbb{N}$. Eventually $\alpha - n \in [0, 1)$ and the geodesic will not leave the fundamental domain via the right-hand boundary, but instead via the bottom boundary. We ignore for the moment some other exceptional possibilities, for example a path that eventually meets a corner of F. At this point we will have to apply the isometry

$$\begin{pmatrix} & -1 \\ 1 & \end{pmatrix} : z \longmapsto -\frac{1}{z},$$

which then replaces the forward end point $\alpha - n$ by $-\frac{1}{\alpha - n}$.

Apart from the switch in sign, the procedure outlined above is reminiscent of the Gauss map

$$T : [0, 1] \setminus \mathbb{Q} \ni s \longmapsto \left\{\frac{1}{s}\right\} \in [0, 1] \setminus \mathbb{Q},$$

where $\{\cdot\}$ denotes the fractional part of a real number. In fact, iterating T involves successively taking reciprocals and fractional parts (and the latter process is subtraction of some $n \in \mathbb{N}_0$). Following the geodesic flow, we similarly will see successively fractional parts and negative reciprocals. Making this more precise reveals a concrete connection between the geodesic flow and the Gauss map in the following sense. There is a Poincaré cross-section (that is, a surface in the three-manifold $\mathrm{PSL}_2(\mathbb{Z}) \backslash \mathrm{PSL}_2(\mathbb{R})$ that is transverse to the geodesic flow) with the property that in a convenient coordinate system the first return map for the geodesic flow to this cross-section is essentially the invertible extension of the Gauss map.

As the Gauss map is, in turn, intimately connected to the continued fraction expansion of real numbers, this indeed links the dynamics of the geodesic flow to the theory of Diophantine approximation for real numbers. Thus it should not be surprising that various Diophantine properties of real numbers can be understood in terms of the behaviour of orbits with respect to the geodesic flow. For now we will simply give an example of this phenomenon.

Definition 3.2.5 A number $s \in \mathbb{R}$ is *badly approximable* if

$$\inf_{n \in \mathbb{N}} n |ns|_\mathbb{Z} > 0,$$

where $|s|_\mathbb{Z} = \min_{m \in \mathbb{Z}} |s - m|$ denotes the distance to the nearest integer.

Lemma 3.2.6 (Artin–Dani for Badly Approximable Numbers) *A number $s \in \mathbb{R}$ is badly approximable if and only if the forward orbit*

$$\left\{ \mathrm{PSL}_2(\mathbb{Z}) \begin{pmatrix} 1 & \\ s & 1 \end{pmatrix} a_t^{-1} \,\bigg|\, t \geq 0 \right\}$$

has compact closure in $\mathrm{PSL}_2(\mathbb{Z}) \backslash \mathrm{PSL}_2(\mathbb{R})$.

Exercise 3.2.7 We will prove the correspondence above in greater generality later, but the reader may find it instructive to attempt to prove it now.

3.3 The Moduli Space of d-Dimensional Lattices

In this section we wish to discuss some examples of the following generalization of Definition 3.2.3 that will be crucial for all of the remaining material.

Definition 3.3.1 Let G be a metrizable locally compact group. A discrete subgroup $\Gamma < G$ is called a *lattice* if the quotient G/Γ supports a G-invariant Borel probability measure, and is called a *uniform lattice* if G/Γ is compact.

Notice that G/Γ inherits the quotient topology from the topology on G.

3.3.1 The Modular Curve

Exercise 3.3.2 Show that for a discrete subgroup $\Lambda < \mathbb{R}^2$ exactly one of the following holds:

- $\Lambda = \{0\}$;
- $\Lambda = \mathbb{Z}\mathbf{v}$ for some $\mathbf{v} \in \mathbb{R}^2 \smallsetminus \{0\}$; or
- $\Lambda = \mathbb{Z}\mathbf{v}_1 + \mathbb{Z}\mathbf{v}_2$ for two vectors $\mathbf{v}_1, \mathbf{v}_2 \in \mathbb{R}^2$ that are linearly independent over $\mathbb{R}$.

Also show that Λ is a (automatically uniform) lattice in $\mathbb{R}^2$ if and only if Λ is of the third type.

For a lattice $\Lambda = \mathbb{Z}\mathbf{v}_1 + \mathbb{Z}\mathbf{v}_2 < \mathbb{R}^2$ as in the above exercise, we say that Λ is *unimodular* if $|\det(\mathbf{v}_1, \mathbf{v}_2)| = 1$ (where we are writing $\mathbf{v}_1, \mathbf{v}_2$ as column vectors). Notice that we may assume that $\det(\mathbf{v}_1, \mathbf{v}_2) > 0$ as we may replace $\mathbf{v}_1$ by $-\mathbf{v}_1$ to ensure this. It follows that the *moduli space of* (in other words, the space of all) *unimodular lattices* in $\mathbb{R}^2$ is given by

$$\mathsf{X}_2 = \left\{ g\mathbb{Z}^2 \mid g \in \mathrm{SL}_2(\mathbb{R}) \right\} \cong \mathrm{SL}_2(\mathbb{R})/\mathrm{SL}_2(\mathbb{Z}).$$

Using the map

$$g\,\mathrm{SL}_2(\mathbb{Z}) \longmapsto \mathrm{PSL}_2(\mathbb{Z})g^{-1},$$

this space is also isomorphic to the unit tangent bundle

$$\mathsf{X}_2 \cong \mathrm{SL}_2(\mathbb{R})/\mathrm{SL}_2(\mathbb{Z}) \cong \mathrm{PSL}_2(\mathbb{Z})\backslash\mathrm{PSL}_2(\mathbb{R})$$

of the surface considered in Chap. 3.2. The *modular curve* Y_2 is the set of unimodular lattices in $\mathbb{R}^2$ considered up to rotations. For this space we then precisely get an isomorphism to the hyperbolic surface

$$\begin{aligned}\mathsf{Y}_2 &\cong \mathrm{SO}_2(\mathbb{R})\backslash\mathrm{SL}_2(\mathbb{R})/\mathrm{SL}_2(\mathbb{Z}) \\ &\cong \mathrm{PSL}_2(\mathbb{Z})\backslash\mathrm{PSL}_2(\mathbb{R})/\mathrm{PSO}_2(\mathbb{R}) \\ &\cong \mathrm{PSL}_2(\mathbb{Z})\backslash\mathbb{H}.\end{aligned}$$

Exercise 3.3.3 Try to understand the isomorphism above more concretely, by finding a formula relating the y-coordinate of a representative in the fundamental domain $F \subseteq \mathbb{H}$ to the length of the shortest vector in the corresponding unimodular lattice $\Lambda < \mathbb{R}^2$.

3.3.2 Minkowski's Geometry of Numbers and Successive Minima

Exercise 3.3.4 Show that a discrete subgroup $\Lambda < \mathbb{R}^d$ is a lattice if and only if we have $\Lambda = g\mathbb{Z}^d$ for some $g \in \mathrm{GL}_d(\mathbb{R})$.

We will refer to $|\det g|$ as the *covolume* $\mathrm{covol}(\Lambda)$ of the lattice $\Lambda = g\mathbb{Z}^d$ with $g \in \mathrm{GL}_d(\mathbb{R})$, since $g[0, 1)^d \subseteq \mathbb{R}^d$ is a fundamental domain for Λ with Lebesgue measure $|\det g|$. Note that we may assume that $\det g > 0$ by switching the sign of one of its column vectors. Once again we say that a lattice $\Lambda < \mathbb{R}^d$ is unimodular it its covolume is 1.

The following theorems of Minkowski [11] are also important ingredients in (algebraic) number theory, forming the basis of the 'geometry of numbers'.

Theorem 3.3.5 (Minkowski's First Theorem) *Let $d \geq 2$ and let $\Lambda < \mathbb{R}^d$ be a unimodular lattice. Then $|\Lambda \cap [-1, 1]^d| > 1$.*

Morally, the proof will reveal that this is a form of pigeonhole principle for lattices.

Proof of Theorem 3.3.5 Let $\varepsilon > 0$ and define the slightly expanded d-dimensional box

$$K_\varepsilon = \left[-\frac{1+\varepsilon}{2}, \frac{1+\varepsilon}{2}\right]^d.$$

As the Lebesgue measure of K_ε is strictly bigger than 1, the canonical projection

$$\pi : \mathbb{R}^d \ni \mathbf{x} \longmapsto \mathbf{x} + \Lambda \in \mathbb{R}^d/\Lambda$$

cannot be injective when restricted to K_ε. Hence there exists some $\mathbf{x} \neq \mathbf{y}$ in K_ε with $\pi(\mathbf{x}) = \pi(\mathbf{y})$, or equivalently with $0 \neq \mathbf{x} - \mathbf{y} \in \Lambda \cap (2K_\varepsilon)$. Letting ε decrease to 0 and recalling that Λ is discrete (and so is also closed), we obtain the result. □

Exercise 3.3.6 Which properties of the set $[-1, 1]^d$ were really used in this argument? Generalize the statement of the theorem accordingly.

The second theorem of Minkowski needs the following definitions.

Definition 3.3.7 (Successive Minima) Let $d \geqslant 2$ and let $\Lambda < \mathbb{R}^d$ be a lattice. For each $k \in \{1, \ldots, d\}$ we define[4]

$$\lambda_k(\Lambda) = \min\left\{r > 0 \;\middle|\; \Lambda \cap \overline{B_r(0)} \text{ contains } k \text{ linearly independent vectors}\right\}.$$

In the following we will write $A \ll_d B$ for two positive quantities A, B if $\frac{A}{B}$ is bounded from above by a constant depending only on d, and write $A \asymp_d B$ if $A \ll_d B$ and $B \ll_d A$.

Theorem 3.3.8 (Minkowski's Second Theorem) *Let $\Lambda < \mathbb{R}^d$ be a lattice. Then*

$$\lambda_1(\Lambda) \cdots \lambda_d(\Lambda) \asymp_d \operatorname{covol}(\Lambda),$$

and there exists a $\mathbb{Z}$-basis $\mathbf{v}_1, \ldots, \mathbf{v}_d \in \Lambda$ of Λ (meaning that $\Lambda = \mathbb{Z}\mathbf{v}_1 + \cdots + \mathbb{Z}\mathbf{v}_d$) so that $\|\mathbf{v}_k\| \asymp_d \lambda_k(\Lambda)$ for $k = 1, \ldots, d$.

Morally, the theorem also says that for the well-chosen $\mathbb{Z}$-basis $\mathbf{v}_1, \ldots, \mathbf{v}_d$ of Λ the angles between the vectors are almost irrelevant for determining the covolume of Λ.

Sketch of Proof of Theorem 3.3.8 We prove the theorem by induction on the dimension d, and always choose $\mathbf{v}_1 \in \Lambda \smallsetminus \{0\}$ to be a vector of minimal length (so that $\|\mathbf{v}_1\| = \lambda_1(\Lambda)$).

In the case $d = 1$ one has $\Lambda = \mathbb{Z}\mathbf{v}_1$ (see also the argument below) and the claims in the theorem are trivial.

For the inductive step, we define W to be the orthogonal complement of $\mathbb{R}\mathbf{v}_1$. Let

$$\pi: \mathbb{R}^d \longrightarrow W \cong \mathbb{R}^{d-1}$$

be the orthogonal projection, and define $\Lambda^\perp = \pi(\Lambda)$. We next claim that for any $\mathbf{w} \in \pi(\Lambda)$ there exists some $\mathbf{v} \in \Lambda$ with $\pi(\mathbf{v}) = \mathbf{w}$ and

$$\|\mathbf{v}\| \leqslant \sqrt{\|\mathbf{w}\|^2 + \tfrac{1}{4}\lambda_1(\Lambda)^2}.$$

In particular, $\Lambda^\perp$ is discrete and all of its non-zero vectors have length at least $\frac{\sqrt{3}}{2}\lambda_1(\Lambda)$.

To prove the claim, suppose $\mathbf{w} = \pi(\widetilde{\mathbf{v}})$ for some $\widetilde{\mathbf{v}} \in \Lambda \smallsetminus \mathbb{R}\mathbf{v}_1$. Using 'division with remainder' for the projection of $\widetilde{\mathbf{v}}$ to $\mathbb{R}\mathbf{v}_1$ with respect to $\mathbf{v}_1$, we can find some $\mathbf{v} \in \Lambda$ with $\pi(\mathbf{v}) = \pi(\widetilde{\mathbf{v}}) = \mathbf{w}$ so that $\mathbf{v} = t\mathbf{v}_1 + \mathbf{w}$ with some $t \in [-\tfrac{1}{2}, \tfrac{1}{2}]$, as illustrated in Fig. 3.5.

---

[4] Because of the equivalence of norms on $\mathbb{R}^d$, it does not matter much which norm we use to define the closed balls $\overline{B_r(0)}$. However, we will use the Euclidean norm as this will be useful for the inductive arguments used later.

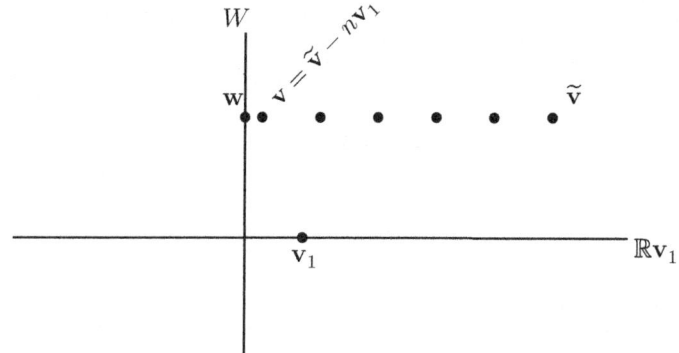

Fig. 3.5 Adding or subtracting integer multiples of v_1 to $\tilde{v} = sv_1 + w$ for some $s \in \mathbb{R}$, we find a vector $v = tv_1 + w$ for some $t \in [-\frac{1}{2}, \frac{1}{2}]$ whose projection to $\mathbb{R}v_1$ has length at most $\frac{1}{2}\|v_1\| = \frac{1}{2}\lambda_1(\Lambda)$

Taking the norm, we obtain

$$\|v\|^2 = \|tv_1\|^2 + \|w\|^2 \leq \tfrac{1}{4}\|v_1\|^2 + \|w\|^2 = \tfrac{1}{4}\lambda_1(\Lambda)^2 + \|w\|^2,$$

which gives the first part of the claim. For the second part, suppose for the moment that $w \neq 0$ and $\|w\| < \frac{\sqrt{3}}{2}\lambda_1(\Lambda)$. Then, by applying the argument above, we find a non-zero vector $v \in \Lambda$ with

$$\|v\|^2 \leq \|w\|^2 + \tfrac{1}{4}\lambda_1(\Lambda)^2 < \tfrac{3}{4}\lambda_1(\Lambda)^2 + \tfrac{1}{4}\lambda_1(\Lambda)^2 = \lambda_1(\Lambda)^2.$$

However, this contradicts the definition of $\lambda_1(\Lambda)$, and the claim follows.

As Λ contains d linearly independent vectors and $\ker \pi = \mathbb{R}v_1$ is a one-dimensional space, $\pi(\Lambda)$ must contain $d-1$ linearly independent vectors. Together with the above discreteness, it follows that $\pi(\Lambda)$ is a lattice in W. The claim above shows that

$$\lambda_1(\pi(\Lambda)) \geq \tfrac{\sqrt{3}}{2}\lambda_1(\Lambda). \tag{3.1}$$

Moreover, applying the claim to k linearly independent vectors in $\pi(\Lambda)$ of length at most $\lambda_k(\pi(\Lambda))$, we also obtain

$$\lambda_{k+1}(\Lambda) \leq \max\left(\lambda_1(\Lambda), \sqrt{\lambda_k(\pi(\Lambda))^2 + \tfrac{1}{4}\lambda_1(\Lambda)^2}\right).$$

Using (3.1) twice in the form

$$\lambda_1(\Lambda) \ll_d \lambda_1(\pi(\Lambda)) \leq \lambda_k(\pi(\Lambda)),$$

we also obtain
$$\lambda_{k+1}(\Lambda) \ll_d \lambda_k(\pi(\Lambda)).$$

As π has one-dimensional kernel and has Lipschitz constant 1, the opposite inequality
$$\lambda_k(\pi(\Lambda)) \leq \lambda_{k+1}(\Lambda)$$
holds trivially for $k, \ldots, d-1$.

Suppose now that $F_W \subseteq W$ is a fundamental domain for $\pi(\Lambda)$. The inductive hypothesis now implies that the Lebesgue measure m_W on W satisfies
$$\mathrm{covol}(\pi(\Lambda)) = m_W(F_W) \asymp_d \lambda_1(\pi(\Lambda)) \cdots \lambda_{d-1}(\pi(\Lambda)).$$

From this we obtain that
$$F = F_W + [0,1]\mathbf{v}_1$$
is a fundamental domain for Λ, and that
$$\begin{aligned}
\mathrm{covol}(\Lambda) &= m_{\mathbb{R}^d}(F_W + [0,1]\mathbf{v}_1) \\
&= m_W(F_W) \cdot \|\mathbf{v}_1\| \\
&\asymp_d \lambda_1(\pi(\Lambda)) \cdots \lambda_{d-1}(\pi(\Lambda))\lambda_1(\Lambda) \\
&\asymp_d \lambda_1(\Lambda) \cdots \lambda_d(\Lambda)
\end{aligned}$$
as claimed.

The existence of the $\mathbb{Z}$-basis as claimed in the theorem follows by the same type of inductive argument. □

The following exercise shows why we need the implicit constants in the norm of the elements of the $\mathbb{Z}$-basis in the result above.

Exercise 3.3.9 Define a lattice in $\mathbb{R}^d$ by
$$\Lambda = \mathbb{Z}^d + \mathbb{Z}\begin{pmatrix} \frac{1}{2} \\ \vdots \\ \frac{1}{2} \end{pmatrix}.$$

Show that $\mathrm{covol}(\Lambda) = \frac{1}{2}$ and $\lambda_1(\Lambda) = \cdots = \lambda_d(\Lambda) = 1$ for $d \geq 4$, but any $\mathbb{Z}$-basis of Λ contains a vector of length at least $\frac{\sqrt{d}}{2}$. Hence for $d \geq 5$ we cannot guarantee the existence of a $\mathbb{Z}$-basis $\mathbf{v}_1, \ldots, \mathbf{v}_k$ of a given lattice with the desired property that $\|\mathbf{v}_k\| \leq \lambda_k(\Lambda)$ for $k = 1, \ldots, d$.

3.3.3 The Moduli Space of Lattices

We define the *moduli space of unimodular lattices* by

$$X_d = \{\Lambda < \mathbb{R}^d \mid \Lambda \text{ is a lattice with } \operatorname{covol}(\Lambda) = 1\}$$
$$= \{g\mathbb{Z}^d \mid g \in \operatorname{SL}_d(\mathbb{R})\}$$
$$\cong \operatorname{SL}_d(\mathbb{R}) / \operatorname{Stab}_{\operatorname{SL}_d(\mathbb{R})}(\mathbb{Z}^d)$$
$$= \operatorname{SL}_d(\mathbb{R}) / \operatorname{SL}_d(\mathbb{Z}).$$

We will often switch between the two viewpoints of a lattice $\Lambda = g\mathbb{Z}^d$ and a coset $x = g \operatorname{SL}_d(\mathbb{R})$ without pointing this out explicitly. Thus, for example, we may speak of the quantity $\lambda_1(x)$ for such a coset. One important goal for any topological space is to understand its compact subsets, and this is achieved by Theorem 3.3.11 below.

Example 3.3.10 We note that for any $d \geq 2$ the space X_d is not compact. One way to see this is to notice that the sequence $(\Lambda_k)_{k \geq 1}$ of lattices, where

$$\Lambda_k = \left\{ \left(kn_1, \tfrac{1}{k}n_2, n_3, \ldots, n_d\right) \,\Big|\, (n_1, \ldots, n_d) \in \mathbb{Z}^d \right\}$$

for $k \geq 1$, does not have a subsequence that converges to an element of X_d.

The following theorem shows that the type of 'divergence' in X_d seen in Example 3.3.10 is essentially the only way for a sequence of elements of X_d to fail to have a convergent subsequence.

Theorem 3.3.11 (Mahler's Compactness Criterion) *A subset $K \subseteq X_d$ has compact closure if and only if $\inf\{\lambda_1(x) \mid x \in K\} > 0$.*

For any $\varepsilon > 0$ we define

$$X_d(\varepsilon) = \{x \in X \mid \lambda_1(x) \geq \varepsilon\}$$

so that

$$X_d = \bigcup_{\varepsilon > 0} X_d(\varepsilon).$$

Using these sets, Theorem 3.3.11 can also be phrased as saying that a set $K \subseteq X_d$ is compact if and only if it is closed and $K \subseteq X_d(\varepsilon)$ for some $\varepsilon > 0$.

Exercise 3.3.12 Show that $\lambda_1 \colon X_d \to (0, \infty)$ is continuous. Conclude that $X_d(\varepsilon)$ is closed for any $\varepsilon > 0$, and that any set with compact closure is contained in $X_d(\varepsilon)$ for some $\varepsilon > 0$.

The heart of Theorem 3.3.11 is the converse of the exercise above.

Proof of Theorem 3.3.11 Because of Exercise 3.3.12, it is enough to prove that $X_d(\varepsilon)$ is compact for all $\varepsilon > 0$. So let us fix some $\varepsilon > 0$ and $x = \Lambda \in X_d(\varepsilon)$. Applying Minkowski's successive minima theorem (Theorem 3.3.8), we find a $\mathbb{Z}$-basis $\mathbf{v}_1, \ldots, \mathbf{v}_d \in \Lambda$ for Λ so that $\|\mathbf{v}_k\| \ll_d \lambda_k(\Lambda)$ for $k = 1, \ldots, d$. Moreover,

$$\varepsilon \leqslant \lambda_1(\Lambda) \leqslant \lambda_2(\Lambda) \leqslant \cdots \leqslant \lambda_d(\Lambda)$$

and

$$\lambda_1(\Lambda) \cdots \lambda_d(\Lambda) \asymp_d 1,$$

which together imply that $\lambda_d(\Lambda) \ll_d \varepsilon^{-(d-1)}$. Switching the sign of $\mathbf{v}_d$ if necessary, we may use $\mathbf{v}_1, \ldots, \mathbf{v}_d$ (written as column vectors) as the columns of a matrix $g \in \mathrm{SL}_d(\mathbb{R})$ with $x = \Lambda = g\mathbb{Z}^d$. By the uniform estimates above, g belongs to a compact subset of $\mathrm{SL}_d(\mathbb{R}) \subseteq \mathrm{Mat}_{d,d}(\mathbb{R})$ (depending only on $\varepsilon > 0$). This proves that $X_d(\varepsilon)$ belongs to the compact image of a compact set which, together with Exercise 3.3.12, shows that $X_d(\varepsilon)$ is compact, and so gives the theorem. □

Using Minkowski's successive minima and a decomposition of $\mathrm{SL}_d(\mathbb{R})$ and of its Haar measure (specifically, the Iwasawa or 'KAN' decomposition of $\mathrm{SL}_d(\mathbb{R})$ that can be obtained from the Gram–Schmidt orthonormalization procedure), one can prove the following by a direct calculation. We will skip this step, referring to [7] for the details.

Theorem 3.3.13 (Finite Volume of Moduli Space) *The moduli space of unimodular lattices*

$$X_d = \mathrm{SL}_d(\mathbb{R})/\mathrm{SL}_d(\mathbb{Z})$$

has finite volume. That is, it supports a finite $\mathrm{SL}_d(\mathbb{R})$-invariant Borel measure m_{X_d}. In other words, $\mathrm{SL}_d(\mathbb{Z}) < \mathrm{SL}_d(\mathbb{R})$ is a non-uniform lattice. Moreover, there exists some $\kappa > 0$ (depending on d only) so that

$$m_{X_d}(X_d \smallsetminus X_d(\varepsilon)) \ll_d \varepsilon^{\kappa}$$

for all $\varepsilon > 0$.

3.4 Ergodicity and Mixing

Having found an interesting finite volume quotient of $\mathrm{SL}_d(\mathbb{R})$ in the previous section, we now turn to dynamical questions concerning the actions of various subgroups of $\mathrm{SL}_d(\mathbb{R})$ on this space. We start this line of thought with the ergodic

3 Homogeneous Dynamics and its Connection to Diophantine Approximation

theory of one-parameter subgroups of the abelian diagonal subgroup

$$A = \left\{ \begin{pmatrix} a_1 & & \\ & \ddots & \\ & & a_d \end{pmatrix} \,\Bigg|\, a_1, \ldots, a_d > 0 \text{ and } a_1 \cdots a_d = 1 \right\}$$

of $\mathrm{SL}_d(\mathbb{R})$.

Theorem 3.4.1 (Ergodicity) *Suppose $d \geq 2$ and $\mathrm{SL}_d(\mathbb{R})$ acts continuously on a locally compact metric space, preserving a finite Borel measure μ. If the action of $\mathrm{SL}_d(\mathbb{R})$ is ergodic, then any non-trivial element of A also acts ergodically.*

Sketch of Proof We consider the Hilbert space $\mathcal{H} = L^2_\mu(X)$, and recall that the measure-preserving action of $\mathrm{SL}_d(\mathbb{R})$ induces a unitary representation of $\mathrm{SL}_d(\mathbb{R})$ on $\mathcal{H}$. In fact for $g \in \mathrm{SL}_d(\mathbb{R})$ the map $\pi_g : \mathcal{H} \to \mathcal{H}$ defined by

$$\pi_g(f)(x) = f(g^{-1} \cdot x)$$

for $f \in \mathcal{H}$ and $x \in X$ is unitary (that is, $\pi_g^* = \pi_g^{-1}$). If $g_n \to g$ as $n \to \infty$ and $f \in C_c(X)$, then $\pi_{g_n}(f) \to \pi_g(f)$ in $L^2_\mu(X)$ by dominated convergence. Using the density of $C_c(X)$ in $L^2_\mu(X)$, this extends to any $f \in L^2_\mu(X)$. This continuity in π_g as $g \in \mathrm{SL}_d(\mathbb{R})$ changes, together with the fact that each π_g is a unitary operator, means that π defines a *unitary representation* of $\mathrm{SL}_d(\mathbb{R})$.

For concreteness, we assume that our non-trivial element of A is

$$a = \begin{pmatrix} e^{-(d-1)} & \\ & eI_{d-1} \end{pmatrix} \in A,$$

where I_{d-1} denotes the identity matrix in dimension $(d-1)$. We note that it is not difficult to generalize the argument to other elements of A. Suppose therefore that $f \in L^2_\mu(X)$ is invariant under a, in the sense that $\pi_a(f) = f$. To prove ergodicity of a (with respect to μ), we have to show that f is constant μ-almost surely. We will achieve this by showing that f satisfies $\pi_g(f) = f$ for all $g \in \mathrm{SL}_d(\mathbb{R})$, that is, f is invariant under the action of the whole group $\mathrm{SL}_d(\mathbb{R})$. The hypothesis of ergodicity for the action of the whole group then shows that f is μ-almost surely constant.

Let $\mathbf{v} \in \mathbb{R}^{d-1}$ and define

$$u_\mathbf{v} = \begin{pmatrix} 1 & \mathbf{v}^t \\ & I_{d-1} \end{pmatrix}.$$

Note that $a^n u_\mathbf{v} a^{-n} = u_{e^{-nd}\mathbf{v}} \to I$ as $n \to \infty$ (by a simple matrix calculation). Using the fact that $\pi_a(f) = f$, that π_a is unitary, and the continuity of the unitary representation, this gives

$$\|\pi_{u_{\mathbf{v}}}(f) - f\|_2 = \|\pi_{a^n}(\pi_{u_{\mathbf{v}}}(\pi_{a^{-n}}(f))) - \pi_{a^n}(f)\|_2$$
$$= \|\pi_{u_{e^{-n}d_{\mathbf{v}}}}(f) - f\|_2 \longrightarrow 0$$

as $n \to \infty$. This is somewhat unexpected, as the initial expression does not depend on n, and so gives $\pi_{u_{\mathbf{v}}}(f) = f$ for all $\mathbf{v} \in \mathbb{R}^{d-1}$.

Using $n \to -\infty$ instead gives $\pi_{u'_{\mathbf{v}}}(f) = f$ for all matrices

$$u'_{\mathbf{v}} = \begin{pmatrix} 1 & \\ \mathbf{v} & I_{d-1} \end{pmatrix}$$

with $\mathbf{v} \in \mathbb{R}^{d-1}$, by the same argument.

Now let $2 \leqslant i \neq j \leqslant d$, $s \in \mathbb{R} \setminus \{0\}$, and define

$$u = I_d + sE_{i,j}$$

to be the matrix with 1s along the diagonal and a single non-zero entry s in the ith row and jth column. Also define $g = I + sE_{1,j}$ and $h = I + E_{i,1}$. By the argument above we already know that $\pi_g(f) = f = \pi_h(f)$. Using this and the relation

$$hgh^{-1}g^{-1} = (I + E_{i,1})(I + sE_{1,j})(I - E_{i,1})(I - sE_{1,j}) = (I + sE_{i,j}) = u,$$

we also obtain $\pi_u(f) = f$.

Putting the above together, we have shown that $\pi_u(f) = f$ whenever

$$u = I + sE_{i,j}$$

for $s \in \mathbb{R}$ and $1 \leqslant i \neq j \leqslant d$. Using (a slightly restricted version of) the Gauss elimination algorithm from first year linear algebra, one can show that these matrices actually generate all of $\mathrm{SL}_d(\mathbb{R})$. It follows that $\pi_g(f) = f$ for all $g \in \mathrm{SL}_d(\mathbb{R})$. Using the assumed ergodicity for the action of $\mathrm{SL}_d(\mathbb{R})$ it follows that f is constant μ-almost surely. As $f \in L^2_\mu(X)$ was only assumed to be invariant under the action of a, we see that a acts ergodically with respect to μ also. □

Together with the rather trivial fact that the transitive action of $\mathrm{SL}_d(\mathbb{R})$ on the quotient space $X_d = \mathrm{SL}_d(\mathbb{R})/\mathrm{SL}_d(\mathbb{Z})$ in Theorem 3.3.13 is ergodic, Theorem 3.4.1 gives the following.

Corollary 3.4.2 *Non-trivial elements of the diagonal subgroup* $A < \mathrm{SL}_d(\mathbb{R})$ *act ergodically on* $X_d = \mathrm{SL}_d(\mathbb{R})/\mathrm{SL}_d(\mathbb{Z})$.

By working a little harder one can show that many other elements of $\mathrm{SL}_d(\mathbb{R})$ act ergodically. In fact even more is true, due to the following result (see [5, Ch. 9 & 11] or [7]).

Theorem 3.4.3 (Howe–Moore) *Using the same assumptions as in Theorem 3.4.1, the* $\mathrm{SL}_d(\mathbb{R})$*-action is strong mixing. Assuming* $\mu(X) = 1$, *this means that*

$$\int_X \pi_g(f_1) f_2 \, d\mu \longrightarrow \int_X f_1 \, d\mu \int_X f_2 \, d\mu$$

as[5] $g \to \infty$ *for any* $f_1, f_2 \in L^2_\mu(X)$.

3.5 Unipotent Dynamics

The material discussed so far has been of a classical and fundamental nature. However, we now come to one of the seminal ideas in the more recent theory of homogeneous dynamics.

An element $u \in \mathrm{SL}_d(\mathbb{R})$ is called *unipotent* if 1 is its only eigenvalue (over $\mathbb{C}$) or, equivalently, if $u - I_d$ is nilpotent (that is, if $(u - I_d)^d = 0$). A subgroup is called *unipotent* if all its elements are unipotent.

While the dynamics of one-parameter subgroups of A are 'chaotic', the dynamics of one-parameter unipotent subgroups exhibit a 'rigid' structure. We will see some of this difference in the results of this section.

3.5.1 Non-divergence

One of the first instances of the less chaotic behaviour of unipotent dynamics, and an important first step towards 'rigidity' phenomenon, is the non-divergence result of Margulis [9] and Dani [3].

Theorem 3.5.1 (Margulis–Dani Quantitative Non-divergence) *Let* $d \geq 2$, *let* I *be a compact interval in* $\mathbb{R}$, *and let* $p \colon I \to \mathrm{SL}_d(\mathbb{R})$ *be a polynomial. Suppose that there exists a constant* $\eta \in (0, 1]$ *such that*

$$\sup_{t \in I} \mathrm{covol}(p(t), V) \geq \eta^{\dim V}$$

for any rational subspace V, *where we write* $\mathrm{covol}(p(t), V)$ *to denote the covolume of the lattice* $p(t)(\mathbb{Z}^d \cap V)$ *in the subspace* $p(t)V$. *Then there exists a constant* $\kappa > 0$

---

[5] Here we write $g \to \infty$ to mean that g eventually leaves any compact subset of $\mathrm{SL}_d(\mathbb{R})$.

(depending on d and the degree of p only) so that

$$m_{\mathbb{R}}\left(\{t \in I \mid \lambda_1(p(t)\operatorname{SL}_d(\mathbb{Z})) < \varepsilon\}\right) \ll \left(\frac{\varepsilon}{\eta}\right)^{\kappa} m_{\mathbb{R}}(I),$$

where $m_{\mathbb{R}}$ denotes the Lebesgue measure on $\mathbb{R}$.

We note that any one-parameter unipotent subgroup U of $\operatorname{SL}_d(\mathbb{R})$ has the form $U = \{\exp(tw) : t \in \mathbb{R}\}$ for some nilpotent matrix w. From this it follows that the exponential map (defined by the usual power series) can be used to define the polynomial

$$\mathbb{R} \ni t \longmapsto u_t = \exp(tw),$$

whose degree depends on the nilpotentcy degree of w. Hence the theorem above applies, for instance, to the polynomial

$$\mathbb{R} \ni t \longmapsto p(t) = \exp(tw)g_0$$

and so to pieces of trajectories of one-parameter unipotent subgroups.

Even though Theorem 3.5.1 and its proof are highly geometric (relying on the geometry of numbers discussed in Sect. 3.3.2), it is more or less impossible to illustrate what is going on except in the minimal dimensional case $d = 2$ (see Fig. 3.6).

Recall that the space of Borel probability measures on a *compact* metric space is compact in the weak* topology (see [6, Sec. 8.2.1]). It follows that any weak* limit of Borel probability measures is also a Borel probability measure. However,

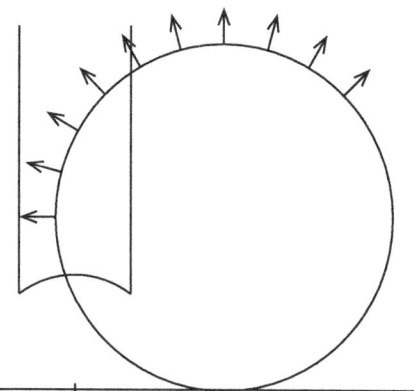

Fig. 3.6 A horocycle orbit in X_2 may reach high up into the cusp (which is the name for a neighbourhood of infinity in X_2), but will always come back. The non-divergence result makes this quantitative, and holds in any dimension

on a non-compact space X this fails as a weak* limit of a sequence of probability measures (defined using elements of $C_c(X)$) may be identically zero. This may also happen in dynamically relevant situations, as illustrated in Exercise 3.5.4.

Corollary 3.5.2 *Let $U = \{u_t \mid t \in \mathbb{R}\}$ be a one-parameter unipotent subgroup of $\mathrm{SL}_d(\mathbb{R})$ and let $x_0 \in \mathsf{X}_d = \mathrm{SL}_d(\mathbb{R})/\mathrm{SL}_d(\mathbb{Z})$. Then any weak* limit of*

$$\frac{1}{T}\int_0^T \delta_{u_t x_0}\, dt$$

for $T \to \infty$ is a probability measure.

Proof Choose $g_0 \in \mathrm{SL}_d(\mathbb{R})$ with $x_0 = g_0\,\mathrm{SL}_d(\mathbb{Z})$; we will apply Theorem 3.5.1 to the polynomial $p(t) = u_t g_0$. Let

$$\eta = \min_V \mathrm{covol}(g_0, V)^{1/\dim(V)}, \tag{3.2}$$

where the minimum is taken over all rational subspaces $V < \mathbb{R}^d$. We note that the minimum exists even though there are infinitely many such subspaces. To see this, we fix a parameter T and let V be a rational subspace with

$$\mathrm{covol}(g_0, V) < T.$$

Applying Theorem 3.3.11 to the lattice $g_0(V \cap \mathbb{Z}^d) < g_0 V$ of covolume less than T, we find a $\mathbb{Z}$-basis consisting of vectors of length $\ll_d T\lambda_1(g_0 \mathbb{Z}^d)^{-(d-1)}$. As $g_0 \mathbb{Z}^d$ is discrete, it follows that there are only finitely many possibilities for these vectors, and hence also for the subspace V. This shows that η is well-defined by (3.2).

The definition of η ensures that η satisfies the assumption of the theorem for the interval $I = [0, T]$ and any $T > 0$. Now let $\varepsilon > 0$ and choose some $f \in C_c(\mathsf{X}_d)$ so that $0 \leqslant f \leqslant 1$ and $f|_{\mathsf{X}_d(\varepsilon)} \equiv 1$. Then

$$\frac{1}{T}\int_0^T f(u_t x)\, dt \geqslant 1 - c\left(\frac{\varepsilon}{\eta}\right)^\kappa$$

for some constants $c > 0$ and $\kappa > 0$ by Theorem 3.5.1. If now μ is a weak* limit of

$$\frac{1}{T}\int_0^T \delta_{u_t x_0}\, dt$$

as $T \to \infty$, then we obtain

$$\int_{\mathsf{X}_d} f\, d\mu \geqslant 1 - c\left(\frac{\varepsilon}{\eta}\right)^\kappa.$$

This gives $\mu(X_d) \geq \mu(\mathrm{Supp}(f)) \geq 1 - c\left(\frac{\varepsilon}{\eta}\right)^\kappa$, and as $\varepsilon > 0$ was arbitrary we obtain the corollary. □

Exercise 3.5.3 Estimate η as in (3.2) in terms of $\lambda_1(x_0)$.

Exercise 3.5.4 Find $x_0 \in X_2$ so that the conclusion of Corollary 3.5.2 fails (spectacularly, in the sense that the limit measure gives the whole space zero mass) for the action of the diagonal subgroup corresponding to the geodesic flow.

3.5.2 Ratner's Rigidity Theorems

While the non-divergence result above could have been proved within the constraints of the Summer school (at the cost of dropping most of the other material), the following two theorems would, in full generality, need about a year-long course to prove (or a semester if sufficient background is assumed).

Theorem 3.5.5 (Measure Rigidity) *Let $X = G/\Gamma$ be the quotient of a closed linear group G by a discrete subgroup $\Gamma < G$. Let U be a connected unipotent subgroup. Then any U-invariant and ergodic Borel probability measure μ on X is homogeneous. That is, there exists a closed connected subgroup $L < G$ containing U with the property that μ is the unique L-invariant Borel probability m_{Lx_0} on a single (automatically closed) L-orbit Lx_0 for some $x_0 \in X$.*

Theorem 3.5.6 (Equidistribution and Orbit Closure) *Let $X = G/\Gamma$ be the quotient of a closed linear group by a lattice, let $U = \{u_t \mid t \in \mathbb{R}\}$ be a one-parameter unipotent subgroup of G, and let $x_0 \in X$. Then there exists a closed connected subgroup $L < G$ containing U with the property that*

$$\overline{Ux_0} = Lx_0$$

and

$$\frac{1}{T}\int_0^T f(u_t x_0)\, dt \longrightarrow \int f\, dm_{Lx_0}$$

as $T \to \infty$ for every $f \in C_c(X)$, where m_{Lx_0} is the unique L-invariant Borel probability measure on the closed orbit Lx_0.

Both of these theorems fail spectacularly for the geodesic flow.

3.5.3 Rational Structures

The results of this and the following subsection will need more background. We recommend that the reader makes a serious attempt at this nonetheless. If the material turns out to be difficult to follow, the reader should skip them and continue in Sect. 3.6, which will only need Theorem 3.5.9 from Sect. 3.5.4 as a 'black box'.

Exercise 3.5.7 One easy special case of Ratner's theorems is given by translation on a d-torus. Let $d \geqslant 2$ and let $U < \mathbb{R}^d$ be a non-trivial subspace. Show that

$$\overline{U + \mathbb{Z}^d} = L/(L \cap \mathbb{Z}^d) \subseteq \mathbb{T}^d = \mathbb{R}^d/\mathbb{Z}^d$$

for some linear subspace L that can be defined by linear equations with coefficients over $\mathbb{Q}$.

The conclusion in the exercise above can also be phrased by saying that 'L is defined over $\mathbb{Q}$'.

To better understand the subgroups L appearing in the theorems of Ratner in Sect. 3.5.2, Borel and Prasad [2] proved the following result.

Theorem 3.5.8 (Borel–Prasad) *Let $d \geqslant 2$ and*

$$x_0 = g_0 \operatorname{SL}_d(\mathbb{Z}) \in X_d = \operatorname{SL}_d(\mathbb{R})/\operatorname{SL}_d(\mathbb{Z}).$$

The subgroups L appearing in Theorems 3.5.5 and 3.5.6 have the property that $g_0^{-1} L g_0$ is the connected component of the $\mathbb{R}$-points of an algebraic subgroup $\mathbb{L}$ defined over $\mathbb{Q}$. Moreover, the radical of $\mathbb{L}$ is unipotent.

Theorem 3.5.8 uses the language of algebraic groups which may initially look intimidating. However, it should really be read as simply saying that $g_0^{-1} L g_0$ is a 'rational subgroup' that can be obtained from a semisimple group and a unipotent group.

3.5.4 A Unique Ergodicity Result

As an example of how we could work with the results above, we will prove the following unique ergodicity result.

Theorem 3.5.9 *Let $X_3 = \operatorname{SL}_3(\mathbb{R})/\operatorname{SL}_3(\mathbb{Z})$, and define the subgroup*

$$H = \left\{ \begin{pmatrix} e^{-t} & & \\ & e^{t/2} & \\ & & e^{t/2} \end{pmatrix} \begin{pmatrix} 1 & & \\ s & 1 & \\ & & 1 \end{pmatrix} \, \middle| \, s, t \in \mathbb{R} \right\} < \operatorname{SL}_3(\mathbb{R}).$$

Then the action of H on X_3 is uniquely ergodic. That is, the normalized Haar measure m_{X_3} is the only H-invariant Borel probability measure on X_3.

Roughly speaking, the proof consists of the following steps.

- By general results in the ergodic theory of group actions, it is enough to show that an ergodic H-invariant Borel probability measure μ must be m_{X_3}.
- As H is not unipotent, it is unclear how to use Theorem 3.5.5 in this context. Because of this, we will use the unipotent subgroup

$$U = \left\{ u_s = \begin{pmatrix} 1 & & \\ s & 1 & \\ & & 1 \end{pmatrix} \;\middle|\; s \in \mathbb{R} \right\} < H.$$

- As μ might not be ergodic with respect to U, we need to consider the decomposition of μ into its U-ergodic components $\mu_x^{\mathcal{E}}$ for μ-almost every x in X, where $\mathcal{E} = \{B \in \mathcal{B}_{X_3} \mid B \text{ is invariant under } U\}$ (see [5, Sec. 8.7] for these notions and the required results).
- Now we can apply Theorem 3.5.5 and obtain for almost every $x \in X$ a connected subgroup L_x so that the U-invariant and ergodic probability measure $\mu_x^{\mathcal{E}}$ is the L_x-invariant probability measure on $L_x x$.
- Using the fact that

$$A = \left\{ a_t = \begin{pmatrix} e^{-t} & & \\ & e^{t/2} & \\ & & e^{t/2} \end{pmatrix} \;\middle|\; t \in \mathbb{R} \right\} < H$$

normalizes U, we obtain $(a_t)_* \mu_x^{\mathcal{E}} = \mu_{a_t x}^{\mathcal{E}}$ and $a_t L_x a_t^{-1} = L_{a_t x}$.
- By analyzing the action of a_t on subspaces of the Lie algebra $\mathfrak{sl}_3(\mathbb{R})$, it follows that the Lie algebra of $a_t L_x a_t^{-1}$ is either equal to the Lie algebra of L_x or will converge in a natural sense to a different subspace as $t \to \infty$. Together with Poincaré recurrence, this gives $a_t L_x a_t^{-1} = L_x$ for μ-almost every $x \in X$. Together with ergodicity of μ with respect to $H = AU$, it follows that there exists one connected subgroup L so that $\mu_x^{\mathcal{E}}$ is the L-invariant probability measure on Lx.
- Next we analyze what the possibilities are for connected subgroups L in $\mathrm{SL}_3(\mathbb{R})$ that are normalized by H.
- Finally, we can use Theorem 3.5.8 and Poincaré recurrence to rule out all of the possibilities for L except for $L = \mathrm{SL}_3(\mathbb{R})$ itself. However, this implies that $\mu = m_{X_3}$ as desired.

We note that part of this argument follows the ideas contained in a paper of Margulis and Tomanov [10]. We will make use of Poincaré recurrence in the following form.

Exercise 3.5.10 If $T: X \to X$ is a measure-preserving map, and $f: X \to \mathbb{R}$ is measurable, then for almost every $x \in X$ there exists a sequence (k_n) so that $\lim_{k \to \infty} f(T^{k_n} x) = f(x)$.

Proof of Theorem 3.5.9 We first show that it suffices to show that there is only one H-invariant and ergodic probability measure (namely m_{X_3}) on X_3. Indeed, for a general H-invariant measure μ on X_3 we can define for the σ-algebra

$$\mathcal{E}(H) = \{B \in \mathcal{B}_{\mathsf{X}_3} \mid B \text{ is } H\text{-invariant}\}$$

the conditional measures $\mu_x^{\mathcal{E}(H)}$ for $x \in \mathsf{X}_3$. The latter are almost surely H-invariant and ergodic probability measures on X_3 with

$$\mu = \int \mu_x^{\mathcal{E}(H)} \, d\mu(x),$$

see [5, Sec. 5.3 & 8.7]. Assuming now that m_{X_3} is the only H-invariant and ergodic probability measure, we obtain $\mu_x^{\mathcal{E}(H)} = m_{\mathsf{X}_3}$ almost surely, and so can conclude that $\mu = m_{\mathsf{X}_3}$ as desired.

From now on we suppose that μ is an H-invariant and ergodic probability measure on X_3. Our goal is to show that $\mu = m_{\mathsf{X}_3}$. For this, we wish to use Theorem 3.5.5 concerning probability measures invariant and ergodic under a unipotent subgroup U. To this end we define

$$U = \left\{ u_s = \begin{pmatrix} 1 & & \\ s & 1 & \\ & & 1 \end{pmatrix} \;\middle|\; s \in \mathbb{R} \right\} < H,$$

which implies that μ is U-invariant. Unfortunately, ergodicity does not descend[6] automatically from the H-action to the action of the subgroup U. Hence we define the σ-algebra

$$\mathcal{E} = \mathcal{E}(U) = \{B \in \mathcal{B}_{\mathsf{X}_3} \mid B \text{ is } U\text{-invariant}\}$$

and once more decompose μ into conditional measures $\mu_x^{\mathcal{E}}$ for $x \in \mathsf{X}_3$. By Einsiedler and Ward [5, Sec. 5.3 & 8.7] we have

$$\mu = \int \mu_x^{\mathcal{E}} \, d\mu(x) \tag{3.3}$$

and that almost surely $\mu_x^{\mathcal{E}}$ is U-invariant and ergodic.

---

[6] This happened in the proof of Theorem 3.4.1, but does not work for U here.

We claim that

$$x \in \operatorname{Supp} \mu_x^{\mathcal{E}} \text{ almost surely.} \tag{3.4}$$

To see this note first that X_3 satisfies the second countability axiom in topology. For a given open set $O \subseteq X_3$ we have

$$\mu(\{x \in O \mid \mu_x^{\mathcal{E}}(O) = 0\}) = \int_{\{x \in O \mid \mu_x^{\mathcal{E}}(O)=0\}} \mathbb{1}_O \, d\mu$$

$$= \int_{\{x \in O \mid \mu_x^{\mathcal{E}}(O)=0\}} \mu_y^{\mathcal{E}}(O) \, d\mu(y) = 0$$

by the characterising properties of conditional measures as

$$\{x \in O \mid \mu_x^{\mathcal{E}}(O) = 0\}$$

is $\mathcal{E}$-measurable (see [5, Sec. 5.3]). Varying O within a countable base of the topology and taking the union N of the null sets $\{x \in O \mid \mu_x^{\mathcal{E}}(O) = 0\}$ we obtain (3.4) outside of N.

Applying Theorem 3.5.5 to $\mu_x^{\mathcal{E}}$ (for a typical $x \in X_3$), we see that there exists a connected subgroup $L_x < \operatorname{SL}_3(\mathbb{R})$ (possibly depending on x) and a point x_0 (possibly depending on x) so that $\mu_x^{\mathcal{E}}$ is the L_x-invariant probability measure $m_{L_x x_0}$ on the closed orbit $L_x x_0$. As

$$x \in \operatorname{Supp} \mu_x^{\mathcal{E}} = \operatorname{Supp} m_{L_x x_0} = L_x x_0$$

by (3.4) we obtain $L_x x = L_x x_0$ and may simply use x for the starting point of the L_x-orbit. To summarize, for almost every $x \in X_3$ there exists a connected subgroup $L_x < \operatorname{SL}_3(\mathbb{R})$ so that the conditional measure $\mu_x^{\mathcal{E}}$ is actually the L_x-invariant probability measure on the closed orbit $L_x x$.

Our goal is to show that almost surely we have $L_x = \operatorname{SL}_3(\mathbb{R})$, which then implies that $\mu_x^{\mathcal{E}} = m_{\operatorname{SL}_3(\mathbb{R})x} = m_{X_3}$ almost surely, and so $\mu = m_{X_3}$ by (3.3). To achieve this, we will first study how L_x varies as x changes in X_3.

Let $h \in H$. Since $U \triangleleft H$, we see that hB is still U-invariant for any set B in $\mathcal{E} = \mathcal{E}(U)$, that is, $h\mathcal{E} = \mathcal{E}$. By Einsiedler and Ward [5, Cor. 5.24] this implies that $(h)_* \mu_x^{\mathcal{E}} = \mu_{hx}^{\mathcal{E}}$ almost surely. Recalling that $\mu_x^{\mathcal{E}} = m_{L_x x}$ almost surely, we obtain that

$$m_{L_{hx} hx} = \mu_{hx}^{\mathcal{E}} = (h)_* \mu_x^{\mathcal{E}} = (h)_* m_{L_x x} = m_{hL_x x} = m_{hL_x h^{-1} hx} \tag{3.5}$$

almost surely. However, as the connected subgroup giving rise to a closed orbit is uniquely determined by the Haar measure on this orbit (namely, as the connected component of its invariance group), we obtain from (3.5) that

$$L_{hx} = hL_x h^{-1} \tag{3.6}$$

almost surely and for any $h \in H$.

For $h = u_s \in U$ we know (by construction) that $m_{L_x x}$ is U-invariant, which implies that $U < L_x$ and hence by (3.6) also that $L_{u_s x} = L_x$ almost surely.

We now define

$$A = \left\{ a_t = \begin{pmatrix} e^{-t} & & \\ & e^{t/2} & \\ & & e^{t/2} \end{pmatrix} \;\middle|\; t \in \mathbb{R} \right\} < H$$

and note that $H = AU$. As our next step, we wish to understand how L_x changes along $a_t \in A$.

As a connected subgroup $L < \mathrm{SL}_3(\mathbb{R})$ is uniquely determined by its Lie algebra, we look at the adjoint representation

$$\mathrm{Ad}_{a_t} : \mathfrak{sl}_3(\mathbb{R}) \longrightarrow \mathfrak{sl}_3(\mathbb{R})$$

for $a_t \in A$ (which is simply a notation for conjugation by a_t). Recall that the Lie algebra $\mathfrak{sl}_3(\mathbb{R})$ consists of all 3×3 matrices with trace zero equipped with the Lie bracket operation $[\mathbf{m}, \mathbf{n}] = \mathbf{mn} - \mathbf{nm}$ for $\mathbf{m}, \mathbf{n} \in \mathfrak{sl}_3(\mathbb{R})$. A direct matrix calculation reveals that Ad_{a_t} has

$$\mathfrak{g}^+ = \left\{ \begin{pmatrix} 0 & 0 & 0 \\ * & 0 & 0 \\ * & 0 & 0 \end{pmatrix} \right\} < \mathfrak{sl}_3(\mathbb{R})$$

as a two-dimensional eigenspace for eigenvalue $e^{\frac{3}{2}t}$,

$$\mathfrak{g}^0 = \left\{ \begin{pmatrix} * & 0 & 0 \\ 0 & * & * \\ 0 & * & * \end{pmatrix} \right\} \cap \mathfrak{sl}_3(\mathbb{R}) < \mathfrak{sl}_3(\mathbb{R})$$

as a four-dimensional eigenspace for eigenvalue 1, and

$$\mathfrak{g}^- = \left\{ \begin{pmatrix} 0 & * & * \\ 0 & 0 & 0 \\ 0 & 0 & 0 \end{pmatrix} \right\} < \mathfrak{sl}_3(\mathbb{R})$$

as a two-dimensional eigenspace for eigenvalue $e^{-\frac{3}{2}t}$. Let $\ell < \mathfrak{sl}_3(\mathbb{R})$ be a linear subspace. Applying a_t via the adjoint representation to ℓ, we claim that there are two distinct possibilities:

- Either ℓ is the sum $(\ell \cap \mathfrak{g}^+) \oplus (\ell \cap \mathfrak{g}^0) \oplus (\ell \cap \mathfrak{g}^+)$ and is invariant under all $a_t \in A$;
- or $\mathrm{Ad}_{a_t}(\ell)$ converges for $t \to \infty$ to a subspace $\ell_\infty \neq \ell$.

We note that the convergence is defined using the appropriate Grassmannian manifold. To see this, we carefully choose a basis of ℓ. First take a basis of the spaces $\ell \cap \mathfrak{g}^+$, $\ell \cap \mathfrak{g}^0$, and $\ell \cap \mathfrak{g}^-$ respectively. If the union is a basis of ℓ then we are in the first case. Suppose this is not the case, and let $\pi_+ \colon \mathfrak{sl}_3(\mathbb{R}) \to \mathfrak{g}^+$ be the projection with kernel $\mathfrak{g}^0 \oplus \mathfrak{g}^-$ and $\pi_0 \colon \mathfrak{g}^0 \oplus \mathfrak{g}^- \to \mathfrak{g}^0$ the projection with kernel $\mathfrak{g}^-$. We first extend the above basis vectors by vectors in ℓ so that the images under π^+ together with the basis of $\ell \cap \mathfrak{g}^+$ give a basis of $\pi^+(\ell)$. Similarly, we extend the above basis by vectors in $\ell \cap (\mathfrak{g}^0 + \mathfrak{g}^-)$ so that their images under π_0 together with the basis of $\ell \cap \mathfrak{g}^0$ give a basis of $\pi_0(\ell \cap (\mathfrak{g}^0 + \mathfrak{g}^-))$. Together it follows from basic linear algebra that the original basis vectors for $\ell \cap \mathfrak{g}^+$, $\ell \cap \mathfrak{g}^0$, and $\ell \cap \mathfrak{g}^-$ and the new vectors give a basis of ℓ. As we assumed that

$$\ell \neq (\ell \cap \mathfrak{g}^+) \oplus (\ell \cap \mathfrak{g}^0) \oplus (\ell \cap \mathfrak{g}^-),$$

we added new vectors either in the first step or in the second step.

Suppose $v \in \ell$ has the property that $\pi^+(v)$ is linearly independent of $\ell \cap \mathfrak{g}^+$. Applying Ad_{a_t} to v, it follows that $\pi^+(v)$ is expanded at the rate $e^{\frac{3}{2}t}$. Normalizing the length of the image vectors, this squashes the components corresponding to $\mathfrak{g}^0$ (which had not changed) and to $\mathfrak{g}^-$ (which in any case decrease at a rate of $e^{-\frac{3}{2}t}$). In other words, projectively $\mathrm{Ad}_{a_t} v$ converges to $\pi^+(v) \notin \ell \cap \mathfrak{g}^+$. Pushing this argument through in the analogous case of $v \in \ell \cap (\mathfrak{g}^0 + \mathfrak{g}^-)$ with $\pi_0(v)$ linearly independent of $\ell \cap \mathfrak{g}^0$, it follows that $\mathrm{Ad}_{a_t}(\ell)$ converges in the Grassmannian to

$$\ell_\infty = \pi^+(\ell) \oplus \pi_0(\ell \cap (\mathfrak{g}^0 + \mathfrak{g}^-)) \oplus (\ell \cap \mathfrak{g}^-) \neq \ell.$$

However, the second possibility gives a contradiction to Poincaré recurrence as follows. Almost surely as $t \to \infty$ the point $a_t x$ must return close to x infinitely often (and this must also be true with respect to other measurable functions, as in Exercise 3.5.10). We note that the map that sends x to the Lie algebra $\ell_x = \mathrm{Lie}(L_x)$ of L_x is (where defined) measurable.[7] Using this measurability and Poincaré recurrence, it follows that $\mathrm{Ad}_{a_t}(\ell_x)$ converges to ℓ_x, at least along some subsequence. However, our dichotomy above says that either $\mathrm{Ad}_{a_t}(\ell_x) = \ell_x$

---

[7] This is a (slightly tedious) exercise in measure theory: Morally speaking it is much more difficult to construct a non-measurable function (and always requires a rather explicit use of the axiom of choice). Our construction of L_x, while certainly technical, was natural and hence must produce a measurable function.

or $\mathrm{Ad}_{a_t}(\ell_x)$ converges as $t \to \infty$ to a subspace $\ell_{x,\infty} \neq \ell_x$. Hence we must have $\mathrm{Ad}_{a_t}(\ell_x) = \ell_x$ almost surely and for all $a_t \in A$.

Let us summarize:

- The Lie algebra ℓ_x of L_x uniquely determines the connected subgroup L_x,
- $\mathrm{Ad}_{a_t}(\ell_x) = \ell_x$ almost surely for $a_t \in A$, and
- $L_{a_t x} = a_t L_x a_t^{-1}$ has Lie algebra $\mathrm{Ad}_{a_t}(\ell_x)$ almost surely.

Recalling moreover that $L_{u_s x} = L_x$ almost surely for $u_s \in U$ and $H = AU$, we see that $L_{hx} = L_x$ almost surely for $h \in H$. However, this shows that $x \mapsto \ell_x$ is an H-invariant measurable function on X_3. By the assumed ergodicity of μ with respect to H, we conclude that there exists one connected subgroup L of $\mathrm{SL}_3(\mathbb{R})$ so that almost surely $\mu_x^{\mathcal{E}} = m_{Lx}$ is the L-invariant probability measure on a single closed L-orbit.

Let ℓ be the Lie algebra of L. By the argument above, we have $\mathrm{Ad}_{a_t}(\ell) = \ell$ for all $a_t \in A$, and so $\ell = \ell^+ \oplus \ell^0 \oplus \ell^-$ for $\ell^+ = \ell \cap \mathfrak{g}^+$, $\ell^0 = \ell \cap \mathfrak{g}^0$, and $\ell^- = \ell \cap \mathfrak{g}^-$. As $U < L$, we also know that

$$\left\{ \begin{pmatrix} 0 & 0 & 0 \\ * & 0 & 0 \\ 0 & 0 & 0 \end{pmatrix} \right\} < \ell^+ < \ell. \tag{3.7}$$

We claim that

$$\dim \ell^+ = \dim \ell^- \in \{1, 2\}. \tag{3.8}$$

To see this, we will again use Poincaré recurrence, this time for the function

$$x \longmapsto \mathrm{vol}(Lx).$$

More precisely, we fix a Haar measure m_L on the group L and, for a given orbit, use a measurable set $F_x \subseteq L$ for which $F_x \ni g \mapsto gx \in Lx$ is bijective to then define $\mathrm{vol}(Lx) = m_L(F_x)$ (which does not depend on the choice of F_x, see [7, Ch. 1] for the details). By construction of L we have $\mathrm{vol}(Lx) < \infty$ for μ-almost every $x \in \mathsf{X}_3$.

For $a_t \in A$ and $x \in \mathsf{X}_3$ and $F_x \subseteq L$ as above, we have that

$$a_t F_x a_t^{-1} \ni g \longmapsto g a_t x \in L_{a_t} x$$

is again bijective, which implies that

$$\mathrm{vol}(L_{a_t} x) = m_L(a_t F_x a_t^{-1}).$$

This composition of the Haar measure m_L by the automorphism defined by a_t again gives a Haar measure on L. As Haar measures are unique up to a scalar multiple, we obtain

$$\operatorname{vol}(La_t x) = m_L(a_t F_x a_t^{-1}) = c_t m_L(a_t F_x) = c_t \operatorname{vol}(Lx)$$

for some constant $c_t > 0$ depending on a_t. In fact one can calculate c_t concretely as the determinant of the restriction of Ad_{a_t} to ℓ. As a_t contracts ℓ^- by $e^{-\frac{3}{2}t}$ and expands ℓ^+ by $e^{\frac{3}{2}t}$, we either have

- that (3.8) holds, that these effects cancel each other out, and that $c_t = 1$ for all $a_t \in A$, or
- that (3.8) fails, and we have $c_t = e^{\delta t}$ for $a_t \in A \setminus \{I\}$ and some $\delta \neq 0$.

However, in the latter case we have that

$$\operatorname{vol}(La_t x) = e^{\delta t} \operatorname{vol}(Lx)$$

almost surely converges to 0 or to ∞ as $t \to \infty$, and so fails Poincaré recurrence (as in Exercise 3.5.10 for the measurable function $f(x) = \operatorname{vol}(Lx)$). This contradiction shows that (3.8) must hold.

If $\dim \ell^+ = \dim \ell^- = 2$, then ℓ contains both

$$\ell^+ = \mathfrak{g}^+ = \left\{ \begin{pmatrix} 0 & 0 & 0 \\ * & 0 & 0 \\ * & 0 & 0 \end{pmatrix} \right\}$$

and

$$\ell^- = \mathfrak{g}^- = \left\{ \begin{pmatrix} 0 & * & * \\ 0 & 0 & 0 \\ 0 & 0 & 0 \end{pmatrix} \right\}.$$

Taking commutators, we also see that

(a) $\left[\begin{pmatrix} 0 & 0 & 0 \\ 1 & 0 & 0 \\ 0 & 0 & 0 \end{pmatrix}, \begin{pmatrix} 0 & 1 & 0 \\ 0 & 0 & 0 \\ 0 & 0 & 0 \end{pmatrix} \right] = \begin{pmatrix} -1 & 0 & 0 \\ 0 & 1 & 0 \\ 0 & 0 & 0 \end{pmatrix} \in \ell,$

(b) $\left[\begin{pmatrix} 0 & 0 & 0 \\ 1 & 0 & 0 \\ 0 & 0 & 0 \end{pmatrix}, \begin{pmatrix} 0 & 0 & 1 \\ 0 & 0 & 0 \\ 0 & 0 & 0 \end{pmatrix} \right] = \begin{pmatrix} 0 & 0 & 0 \\ 0 & 0 & 1 \\ 0 & 0 & 0 \end{pmatrix} \in \ell,$

(c) $\left[\begin{pmatrix} 0 & 0 & 0 \\ 0 & 0 & 0 \\ 1 & 0 & 0 \end{pmatrix}, \begin{pmatrix} 0 & 0 & 1 \\ 0 & 0 & 0 \\ 0 & 0 & 0 \end{pmatrix} \right] = \begin{pmatrix} -1 & 0 & 0 \\ 0 & 0 & 0 \\ 0 & 0 & 1 \end{pmatrix} \in \ell,$ and

(d) $\left[\begin{pmatrix} 0 & 0 & 0 \\ 0 & 0 & 0 \\ 1 & 0 & 0 \end{pmatrix}, \begin{pmatrix} 0 & 1 & 0 \\ 0 & 0 & 0 \\ 0 & 0 & 0 \end{pmatrix} \right] = \begin{pmatrix} 0 & 0 & 0 \\ 0 & 0 & 0 \\ 0 & 1 & 0 \end{pmatrix} \in \ell.$

Together with $\mathfrak{g}^+$ and $\mathfrak{g}^-$, these span $\ell = \mathfrak{sl}_3(\mathbb{R})$, which implies that L is all of $\mathrm{SL}_3(\mathbb{R})$. This implies that $\mu = m_{X_3}$ as explained earlier.

Hence it remains to show that $\dim \ell^+ = \dim \ell^- = 1$ is not possible, which we will achieve by using Poincaré recurrence one last time. To define the function we are going to use Theorem 3.5.8.

As we are assuming that $\dim \ell^+ = 1$, we obtain from (3.7) that

$$\ell^+ = \left\{ \begin{pmatrix} 0 & 0 & 0 \\ * & 0 & 0 \\ 0 & 0 & 0 \end{pmatrix} \right\}.$$

However, for ℓ^- there are essentially two different cases to consider. In any case,

$$\ell^- = \left\{ \begin{pmatrix} 0 & t\alpha & t\beta \\ 0 & 0 & 0 \\ 0 & 0 & 0 \end{pmatrix} \,\middle|\, t \in \mathbb{R} \right\}$$

for some $(\alpha, \beta) \in \mathbb{R}^2 \setminus \{0\}$ and we will distinguish between the cases $\alpha \neq 0$ and $\alpha = 0$.

Suppose first that $\alpha \neq 0$. Applying push-forward under

$$g = \begin{pmatrix} 1 & 0 & 0 \\ 0 & 1 & s \\ 0 & 0 & 1 \end{pmatrix}$$

to μ, we obtain a measure $\mu' = g_* \mu$ invariant under $gHg^{-1} = H$ (since g commutes with H). The subgroup L gets replaced in this case by $L' = gLg^{-1}$ whose Lie algebra now contains

$$\mathrm{Ad}_g \left(\begin{pmatrix} 0 & \alpha & \beta \\ 0 & 0 & 0 \\ 0 & 0 & 0 \end{pmatrix} \right) = \begin{pmatrix} 1 & & \\ & 1 & s \\ & & 1 \end{pmatrix} \begin{pmatrix} 0 & \alpha & \beta \\ 0 & 0 & 0 \\ 0 & 0 & 0 \end{pmatrix} \begin{pmatrix} 1 & & \\ & 1 & -s \\ & & 1 \end{pmatrix}$$

$$= \begin{pmatrix} 0 & \alpha & \beta - \alpha s \\ 0 & 0 & 0 \\ 0 & 0 & 0 \end{pmatrix}.$$

For $s = \frac{\beta}{\alpha}$ this simplifies our discussion a little (in the case $\alpha \neq 0$), and shows that we may assume that

$$\ell^- = \left\{ \begin{pmatrix} 0 & * & 0 \\ 0 & 0 & 0 \\ 0 & 0 & 0 \end{pmatrix} \right\}.$$

In this case ℓ^+ and ℓ^- together generate

$$\langle \ell^+, \ell^- \rangle = \left\{ \begin{pmatrix} * & * & 0 \\ * & * & 0 \\ 0 & 0 & 0 \end{pmatrix} \right\} \cap \mathfrak{sl}_3(\mathbb{R}) < \ell. \tag{3.9}$$

We claim that we actually have $\ell = \langle \ell^+, \ell^- \rangle$. For this, first note that ℓ contains

$$\mathbf{m} = \begin{pmatrix} -1 & 0 & 0 \\ 0 & 1 & 0 \\ 0 & 0 & 0 \end{pmatrix} = \left[\begin{pmatrix} 0 & 0 & 0 \\ 1 & 0 & 0 \\ 0 & 0 & 0 \end{pmatrix}, \begin{pmatrix} 0 & 1 & 0 \\ 0 & 0 & 0 \\ 0 & 0 & 0 \end{pmatrix} \right]$$

by (3.9). Recall that $\ell = \ell^+ \oplus \ell^0 \oplus \ell^-$ and note that $\mathbf{m} \in \ell^0 \subseteq \mathfrak{g}^0$. The adjoint map $\mathrm{ad}_\mathbf{m}$ restricted to $\mathfrak{g}^0$ has the diagonal sub-algebra as eigenspace for eigenvalue 0 and the two subspaces

$$\left\{ \begin{pmatrix} 0 & 0 & 0 \\ 0 & 0 & * \\ 0 & 0 & 0 \end{pmatrix} \right\}, \left\{ \begin{pmatrix} 0 & 0 & 0 \\ 0 & 0 & 0 \\ 0 & * & 0 \end{pmatrix} \right\}$$

as eigenspaces for eigenvalue 1 and -1 respectively. If ℓ^0 were to contain either of these two subspaces, then we could obtain

$$\begin{pmatrix} 0 & 0 & 1 \\ 0 & 0 & 0 \\ 0 & 0 & 0 \end{pmatrix} \notin \ell$$

or

$$\begin{pmatrix} 0 & 0 & 0 \\ 0 & 0 & 0 \\ 1 & 0 & 0 \end{pmatrix} \notin \ell$$

as a commutator of elements in $\ell^\pm$ and of ℓ^0. Therefore

$$\ell^0 \subseteq \left\{ \begin{pmatrix} * & 0 & 0 \\ 0 & * & 0 \\ 0 & 0 & * \end{pmatrix} \right\} \cap \mathfrak{sl}_3(\mathbb{R}),$$

and either $\ell^0 = \mathbb{R}h$ or ℓ^0 is the full two-dimensional diagonal sub-algebra of $\mathfrak{sl}_3(\mathbb{R})$. In the first case, we obtain

$$L = \left\{ \begin{pmatrix} g & \\ & 1 \end{pmatrix} \,\bigg|\, g \in \mathrm{SL}_2(\mathbb{R}) \right\} \cong \mathrm{SL}_2(\mathbb{R}). \tag{3.10}$$

In the second, we obtain

$$L = \left\{ \begin{pmatrix} g & \\ & (\det g)^{-1} \end{pmatrix} \,\bigg|\, \det g > 0 \right\} \tag{3.11}$$

and the radical (and simultaneously the centre of) L is given by the diagonal subgroup

$$\left\{ \begin{pmatrix} e^s & & \\ & e^s & \\ & & e^{-2s} \end{pmatrix} \,\bigg|\, s \in \mathbb{R} \right\}. \tag{3.12}$$

However, Theorem 3.5.8 actually stated that the radical of L is unipotent. As this is not the case for (3.12), this rules out the second possibility. Therefore we must have (3.10), which is equivalent to our earlier claim that we have equality in (3.9).

Let now $x = g\,\mathrm{SL}_3(\mathbb{Z})$ be such that for L as in (3.10) the orbit Lx is closed with finite volume. Then by Theorem 3.5.8 the group $g^{-1}Lg$ is defined over $\mathbb{Q}$. Without going into the details of what this means, this should also mean that every reasonable unique object attached to $g^{-1}Lg$ should also be defined over $\mathbb{Q}$. We apply this principle to the common eigenspace[8] for eigenvalue 1. For L this is obviously given by $\mathbb{R}e_3$, and for $g^{-1}Lg$ the above rationality implies that $\mathbb{R}g^{-1}e_3$ is rational and so contains some integer vector $\mathbf{p} \in (\mathbb{R}g^{-1}e_3) \cap \mathbb{Z}^3 \setminus \{0\}$. We may choose $\mathbf{p}$ of minimal length (which makes it unique up to a choice of sign). Applying g, we see that the intersection of $\mathbb{R}e_3$ with the lattice $\Lambda_x = g\mathbb{Z}^3$ corresponding to $x = g\mathbb{Z}^3$ is given by $\mathbb{Z}g\mathbf{p}$. We now define the function

$$f(x) = \mathrm{vol}\,(\mathbb{R}e_3/(\mathbb{R}e_3 \cap \Lambda_x)) = \|g\mathbf{p}\|$$

---

[8] As $g^{-1}Lg$ is defined over $\mathbb{Q}$, the same holds for its Lie algebra $\mathrm{Ad}_{g^{-1}}(\ell)$. Now take a basis $\mathbf{b}_1, \mathbf{b}_2, \mathbf{b}_3$ for $\mathrm{Ad}_{g^{-1}}(\ell)$ over $\mathbb{Q}$ to see that this common eigenspace of $g^{-1}Lg$ is the intersection of the kernels of $\mathbf{b}_1, \mathbf{b}_2$, and $\mathbf{b}_3$.

which, by our discussion, is finite for any x with $\text{vol}(Lx) < \infty$. Applying $a_t \in A$ to x, we note that $g^{-1}a_t^{-1}La_t g = g^{-1}Lg$ is unchanged, and so is $\mathbf{p} \in \mathbb{Z}^3$, hence

$$f(a_t x) = \|a_t g\mathbf{p}\| = e^{\frac{t}{2}} \|g\mathbf{p}\|$$

as $g\mathbf{p} \in \mathbb{R}\mathbf{e}_3$ and a_t is a diagonal matrix. However, this once more contradicts Poincaré recurrence for the function f, μ-almost everywhere as $t \to \infty$.

So suppose now that $\alpha = 0$ or, equivalently,

$$\ell^- = \left\{ \begin{pmatrix} 0 & 0 & * \\ 0 & 0 & 0 \\ 0 & 0 & 0 \end{pmatrix} \right\}.$$

Taking the commutator of ℓ^+ and ℓ^-, we see that

$$\left\{ \begin{pmatrix} 0 & 0 & 0 \\ 0 & 0 & * \\ 0 & 0 & 0 \end{pmatrix} \right\} < \ell^0.$$

We claim that we actually have equality, so

$$\ell = \left\{ \begin{pmatrix} 0 & 0 & * \\ * & 0 & * \\ 0 & 0 & 0 \end{pmatrix} \right\}. \tag{3.13}$$

To see this, note that $\mathfrak{g}^0$ normalizes $\mathfrak{g}^-$, which implies that $\ell^0 = \mathfrak{g}^0 \cap \ell$ normalizes $\ell^- = \mathfrak{g}^- \cap \ell$. So suppose

$$m = \begin{pmatrix} a & 0 & 0 \\ 0 & b & c \\ 0 & d & e \end{pmatrix} \in \ell^0$$

and calculate

$$\text{ad}_m \left(\begin{pmatrix} 0 & 0 & 1 \\ 0 & 0 & 0 \\ 0 & 0 & 0 \end{pmatrix} \right) = \left[m, \begin{pmatrix} 0 & 0 & 1 \\ 0 & 0 & 0 \\ 0 & 0 & 0 \end{pmatrix} \right]$$

$$= \begin{pmatrix} a & 0 & 0 \\ 0 & b & c \\ 0 & d & e \end{pmatrix} \begin{pmatrix} 0 & 0 & 1 \\ 0 & 0 & 0 \\ 0 & 0 & 0 \end{pmatrix} - \begin{pmatrix} 0 & 0 & 1 \\ 0 & 0 & 0 \\ 0 & 0 & 0 \end{pmatrix} \begin{pmatrix} a & 0 & 0 \\ 0 & b & c \\ 0 & d & e \end{pmatrix}$$

$$\begin{pmatrix} 0 & 0 & a \\ 0 & 0 & 0 \\ 0 & 0 & 0 \end{pmatrix} - \begin{pmatrix} 0 & d & e \\ 0 & 0 & 0 \\ 0 & 0 & 0 \end{pmatrix} = \begin{pmatrix} 0 & d & a-e \\ 0 & 0 & 0 \\ 0 & 0 & 0 \end{pmatrix} \in \ell^-,$$

which shows that $d = 0$. Hence

$$\left\{ \begin{pmatrix} 0 & 0 & 0 \\ 0 & 0 & * \\ 0 & 0 & 0 \end{pmatrix} \right\} < \ell^0 < \left\{ \begin{pmatrix} * & 0 & 0 \\ 0 & * & * \\ 0 & 0 & * \end{pmatrix} \right\} \cap \mathfrak{sl}_3(\mathbb{R}),$$

and this implies that

$$\left\{ \begin{pmatrix} 0 & 0 & * \\ * & 0 & * \\ 0 & 0 & 0 \end{pmatrix} \right\} < \ell < \left\{ \begin{pmatrix} * & 0 & * \\ * & * & * \\ 0 & 0 & * \end{pmatrix} \right\} \cap \mathfrak{sl}_3(\mathbb{R}).$$

Permuting the first and second basis vector, we see that the latter Lie algebra is isomorphic to the upper triangular (Borel) sub-algebra

$$\begin{pmatrix} * & * & * \\ 0 & * & * \\ 0 & 0 & * \end{pmatrix},$$

and so is solvable. However, as the radical of L is unipotent we obtain the claim (3.13) by Theorem 3.5.8.

Suppose $x = g \operatorname{SL}_3(\mathbb{Z})$ has the property that Lx is closed with finite volume. By Theorem 3.5.8, this implies that $g^{-1}Lg$ is defined over $\mathbb{Q}$. Once more the common eigenspace of $g^{-1}Lg$ for eigenvalue 1 must be defined over $\mathbb{Q}$ as well. For L the common eigenspace is $\mathbb{R}\mathbf{e}_2$, and so $\mathbb{R}g^{-1}\mathbf{e}_2$ is a rational subspace. We let $\mathbf{p} \in (\mathbb{R}g^{-1}\mathbf{e}_2) \cap \mathbb{Z}^3 \setminus \{0\}$ be of minimal length and define

$$f(x) = \operatorname{vol}\left(\mathbb{R}\mathbf{e}_2/(\mathbb{R}\mathbf{e}_2 \cap \Lambda_x)\right) = \|g\mathbf{p}\| < \infty.$$

Applying $a_t \in A$ to x, we again obtain

$$f(a_t x) = \|a_t g\mathbf{p}\| = e^{\frac{t}{2}} \|g\mathbf{p}\|$$

and see that any $x \in X_3$ for which $\operatorname{vol}(Lx) < \infty$ cannot satisfy Poincaré recurrence. It follows that this second possibility of a subgroup L with

$$\dim \ell^- = \dim \ell^+ = 1$$

is also not possible, which concludes the proof. $\square$

We hope the reader has enjoyed the tour above, which visited many different areas of mathematics in order to reach a unique ergodicity result. In the remaining section we try to use this result to derive consequences in the theory of Diophantine approximation.

3.6 Diophantine Approximation

In this section we will explain some instances of the Dani correspondence mentioned in Sect. 3.2.3. In fact we will use the results of Sects. 3.3–3.5 to prove theorems in the theory of Diophantine approximation. We will start with elementary results that should help to develop understanding of the fundamental connection,[9] but will also reach more refined recent results.

The following is a natural starting point for higher-dimensional Diophantine approximation, and we will prove it with our measure-theoretic pigeonhole principle from Theorem 3.3.5.

Theorem 3.6.1 (Dirichlet) *Let $d \geqslant 1$, $\mathbf{v} \in \mathbb{R}^d$, and $T > 1$. Then there exists a natural number $q \in \mathbb{N}$ and an integer vector $\mathbf{p} \in \mathbb{Z}^d$ so that*

$$\left. \begin{array}{l} 1 \leqslant q \leqslant T \quad \text{and} \\ \left\| \mathbf{v} - \frac{1}{q}\mathbf{p} \right\|_\infty \leqslant \frac{1}{T^{1/d}q}. \end{array} \right\} \quad (3.14)$$

Proof We set

$$u = \begin{pmatrix} 1 & \\ \mathbf{v} & I_d \end{pmatrix}$$

and

$$a = \begin{pmatrix} T^{-1} & \\ & T^{1/d} I_d \end{pmatrix}$$

so that

$$\Lambda = au\mathbb{Z}^{d+1} \in X_{d+1}$$

is a unimodular lattice.

---

[9] The first few theorems also have easier elementary proofs, but our goal is to explain the connection.

3 Homogeneous Dynamics and its Connection to Diophantine Approximation

By Theorem 3.3.5 there exists some nonzero vector

$$\begin{pmatrix} q \\ -\mathbf{p} \end{pmatrix} \in \mathbb{Z}^{d+1}$$

so that

$$\left\| au \begin{pmatrix} q \\ -\mathbf{p} \end{pmatrix} \right\|_\infty \leq 1. \tag{3.15}$$

If $q = 0$, then $\mathbf{p} \neq 0$ and so

$$au \begin{pmatrix} q \\ -\mathbf{p} \end{pmatrix} = a \begin{pmatrix} 0 \\ -\mathbf{p} \end{pmatrix} = T^{1/d} \begin{pmatrix} 0 \\ -\mathbf{p} \end{pmatrix}$$

cannot have supremum norm less than or equal to 1. It follows that $q \neq 0$, and (by simultaneously switching the signs of $\mathbf{p}$ and q if necessary) we may assume that $q \geq 1$. Note that

$$au \begin{pmatrix} q \\ -\mathbf{p} \end{pmatrix} = a \begin{pmatrix} q \\ q\mathbf{v} - \mathbf{p} \end{pmatrix} = \begin{pmatrix} T^{-1}q \\ T^{1/d}(q\mathbf{v} - \mathbf{p}) \end{pmatrix}$$

and hence (3.15) is equivalent to (3.14) as in the theorem. $\square$

Corollary 3.6.2 (Dirichlet) *Let $d \geq 1$ and $\mathbf{v} \in \mathbb{R}^d$. Then there exist infinitely many $(q, \mathbf{p}) \in \mathbb{N} \times \mathbb{Z}^d$ so that*

$$\left\| \mathbf{v} - \frac{1}{q}\mathbf{p} \right\|_\infty \leq \frac{1}{q^{1+\frac{1}{d}}}. \tag{3.16}$$

Proof If $\mathbf{v} = \frac{1}{q_0}\mathbf{p}_0 \in \mathbb{Q}^d$ with $(q_0, \mathbf{p}_0) \in \mathbb{N} \times \mathbb{Z}^d$, we simply set $q = nq_0$ and $\mathbf{p} = n\mathbf{p}_0$ for any integer $n \geq 1$.

So assume that $\mathbf{v} \notin \mathbb{Q}^d$. Notice that the upper bound for the second statement in (3.14) satisfies

$$\frac{1}{T^{\frac{1}{d}}q} \leq \frac{1}{q^{1+\frac{1}{d}}}$$

for any $(q, \mathbf{p}) \in \mathbb{N} \times \mathbb{Z}^d$ as in Theorem 3.6.1 applied to some $T > 1$. Hence (3.14) always gives a solution to (3.16). Since we now assume that $\mathbf{v} \notin \mathbb{Q}^d$, the left-hand side of (3.16) is always positive. Having already found finitely many solutions to (3.16), we may let $\varepsilon > 0$ be the minimum of the left-hand side of (3.16) for these finitely many $(q, \mathbf{p}) \in \mathbb{N} \times \mathbb{Z}^d$. Now choose $T > 1$ so that $\frac{1}{T^{1/d}} < \varepsilon$ and apply

Theorem 3.6.1. This leads to a new solution of (3.16) with $\|\mathbf{v} - \frac{1}{q}\mathbf{p}\| < \varepsilon$. Iterating this gives infinitely many solutions. □

In the following sections we will ask whether Dirichlet's theorem and its corollary above are optimal in various ways. Using dynamics on X_d, we will partially answer these questions.

3.6.1 Very Well Approximable Vectors

We start ambitiously, by asking whether the exponent in Corollary 3.6.2 can be improved.

Definition 3.6.3 A vector $\mathbf{v} \in \mathbb{R}^d$ is very well approximable (with exponent α) if there exists a constant $\alpha > \frac{1}{d}$ so that

$$\left\| \mathbf{v} - \frac{1}{q}\mathbf{p} \right\|_\infty \leqslant \frac{1}{q^{1+\alpha}} \tag{3.17}$$

has infinitely many solutions $q \in \mathbb{N}$ and $\mathbf{p} \in \mathbb{Z}^d$.

Proposition 3.6.4 (Shrinking Targets and Borel–Cantelli) *Almost every $\mathbf{v} \in \mathbb{R}^d$ with respect to Lebesgue measure is not very well approximable.*

For the proof of this (and other results later) we define the diagonal element

$$a_t = \begin{pmatrix} e^{-t} & \\ & e^{t/d} I_d \end{pmatrix}$$

for $t \in \mathbb{R}$. We briefly explain why Proposition 3.6.4 is called a 'shrinking target' problem. As the proof will show, very well approximable vectors correspond to elements $x \in \mathsf{X}_{d+1}$ that visit a shrinking sequence of sets (B_n) after applying a_n, infinitely often. In other words, we are aiming at the target B_n to have $a_n x$ lying in B_n, and even though the sets B_n are shrinking we are supposed to achieve this goal for infinitely many n.

Proof Let $\alpha > \frac{1}{d}$ be fixed and suppose that $\mathbf{v} \in \mathbb{R}^d$ is very well approximable with exponent α. Let

$$\begin{pmatrix} q \\ \mathbf{p} \end{pmatrix} \in \mathbb{N} \times \mathbb{Z}^d$$

be a solution to (3.17). Let $\beta > 1$ (to be specified later) and set $n = \lfloor \beta \log q \rfloor$ so that

$$e^n \leqslant q^\beta \leqslant e^{n+1}.$$

Then
$$a_n \begin{pmatrix} 1 & \\ \mathbf{v} & I_d \end{pmatrix} \begin{pmatrix} q \\ -\mathbf{p} \end{pmatrix} = \begin{pmatrix} e^{-n}q \\ e^{n/d}(q\mathbf{v} - \mathbf{p}) \end{pmatrix}$$

satisfies
$$e^{-n}q \leq e^{-n+\frac{n}{\beta}+\frac{1}{\beta}} \leq e^{-n+\frac{n}{\beta}+1}$$

and, by (3.17), also
$$e^{n/d}|q\mathbf{v} - \mathbf{p}| \leq e^{n/d}q^{-\alpha} \leq e^{(\frac{1}{d} - \frac{\alpha}{\beta})n}.$$

As $\alpha > \frac{1}{d}$ we may choose $\beta > 1$ sufficiently close to 1 to ensure that $\frac{1}{d} - \frac{\alpha}{\beta} < 0$. Define
$$\gamma = \max\left(-1 + \frac{1}{\beta}, \frac{1}{d} - \frac{\alpha}{\beta}\right) < 0$$

so that the above gives
$$\left\| a_n \begin{pmatrix} 1 & \\ \mathbf{v} & I_d \end{pmatrix} \begin{pmatrix} q \\ -\mathbf{p} \end{pmatrix} \right\|_\infty \leq e e^{\gamma n}.$$

Using the fact that $\mathbf{v}$ is very well approximable with exponent α, this shows that
$$a_n u \mathbb{Z}^{d+1} \notin \mathsf{X}_{d+1}(ce^{\gamma n})$$

for infinitely many $n \geq 1$ and some uniform constant $c > 0$.

Recall from Theorem 3.3.13 that
$$m_{\mathsf{X}_{d+1}}(\mathsf{X}_{d+1} \setminus \mathsf{X}_{d+1}(\varepsilon)) \ll_d \varepsilon^\kappa$$

and hence
$$m_{\mathsf{X}_{d+1}}(\mathsf{X}_{d+1} \setminus \mathsf{X}_{d+1}(ce^{\gamma n})) \ll_d e^{\gamma \kappa n}.$$

Therefore, the Borel–Cantelli lemma implies that $m_{\mathsf{X}_{d+1}}$-almost every lattice x satisfies
$$a_n x \in \mathsf{X}_{d+1} \setminus \mathsf{X}_{d+1}(ce^{\gamma n})$$

only finitely often. This suggests that the set $\{\mathbf{v} \in \mathbb{R}^d \mid \mathbf{v} \text{ has exponent } \alpha\}$ is a null set by the Borel–Cantelli lemma. Varying $\alpha > \frac{1}{d}$ then gives the theorem.

However, this argument still has a gap as we wish to obtain a statement for the d-dimensional Lebesgue measure on $\mathbb{R}^d$, while we have a good estimate for $X_{d+1}\setminus X_{d+1}(ce^{\gamma n})$ for the $((d+1)^2 - 1)$-dimensional Haar measure $m_{X_{d+1}}$.

To deal with this issue, we note that

$$\left\{ \begin{pmatrix} 1 & \\ \mathbf{v} & I_d \end{pmatrix} \;\Big|\; v \in \mathbb{R}^d \right\} \mathrm{SL}_{d+1}(\mathbb{Z})$$

is the full unstable manifold for $I_{d+1}\,\mathrm{SL}_{d+1}(\mathbb{Z})$ and a_t for $t > 0$. This allows the above almost everywhere statement with respect to $m_{X_{d+1}}$ to be pushed to the same statement for Lebesgue almost every $\mathbf{v} \in \mathbb{R}^d$ (see Exercise 3.6.5). □

Exercise 3.6.5 (Finishing the Proof)

(a) Let $x \in \mathsf{X}_{d+1}$ and a_t be as above. Let

$$p = \begin{pmatrix} \det(g)^{-1} & \mathbf{w}^{\mathrm{t}} \\ & g \end{pmatrix}$$

for some $g \in \mathrm{GL}_d(\mathbb{R})$ and $\mathbf{w} \in \mathbb{R}^d$.

(b) We assume that the metric on X_{d+1} is derived from a right-invariant metric on $\mathrm{SL}_{d+1}(\mathbb{R})$ that defines its topology.

(c) Calculate $a_t p a_t^{-1}$ and prove that $a_t x$ and $a_t p x$ are a bounded distance from each other for all $t \geqslant 0$.

(d) Show, in particular, that for any compact set Ω of elements as above, there exists a constant $c_\Omega > 0$ so that $a_t x \in \mathsf{X}_{d+1}(\varepsilon)$ implies $a_t p x \in \mathsf{X}_{d+1}(c_\Omega \varepsilon)$ for any $t \geqslant 0$.

(e) Finish the proof of Proposition 3.6.4 by using the fact that Haar measure on $\mathrm{SL}_{d+1}(\mathbb{R})$ is locally the product of the left-invariant Haar measure on

$$P = \left\{ \begin{pmatrix} \det(g)^{-1} & \mathbf{w}^{\mathrm{t}} \\ & g \end{pmatrix} \;\Big|\; g \in \mathrm{GL}_d(\mathbb{R}),\, \mathbf{w} \in \mathbb{R}^d \right\}$$

and the Haar (which, in this case, is Lebesgue) measure on

$$U = \left\{ \begin{pmatrix} 1 & 0 \\ \mathbf{v} & I_d \end{pmatrix} \;\Big|\; \mathbf{v} \in \mathbb{R}^d \right\} \cong \mathbb{R}^d.$$

Let us now turn to a more delicate theorem in the area.

Theorem 3.6.6 (Sprindzuk [13], Kleinbock–Margulis [8]) *For Lebesgue almost every $s \in \mathbb{R}$ the vector $(s, s^2, \ldots, s^d)^{\mathrm{t}}$ is not very well approximable.*

3 Homogeneous Dynamics and its Connection to Diophantine Approximation

We will outline the argument of Kleinbock–Margulis [8], and note that their method works for much more general smooth sub-manifolds of $\mathbb{R}^d$. We define

$$u_s = \begin{pmatrix} 1 & & & \\ s & 1 & & \\ \vdots & & \ddots & \\ s^d & & & 1 \end{pmatrix} \in \mathrm{SL}_{d+1}(\mathbb{R}) \tag{3.18}$$

for $s \in \mathbb{R}$.

Proof of Theorem 3.6.6 The result again follows from the Borel–Cantelli lemma. However, instead of the quite basic volume estimate in Theorem 3.3.13 we are going to use the more powerful non-divergence result in Theorem 3.5.1.

To see this, let $I \subseteq \mathbb{R}$ be a compact interval. We fix $t \geq 0$ and define the polynomial $p(s) = a_t u_s$ for $s \in I$. We suppose for a moment that $\eta = 1$ satisfies the assumption in Theorem 3.5.1 for all rational subspaces $V < \mathbb{R}^{d+1}$ and all sufficiently large $t \geq 0$.

Fix $\alpha > \frac{1}{d}$ and $\gamma < 0$ as in the proof of Proposition 3.6.4. As in that proof, it now follows that if $(s, s^2, \ldots, s^d)^{\mathrm{t}}$ is very well approximable with exponent α, then there are infinitely many $n \in \mathbb{N}$ so that $a_n u_s \mathbb{Z}^{d+1} \notin \mathsf{X}_{d+1}(c \mathrm{e}^{\gamma n})$.

On the other hand, the assumption above and Theorem 3.5.1 show that for a fixed but sufficiently large n we have

$$m_{\mathbb{R}}\left(\left\{s \in I \mid a_n u_s \mathbb{Z}^{d+1} \notin \mathsf{X}_{d+1}(c \mathrm{e}^{\gamma n})\right\}\right) \ll_d \left(\mathrm{e}^{\gamma n}\right)^\kappa m_{\mathbb{R}}(I)$$

for some $\kappa > 0$. Together with the Borel–Cantelli lemma (here applied directly to the Lebesgue measure on I), this now implies that almost every $s \in I$ is not very well approximable with exponent α. Varying $\alpha > \frac{1}{d}$ and $I \subseteq \mathbb{R}$, the theorem follows.

So it remains to show that $\eta = 1$ satisfies the assumptions in Theorem 3.5.1. More precisely, it suffices to show the following.

Main Claim For all sufficiently large t (depending only on d and I) and all rational subspaces $V < \mathbb{R}^{d+1}$ we have

$$\sup_{s \in I} \mathrm{covol}\,(a_t u_s, V) \geq 1. \tag{3.19}$$

Note that for $V = \mathbb{R}^{d+1}$ there is nothing to prove, so we may now assume that $k = \dim V \leq d$. The reader may find it helpful in a first reading of the remaining argument to focus on the cases $k = 1$ or $d = 3$. For the proof of the claim, we are going to use the following statement.

Angle Claim For any subspace $V < \mathbb{R}^{d+1}$ of dimension $k \leq d$ there exists some $s \in I$ so that the angle between $\mathbb{R}\mathbf{e}_1$ and $u_s V$ is positive. In fact, there exists a constant $\omega_0 > 0$ (only depending on d and I) so that we can ensure that the above angle is at least ω_0.

To see this claim, let $\mathbb{G}_{k,d+1}$ be the Grassmannian consisting of k-dimensional subspaces in $\mathbb{R}^{d+1}$ (which for $k = 1$ is just the d-dimensional projective space). We recall that $\mathbb{G}_{k,d+1}$ is a compact manifold of dimension $k(d+1-k)$ with topology defined by saying that two subspaces are close if their intersections with the unit sphere in $\mathbb{R}^{d+1}$ are close in the Hausdorff metric. Next convince yourself that the angle $\omega(V) \in [0, \frac{\pi}{2}]$ between $\mathbb{R}^d \mathbf{e}_1$ and V depends continuously on V, and that

$$\mathbb{R} \times \mathbb{G}_{k,d+1} \ni (s, V) \mapsto u_s V \in \mathbb{G}_{k,d+1}$$

is continuous. From the compactness of $I \times \mathbb{G}_{k,d+1}$ it now follows that

$$\mathbb{G}_{k,d+1} \ni V \mapsto \max_{s \in I} \omega(u_s V)$$

is continuous. Therefore either

$$\omega_0 = \min_{V \in \mathbb{G}_{k,d+1}} \left(\max_{s \in I} \omega(u_s V) \right) > 0,$$

which gives the claim, or there exists some $V \in \mathbb{G}_{k,d+1}$ with $\omega(u_s V) = 0$ for all $s \in I$.

So suppose, for the purpose of a contradiction, the latter. This actually means that $\mathbb{R}\mathbf{e}_1 \in u_s V$ for all $s \in I$, or equivalently that

$$u_s^{-1} \begin{pmatrix} 1 \\ 0 \\ \vdots \\ 0 \end{pmatrix} = \begin{pmatrix} 1 \\ -s \\ \vdots \\ -s^d \end{pmatrix} \in V$$

for all $s \in I$. However if $s_1, \ldots, s_{d+1} \in I$ are pairwise different, then

$$\det \begin{pmatrix} 1 & \cdots & 1 \\ -s_1 & \cdots & -s_{d+1} \\ \vdots & & \vdots \\ -s_1^d & \cdots & -s_{d+1}^d \end{pmatrix} = (-1)^d \det \begin{pmatrix} 1 & \cdots & 1 \\ s_1 & \cdots & s_{d+1} \\ \vdots & & \vdots \\ s_1^d & \cdots & s_{d+1}^d \end{pmatrix} \neq 0$$

by the so-called Vandermonde determinant formula. This now implies that V must be $\mathbb{R}^{d+1}$, and contradicts the assumption that $k = \dim V \leq d$. Therefore the above Angle Claim must hold.

For the proof of the Main Claim from page 145 for rational subspaces V with dim $V = k$, we will use the exterior product $\Lambda^k \mathbb{R}^{d+1}$ and its natural Euclidean norm. The latter may be defined (for example) by declaring the basis

$$\mathbf{e}_{j_1} \wedge \cdots \wedge \mathbf{e}_{j_k} \in \Lambda^k \mathbb{R}^{d+1} \tag{3.20}$$

for $1 \leq j_1 < \cdots < j_k \leq d+1$ to be orthonormal. With this norm the k-dimensional volume of

$$\left\{ \sum_{j=1}^{k} t_j \mathbf{v}_j \;\middle|\; t_1, \ldots, t_k \in [0, 1) \right\}$$

is equal to the Euclidean length of

$$\mathbf{v}_1 \wedge \cdots \wedge \mathbf{v}_k \in \Lambda^k \mathbb{R}^{d+1}$$

for any $\mathbf{v}_1, \ldots, \mathbf{v}_k \in \mathbb{R}^{d+1}$. This allows a convenient reformulation of the covolume $\mathrm{covol}(a_t u_s, V)$ appearing in (3.19). Indeed, for a rational subspace $V < \mathbb{R}^{d+1}$ of dimension k we may choose a $\mathbb{Z}$-basis $\mathbf{v}_1, \ldots, \mathbf{v}_k$ of $V \cap \mathbb{Z}^{d+1}$, which gives

$$\mathrm{covol}(a_t u_s, V) = \|(a_t u_s \mathbf{v}_1) \wedge \cdots \wedge (a_t u_s \mathbf{v}_k)\| = \left\| \Lambda^k(a_t u_s) \mathbf{v}_1 \wedge \cdots \wedge \mathbf{v}_k \right\|$$

for any $s, t \in \mathbb{R}$, where

$$\Lambda^k(a_t u_s) = \Lambda^k(a_t) \Lambda^k(u_s)$$

is the linear map induced by $a_t u_s$ on $\Lambda^k \mathbb{R}^{d+1}$. For us it will be important to understand $\Lambda^k a_t$ more precisely. As a_t is diagonal with only the two eigenvalues e^{-t} and $e^{\frac{t}{d}}$, this is actually quite straightforward. Indeed, $\Lambda^k(a_t)$ is also diagonal with respect to the basis vectors in (3.20). For

$$1 = j_1 < j_2 < \cdots < j_k \leq d+1$$

the eigenvalue is $e^{(-1 + \frac{k-1}{d})t}$, and for

$$2 \leq j_1 < j_2 < \cdots < j_k \leq d+1$$

the eigenvalue is $e^{\frac{k}{d}t}$. Note that the former directions are contracted (and so are unlikely to be helpful in proving a lower bound on the expression $\mathrm{covol}(a_t u_s, V)$), while the latter are uniformly expanded and give a basis of $\Lambda^k(\{0\} \times \mathbb{R}^d)$.

We now fix a k-dimensional rational subspace $V < \mathbb{R}^{d+1}$. Let

$$\mathbf{v}_1, \ldots, \mathbf{v}_k \in V \cap \mathbb{Z}^{d+1}$$

be a $\mathbb{Z}$-basis of $V \cap \mathbb{Z}^{d+1}$. Then $\mathbf{v}_1 \wedge \cdots \wedge \mathbf{v}_k \in \Lambda^k \mathbb{Z}^{d+1}$, where the latter denotes the subgroup spanned by the basis vectors in (3.20). In particular, we see that

$$\|\mathbf{v}_1 \wedge \cdots \wedge \mathbf{v}_k\| \geqslant 1. \tag{3.21}$$

We let $s \in I$ have the property that the angle between $\mathbb{R}\mathbf{e}_1$ and $u_s V$ is at least ω_0 as in the Angle Claim. Since I is a bounded interval and the map $u_s \colon I \to \mathrm{SL}_{d+1}(\mathbb{R})$ is continuous, applying u_s to $\mathbf{v}_1, \ldots, \mathbf{v}_k$ only affects the estimate (3.21) by a uniformly bounded constant. In other words, there is a constant $c_0 > 0$ depending only on d and I so that

$$\|u_s \mathbf{v}_1 \wedge \cdots \wedge u_s \mathbf{v}_k\| \geqslant c_0.$$

Next we define an orthonormal basis of $u_s V$. If $u_s V \subseteq \{0\} \times \mathbb{R}^d$ then we let $\mathbf{w}_1, \cdots, \mathbf{w}_k \in V$ be any orthonormal basis. If, on the other hand,

$$\dim\left((u_s V) \cap (\{0\} \times \mathbb{R}^d)\right) = k - 1$$

then we let $\mathbf{w}_2, \cdots, \mathbf{w}_k$ be an orthonormal basis of $(u_s V) \cap (\{0\} \times \mathbb{R}^d)$ and complete this to an orthonormal basis of V by another vector $\mathbf{w}_1$. This choice also gives a concrete interpretation of our choice of $s \in I$. In fact

$$\mathbf{w}_1 = \pm(\cos \omega)\mathbf{e}_1 + (\sin \omega)\mathbf{w}'_1,$$

where $\omega \in [0, \frac{\pi}{2}]$ is the angle between $u_s V$ and $\mathbb{R}\mathbf{e}_1$, and $\mathbf{w}'_1 \in \{0\} \times \mathbb{R}^d$ has length one and is orthonormal to $\mathbf{w}_2, \cdots, \mathbf{w}_k$. Taking the tensor product, we obtain

$$\mathbf{w}_1 \wedge \cdots \wedge \mathbf{w}_k = \pm(\cos \omega)\mathbf{e}_1 \wedge \mathbf{w}_2 \wedge \cdots \wedge \mathbf{w}_k + (\sin \omega)\mathbf{w}'_1 \wedge \mathbf{w}_2 \wedge \cdots \wedge \mathbf{w}_k,$$

that $\|\mathbf{w}_1 \wedge \cdots \wedge \mathbf{w}_k\| = 1$, and that $\mathbf{w}_1 \wedge \cdots \wedge \mathbf{w}_k$ and $u_s \mathbf{v}_1 \wedge \cdots \wedge u_s \mathbf{v}_k$ are multiples of each other. Combining these, it follows that there exists a uniform constant $c_1 > 0$ (concretely, $c_1 = c_0 \sin \omega_0$) so that

$$(u_s \mathbf{v}_1) \wedge \cdots \wedge (u_s \mathbf{v}_k) = \alpha \mathbf{e}_1 \wedge \mathbf{w}_2 \wedge \cdots \wedge \mathbf{w}_k + \beta \mathbf{w}'_1 \wedge \mathbf{w}_2 \cdots \wedge \mathbf{w}_k$$

for $\alpha, \beta \in \mathbb{R}$ with $|\beta| \geqslant c_1$. Note that the two summands are orthogonal and that $\|\mathbf{w}'_1 \wedge \mathbf{w}_2 \wedge \cdots \wedge \mathbf{w}_k\| = 1$. We now apply a_t and obtain

$$\mathrm{covol}(a_t u_s, V) = \left\|\Lambda^k a_t u_s (\mathbf{v}_1 \wedge \cdots \wedge \mathbf{v}_k)\right\| \geqslant c_1 e^{\frac{k}{d}t}.$$

3 Homogeneous Dynamics and its Connection to Diophantine Approximation

Assuming $t > 0$ is large enough to ensure that $c_1 e^{\frac{k}{d}t} \geq 1$, it follows that

$$\sup_{s \in I} \operatorname{covol}(a_t u_s, V) \geq 1$$

for any rational subspace $V < \mathbb{R}^{d+1}$, as claimed. □

3.6.2 Well and Badly Approximable Vectors

The previous section showed that improving the exponent in Corollary 3.6.2 is usually too much to ask. Hence we now ask whether we can improve the approximation in Corollary 3.6.2 at least by an arbitrarily small multiplicative constant.

Definition 3.6.7 A vector $\mathbf{v} \in \mathbb{R}^d$ is called *well approximable* if, for every $\varepsilon > 0$, one can find $q \in \mathbb{N}$ and $\mathbf{p} \in \mathbb{Z}^d$ with

$$\left\| \mathbf{v} - \frac{\mathbf{p}}{q} \right\|_\infty \leq \frac{\varepsilon}{q^{1+\frac{1}{d}}}. \tag{3.22}$$

Otherwise $\mathbf{v}$ is said to be *badly approximable*.

Exercise 3.6.8 Show that it does not matter in Definition 3.6.7 whether we require that (3.22) has for any $\varepsilon > 0$ one or infinitely many solutions. [Hint: Distinguish the cases $\mathbf{v} \in \mathbb{Q}^d$ and $\mathbf{v} \in \mathbb{R}^d \setminus \mathbb{Q}^d$ and argue as in the proof of Corollary 3.6.2.]

For the notion of badly approximable vectors, the Dani correspondence takes the following form.

Proposition 3.6.9 (Dani Correspondence for Badly Approximable Vectors) *A vector $\mathbf{v} \in \mathbb{R}^d$ is badly approximable if and only if the forward orbit*

$$\left\{ a_t \begin{pmatrix} 1 & \\ \mathbf{v} & I_d \end{pmatrix} \mathbb{Z}^{d+1} \;\middle|\; t \geq 0 \right\} \subseteq \mathsf{X}_d$$

is bounded.

Proof Suppose first that $\mathbf{v} \in \mathbb{R}^d$ has the property that the forward orbit of

$$x = \begin{pmatrix} 1 & \\ \mathbf{v} & I_d \end{pmatrix} \mathbb{Z}^{d+1}$$

is unbounded. By Theorem 3.3.11, this means that the images of x under a_t for $t \geq 0$ (which are lattices in $\mathbb{R}^{d+1}$) must contain arbitrarily small non-zero elements. More precisely, for any $\varepsilon \in (0, 1)$ there exists some $t \geq 0$ so that the lattice $a_t x$ contains a vector with supremum norm less than ε. In other words, there exists

$$\begin{pmatrix} q \\ -\mathbf{p} \end{pmatrix} \in \mathbb{Z}^{d+1}$$

with

$$\left\| a_t \begin{pmatrix} 1 & \\ \mathbf{v} & I_d \end{pmatrix} \begin{pmatrix} q \\ -\mathbf{p} \end{pmatrix} \right\|_\infty \leqslant \varepsilon \qquad (3.23)$$

or, equivalently, with

$$\left. \begin{array}{l} \left| e^{-t} q \right| \leqslant \varepsilon \quad \text{and} \\ \left\| e^{\frac{t}{d}} (q\mathbf{v} - \mathbf{p}) \right\|_\infty \leqslant \varepsilon. \end{array} \right\}$$

Note that $t \geqslant 0$ and $\varepsilon < 1$ forces $q \neq 0$ (by the second estimate) and so we may assume $q \in \mathbb{N}$. Multiplying the second estimate by $e^{-\frac{t}{d}} q^{\frac{1}{d}} \leqslant \varepsilon^{\frac{1}{d}}$ now gives

$$q^{\frac{1}{d}} \|q\mathbf{v} - \mathbf{p}\|_\infty \leqslant \varepsilon^{1+\frac{1}{d}} \leqslant \varepsilon.$$

As $\varepsilon > 0$ was arbitrary, we see that $\mathbf{v}$ is well approximable.

Suppose now for the converse that $\mathbf{v}$ is well approximable. For $\varepsilon > 0$ there must therefore exist $q \in \mathbb{N}$ and $\mathbf{p} \in \mathbb{Z}^d$ so that

$$\left\| \mathbf{v} - \frac{1}{q}\mathbf{p} \right\|_\infty \leqslant \frac{\varepsilon^{1+\frac{1}{d}}}{q^{1+\frac{1}{d}}}$$

or, equivalently,

$$q^{\frac{1}{d}} \|q\mathbf{v} - \mathbf{p}\|_\infty \leqslant \varepsilon^{1+\frac{1}{d}}.$$

We let $t = \log \frac{q}{\varepsilon} \geqslant 0$ so that

$$e^{-t} q = \varepsilon$$

and

$$\left\| e^{\frac{t}{d}} (q\mathbf{v} - \mathbf{p}) \right\|_\infty = \varepsilon^{-\frac{1}{d}} q^{\frac{1}{d}} \|q\mathbf{v} - \mathbf{p}\|_\infty \leqslant \varepsilon,$$

which together are equivalent to (3.23). However, this shows together with Theorem 3.3.11 that the forward orbit

$$\left\{ a_t \begin{pmatrix} 1 & \\ \mathbf{v} & I_d \end{pmatrix} \mathbb{Z}^{d+1} \,\middle|\, t \geq 0 \right\}$$

is unbounded. □

The connection to the notion of badly approximable in Proposition 3.6.9 allows us to use ergodicity (as in Theorem 3.4.1) to prove the following result.

Proposition 3.6.10 *Lebesgue almost every* $\mathbf{v} \in \mathbb{R}^d$ *is well approximable.*

Exercise 3.6.11 Prove Proposition 3.6.10 (by using the ergodicity in Theorem 3.4.1 and the method in Exercise 3.6.5).

However, the following theorem requires deeper results, and we postpone the outline of the proof to Sect. 3.6.4.

Theorem 3.6.12 (Shah) *For Lebesgue almost every* $s \in \mathbb{R}$ *the vector*

$$\gamma(s) = (s, s^2, \ldots, s^d)^{\mathrm{t}} \in \mathbb{R}^d$$

is well approximable.

The method used to prove this once more works in much greater generality.

3.6.3 Dirichlet-Improvable Vectors

Motivated by the positive answer that Corollary 3.6.2 can be improved almost surely, we come to our last question. Can the original formulation in Dirichlet's theorem (Theorem 3.6.1) instead of Corollary 3.6.2 also be improved by an arbitrary multiplicative constant $\lambda < 1$.

Definition 3.6.13 A vector $\mathbf{v} \in \mathbb{R}^d$ is said to be λ-*Dirichlet improvable* for some constant $\lambda \in (0, 1)$ if for any large enough $T > 1$ there exists $q \in \mathbb{N}$ and $\mathbf{p} \in \mathbb{Z}^d$ with

$$\left. \begin{array}{c} 1 \leq q \leq \lambda T \quad \text{and} \\ \left\| \mathbf{v} - \dfrac{\mathbf{p}}{q} \right\|_\infty \leq \dfrac{\lambda}{T^{1/d} q}. \end{array} \right\} \tag{3.24}$$

Moreover, $\mathbf{v}$ is called *Dirichlet improvable* if it is λ-Dirichlet improvable for some constant $\lambda < 1$.

Proposition 3.6.14 (Dynamical Observation) *If $\mathbb{Z}^d$ is a limit point of the forward orbit*

$$\left\{ \begin{pmatrix} e^{-t} & \\ & e^{t/d} I_d \end{pmatrix} \begin{pmatrix} 1 & \\ \mathbf{v} & I_d \end{pmatrix} \mathbb{Z}^{d+1} \,\bigg|\, t \geqslant 0 \right\},$$

then $\mathbf{v}$ is not Dirichlet improvable.

Proof Suppose $\lambda < 1$ and $\mathbf{v}$ is λ-Dirichlet improvable, and $T = e^t$ is sufficiently large (as in the definition). Then (3.24) has a solution, which means that

$$\left\| a_t \begin{pmatrix} 1 & \\ \mathbf{v} & I_d \end{pmatrix} \begin{pmatrix} q \\ -\mathbf{p} \end{pmatrix} \right\|_\infty = \left\| \begin{pmatrix} T^{-1}q \\ T^{1/d}(q\mathbf{v} - \mathbf{p}) \end{pmatrix} \right\|_\infty \leqslant \lambda.$$

This implies that the lattice cannot be close to $\mathbb{Z}^d$ (which does not have non-zero vectors with sup norm bounded by λ). As this holds for any sufficiently large value of $T = e^t$, we see that $\mathbb{Z}^d$ cannot be a limit point of the forward orbit appearing in the proposition. □

Proposition 3.6.15 *Lebesgue almost every $\mathbf{v} \in \mathbb{R}^d$ is not Dirichlet improvable.*

Exercise 3.6.16 Prove Proposition 3.6.15 (using the ergodicity of Theorem 3.4.1 and the method in Exercises 3.6.5 and 3.6.11).

Using more general arguments from homogeneous dynamics, we can once more pass to curves.

Theorem 3.6.17 (Shah) *For Lebesgue almost every $s \in \mathbb{R}$, the vector*

$$\gamma(s) = (s, s^2, \ldots, s^d)^{\mathrm{t}} \in \mathbb{R}^d \tag{3.25}$$

is not Dirichlet improvable.

The method used to prove Theorem 3.6.17 works more generally. We outline the proof in Sect. 3.6.4.

3.6.4 Equidistribution of Curves

The two theorems of Shah above (Theorems 3.6.12 and 3.6.17) follow from the next result.

3 Homogeneous Dynamics and its Connection to Diophantine Approximation

Theorem 3.6.18 (Shah) *For any compact interval $I \subseteq \mathbb{R}$ of positive length, initial point $x_0 \in X_{d+1}$, and u_s as in (3.18) we have*

$$\frac{1}{m_{\mathbb{R}}(I)} \int_I f(a_t u_{\gamma(s)} x_0) \, ds \longrightarrow \int_{X_{d+1}} f \, dm_{X_{d+1}}$$

as $t \to \infty$ for all $f \in C_c(X_{d+1})$.

We now show how this theorem can be used to derive the two results concerning Diophantine properties on curves.

Proof of Theorem 3.6.12 Assuming Theorem 3.6.18 By Proposition 3.6.9 we need to show that

$$\{s \in \mathbb{R} \mid u_s \mathbb{Z}^{d+1} \text{ has bounded forward orbit}\}$$

is a null set. Using $\varepsilon = \frac{1}{n}$ for $n \in \mathbb{N}$ and Theorem 3.3.11 it suffices to show for a fixed $\varepsilon > 0$ that

$$B = \{s \in \mathbb{R} \mid a_t u_s \mathbb{Z}^{d+1} \in X_{d+1}(\varepsilon) \text{ for all } t \geqslant 0\} \tag{3.26}$$

is a null set. We argue indirectly, and suppose this does not hold. Let f in $C_c(X_{d+1})$ be a non-zero function with

$$0 \leqslant f \leqslant 1 \tag{3.27}$$

and

$$\operatorname{Supp} f \subseteq X_{d+1} \setminus X_{d+1}(\varepsilon). \tag{3.28}$$

Let

$$\delta = \int_{X_{d+1}} f \, dm_{X_{d+1}} > 0. \tag{3.29}$$

As we are assuming that B as in (3.26) is not a null set, we can find by Lebesgue density a compact interval $I \subseteq \mathbb{R}$ so that

$$m_{\mathbb{R}}(B \cap I) \geqslant \left(1 - \frac{\delta}{2}\right) m_{\mathbb{R}}(I). \tag{3.30}$$

For $t \geqslant 0$ we now have $f(a_t u_s \mathbb{Z}^{d+1}) = 0$ by (3.28) if $s \in B$ and $f(a_t u_s \mathbb{Z}^{d+1}) \leqslant 1$ otherwise by (3.27). Together with (3.30) this gives

$$\frac{1}{m_{\mathbb{R}}(I)} \int_I f(a_t u_s \mathbb{Z}^{d+1}) \, ds \leqslant \frac{m_{\mathbb{R}}(I \setminus B)}{m_{\mathbb{R}}(I)} \leqslant \frac{\delta}{2}.$$

Letting $t \to \infty$ and applying Theorem 3.6.18 we therefore obtain

$$\int_{X_{d+1}} f \, dm_{X_{d+1}} \leq \frac{\delta}{2},$$

which contradicts our definition of δ in (3.29). This contradiction shows that B as in (3.26) must be a null set, and the theorem follows. □

The argument for Theorem 3.6.17 is actually very similar to the above.

Exercise 3.6.19 Use functions $f \in C_c(X_{d+1})$ with support near $\mathbb{Z}^{d+1}$, Proposition 3.6.14, and the arguments from the proof of Theorem 3.6.12 to prove Theorem 3.6.17.

3.6.5 The Twisting Trick

We will not prove Theorem 3.6.18 as it relies heavily on (extensions of) the linearization technique developed by Ratner, Dani–Margulis, and Shah. Instead we wish to explain one of the ideas of Shah that allows averages over curves as in Theorem 3.6.18 to be connected to unipotent dynamics.

For this we suppose that $I \subseteq \mathbb{R}$ is a compact interval and $\gamma \colon I \to \mathbb{R}^d$ is C^2. We also assume[10] that $\|\gamma'(s)\|_2 = 1$ for $s \in I$ and that $I \ni s \mapsto k_s \in \mathrm{SO}_d(\mathbb{R})$ is continuous so that $k_s \gamma'(s) = \mathbf{e}_1$ for all $s \in I$. We identify $k \in \mathrm{SO}_d(\mathbb{R})$ with

$$\begin{pmatrix} 1 & \\ & k \end{pmatrix} \in \mathrm{SO}_{d+1}(\mathbb{R})$$

to simplify the notation.

Lemma 3.6.20 (Twisting Trick, First Step) *Let $x_0 \in X_{d+1}$. Using the above assumptions and notation, any weak* limit of*

$$\frac{1}{m_{\mathbb{R}}(I)} \int_I \delta_{a_t k_s u_{\gamma(s)} x_0} \, ds \qquad (3.31)$$

for $t \to \infty$ is invariant under the one-parameter subgroup $U = \{u_{r\mathbf{e}_1} \mid r \in \mathbb{R}\}$.

For simplicity we will use the shorthand

$$\fint_I F(s) \, ds = \frac{1}{m_{\mathbb{R}}(I)} \int_I F(s) \, ds$$

for the normalized integral of a function $F \colon I \to \mathbb{R}$.

---

[10] This is not satisfied for γ as in (3.25), but a simple re-parametrization will make it true.

Proof of Lemma 3.6.20 Let $f \in C_c(X_{d+1})$, $r \in \mathbb{R}$, and $t \geqslant 0$. Then we have

$$\fint_I f(u_{re_1} a_t k_s u_{\gamma(s)} x_0) \, ds = \fint_I f(a_t u_{e^{-(1+1/d)t} re_1} k_s u_{\gamma(s)} x_0) \, ds$$

$$= \fint_I f(a_t k_s u_{e^{-(1+1/d)t} r\gamma'(s) + \gamma(s)} x_0) \, ds$$

by the way a_t and k_s interact with u_{e_1}. Fixing r and thinking of a large $t > 0$, we have

$$\gamma(s) + e^{-(1+\frac{1}{d})t} r\gamma'(s) = \gamma\left(s + e^{-(1+\frac{1}{d})t} r\right) + \varepsilon(s)$$

for an error term $\varepsilon(s) = \varepsilon_{r,t}(s) = O\left(e^{-2(1+\frac{1}{d})t}\right)$ as γ is assumed to be C^2. Let

$$I' = I \cap \left(I + e^{-2(1+\frac{1}{d})t} r\right)$$

which (for fixed $r \in \mathbb{R}$ and large $t > 0$) is basically equal to I. We let

$$s' = s + e^{-(1+\frac{1}{d})t} r$$

for $s \in I$, and obtain

$$\fint_I f(u_{re_1} a_t k_s u_{\gamma(s)} x_0) \, ds = \fint_{I'} f(a_t k_s u_{\gamma(s') + \varepsilon(s)} x_0) \, ds' + O\left(re^{-(1+\frac{1}{d})t}\right).$$

We now take the error term $\varepsilon(s)$ and again move it across k_s (which will rotate it) and a_t (which will expand it) to the left. Defining $\varepsilon_{\mathrm{rot}}(s) = k_s \varepsilon(s)$, we obtain

$$\fint_I f(u_{re_1} a_t k_s u_{\gamma(s)} x_0) \, ds = \fint_{I'} f(u_{e^{(1+1/d)t} \varepsilon_{\mathrm{rot}}(s)} a_t k_s u_{\gamma(s')}) \, ds' + O\left(re^{-(1+\frac{1}{d})t}\right).$$

As $\varepsilon(s) = O\left(e^{-2(1+\frac{1}{d})t}\right)$, we know that

$$e^{(1+\frac{1}{d})t} \varepsilon_{\mathrm{rot}}(s) = O\left(e^{-(1+\frac{1}{d})t}\right)$$

is tiny. Using continuity of f, we see that the term $u_{e^{(1+1/d)t} \varepsilon_{\mathrm{rot}}(s)}$ does not change the value of f much. For a weak* limit μ of (3.31) as $t \to \infty$ this shows that

$$\int f(u_{re_1} x) \, d\mu(x) = \int f \, d\mu.$$

As $f \in C_c(X_{d+1})$ was arbitrary, it follows that μ is invariant under U, as desired. □

Suppose now we can use unipotent magic (meaning Theorems 3.5.5 and 3.5.6, and extensions of the arguments needed to prove them) to analyze the weak* limits of (3.31) appearing in Lemma 3.6.20. What does this have to do with the averages appearing in Theorem 3.6.18?

Lemma 3.6.21 (Twisting Trick, Second Step) *Suppose now in addition that for any interval I and any $x_0 \in X_{d+1}$ we have*

$$\frac{1}{m_\mathbb{R}(I)} \int_I \delta_{a_t k_s u_{\gamma(s)} x_0} \, ds \longrightarrow m_{X_{d+1}}.$$

Then for any interval I and $x_0 \in X_{d+1}$ we also have that

$$\frac{1}{m_\mathbb{R}(I)} \int_I \delta_{a_t u_{\gamma(s)} x_0} \, ds \longrightarrow m_{X_{d+1}}.$$

In other words, the orthogonal matrix that was so helpful above to obtain unipotent invariance can simply be forgotten.

Proof of Lemma 3.6.21 Let $\varepsilon > 0$ and $f \in C_c(X_{d+1})$. By the assumed continuity of the map $I \ni s \mapsto k_s$, the function $I \times X_{d+1} \ni (s, x) \mapsto f(k_s^{-1} x)$ is uniformly continuous. Hence there exists some $\delta > 0$ so that

$$\left| f(k_{s_1}^{-1} x) - f(k_{s_2}^{-1} x) \right| < \varepsilon \qquad (3.32)$$

whenever $x \in X_{d+1}$ and $s_1, s_2 \in I$ satisfy $|s_1 - s_2| < \delta$.

We split I into finitely many sub-intervals I_ℓ for $\ell = 1, \ldots, L$ of equal length and length less than δ. Using the assumed equidistribution for each of these intervals, the above continuity property of f, and the fact that k_s commutes with a_t, we can now obtain the desired conclusion up to 2ε. Indeed, fix some $s_\ell \in I_\ell$ for $\ell = 1, \ldots, L$ and apply our assumption to the interval I_ℓ and the function $X_{d+1} \ni x \mapsto f(k_{s_\ell}^{-1} x)$. As $m_{X_{d+1}}$ is invariant under k_{s_ℓ}, we therefore have

$$\left| \fint_{I_\ell} f\left(k_{s_\ell}^{-1} k_s a_t u_{\gamma(s)} x_0 \right) ds - \int_{X_{d+1}} f \, dm_{X_{d+1}} \right| < \varepsilon$$

for $\ell = 1, \ldots, L$ and all sufficiently large t. Now we may use the estimate (3.32) for $x = k_s a_t u_{\gamma(x)} x_0$, $s, s_\ell \in I_\ell$, and $\ell = 1, \ldots, L$ to obtain

$$\left| \fint_I f(a_t u_{\gamma(s)} x_0) \, ds - \int_{X_{d+1}} f \, dm_{X_{d+1}} \right|$$

$$= \left| \frac{1}{L} \sum_{\ell=1}^L \fint_I f(k_s^{-1} k_s a_t u_{\gamma(s)} x_0) \, ds - \int_{X_{d+1}} f \, dm_{X_{d+1}} \right|$$

$$\leqslant \left| \frac{1}{L} \sum_{\ell=1}^{L} \left(\fint_I f(k_{s_\ell}^{-1} k_s a_t u_{\gamma(s)} x_0) \, ds - \int_{X_{d+1}} f \, dm_{X_{d+1}} \right) \right| + \varepsilon$$

$$\leqslant 2\varepsilon$$

for all sufficiently large $t > 0$. As $f \in C_c(X_{d+1})$ and $\varepsilon > 0$ were arbitrary, we obtain the lemma. □

3.6.6 Completing the Proof for Planar Curves

We conclude our excursion into homogeneous dynamics and its connection to Diophantine approximation by using non-divergence, the twisting trick above, and Theorem 3.5.9 to prove the following result.

Proposition 3.6.22 *For any compact interval $I \subseteq \mathbb{R}$ of positive length, the curve $\gamma: s \in \mathbb{R} \mapsto (s, s^2)^t$, $x_0 \in X_3$, and any $f \in C_c(X_3)$ we have*

$$\frac{1}{T m_{\mathbb{R}}(I)} \int_0^T \!\! \fint_I f(a_t u_{\gamma(s)} x_0) \, ds \longrightarrow \int_{X_3} f \, dm_{X_3}$$

as $T \to \infty$. This is enough to prove Theorems 3.6.12 and 3.6.17 for the case $d = 2$.

Sketch of Proof Let $I \subseteq \mathbb{R}$ be a compact interval and $x_0 \in X_3$ as in the proposition. With these, we define the measures

$$\mu_T = \frac{1}{T} \int_0^T \!\! \fint_I \delta_{a_t u_{\gamma(s)} x_0} \, ds \, dt$$

and

$$\mu_T^{\text{twist}} = \frac{1}{T} \int_0^T \!\! \fint_I \delta_{a_t k_s u_{\gamma(s)} x_0} \, ds \, dt.$$

Our goal is to show that μ_T converges in the weak* topology to m_{X_3} as $T \to \infty$. We will achieve this by using the following main steps.

- Any weak* limit of μ_T and of μ_T^{twist} is a probability measure.
- Any weak* limit of μ_T^{twist} is invariant under $A = \{a_t \mid t \in \mathbb{R}\}$ and

$$U = \left\{ \begin{pmatrix} 1 & & \\ * & 1 & \\ & & 1 \end{pmatrix} \right\}.$$

- By Theorem 3.5.9 the latter gives $\mu_T^{\text{twist}} \to m_{X_3}$ as $T \to \infty$.
- Using the twisting trick in Sect. 3.6.5, this also gives the weak* convergence $\mu_T \to m_{X_3}$ as $T \to \infty$.

All of these steps have already been discussed, which is why we only sketch the argument.

Indeed, in Corollary 3.5.2 we already showed how the quantitative non-divergence result in Theorem 3.5.1 can be used to prove that certain limit measures are probability measures. However, the concrete definition of η in (3.2) relied on the fact that all our averages started at the same point x_0, which is not the case for our averages

$$\fint_I \delta_{a_t u_{\gamma(s)} x_0} \, ds.$$

However, in the proof of Theorem 3.6.6 our Main Claim showed that for these averages we can set $\eta = 1$ once $t > 0$ is sufficiently large. As we are averaging over $t \in [0, T]$ with $T \to \infty$, we can certainly restrict to sufficiently large t and hence set $\eta = 1$. This allows us to prove that any weak* limit of the measures μ_T is a probability measure.

The measures μ_T and μ_T^{twist} differ in their definitions only due to the factor $k_s \in SO_2(\mathbb{R})$. Recalling that $a_t k_s = k_s a_t$, it follows for example that any function $f \in C_c(X_3)$ invariant under $SO_2(\mathbb{R})$ will have the same integral with respect to μ_T^{twist} as it does with μ_T. From this it follows that any weak* limit of the measures μ_T^{twist} must also be a probability measure.

Applying the first part of the twisting trick in Lemma 3.6.20 (after verifying that we can replace each average $\fint_I f_I \cdot ds$ in the proof by the average $\frac{1}{T} \int_0^T f_I \cdot ds \, dt$) we see that any weak* limit of the measures μ_T^{twist} is invariant under U.

The invariance of a weak* limit of μ_T^{twist} (or of μ_T) under A is quickly verified since composing some $f \in C_c(X_3)$ with an element $a_{t_0} \in A$ corresponds to shifting the integral $\int_0^T \cdot dt$ by t_0. Hence any weak* limit of μ_T^{twist} is an H-invariant probability measure on X_3 where $H = AU$. However, by Theorem 3.5.9, this shows that m_{X_3} is the only weak* limit of μ_T^{twist} for $T \to \infty$. By compactness of the closed unit ball in the dual of $C_c(X_3)$ (Alaoglu's theorem) this means that $\mu_T^{\text{twist}} \to m_{X_3}$ as $T \to \infty$.

The above holds for any compact interval with positive length. Applying the second part of the twisting trick in Lemma 3.6.21 (after verifying that the additional averaging does not affect the proof) we deduce that $\mu_T \to m_{X_3}$ as $T \to \infty$.

Finally, we go back to the proof of Theorem 3.6.12 in Sect. 3.6.4 (and to Exercise 3.6.19) and verify that the extra averaging in our equidistribution result does not affect the proofs of Theorem 3.6.12 and 3.6.17. □

References

1. Artin, E.: Ein mechanisches system mit quasiergodischen Bahnen. Abh. Math. Sem. Univ. Hamburg **3**(1), 170–175 (1924). https://doi.org/10.1007/BF02954622
2. Borel, A., Prasad, G.: Values of isotropic quadratic forms at S-integral points. Compos. Math. **83**(3), 347–372 (1992). http://www.numdam.org/item?id=CM_1992__83_3_347_0
3. Dani, S.G.: On invariant measures, minimal sets and a lemma of Margulis. Invent. Math. **51**(3), 239–260 (1979). https://doi.org/10.1007/BF01389917
4. Dani, S.G.: Divergent trajectories of flows on homogeneous spaces and Diophantine approximation. J. Reine Angew. Math. **359**, 55–89 (1985). Correction: **360** (1985), 214. https://doi.org/10.1515/crll.1985.359.55
5. Einsiedler, M., Ward, T.: Ergodic Theory with a View Towards Number Theory. Graduate Texts in Mathematics vol. 259. Springer, London (2011). https://doi.org/10.1007/978-0-85729-021-2
6. Einsiedler, M., Ward, T.: Functional Analysis, Spectral Theory, and Applications. Graduate Texts in Mathematics, vol. 276. Springer, London (2017). https://doi.org/10.1007/978-3-319-58540-6
7. Einsiedler, M., Ward, T.: Homogeneous dynamics and applications (in preparation). https://tbward0.wixsite.com/books
8. Kleinbock, D.Y., Margulis, G.A.: Flows on homogeneous spaces and Diophantine approximation on manifolds. Ann. Math. **148**(1), 339–360 (1998). https://doi.org/10.2307/120997
9. Margulis, G.A.: The action of unipotent groups in a lattice space. Mat. Sb. **86**(128), 552–556 (1971). https://doi.org/10.1070/SM1971v015n04ABEH001560
10. Margulis, G.A., Tomanov, G.M.: Measure rigidity for almost linear groups and its applications. J. Anal. Math. **69**, 25–54 (1996). https://doi.org/10.1007/BF02787100
11. Minkowski, H.: Geometrie der Zahlen. In: Bibliotheca Mathematica Teubneriana, Band, vol. 40. Johnson Reprint Corp., New York (1968)
12. Series, C.: The modular surface and continued fractions. J. Lond, Math. Soc. **31**(1), 69–80 (1985). https://doi.org/10.1112/jlms/s2-31.1.69
13. Sprindžuk, V.G.: More on Mahler's conjecture. Dokl. Akad. Nauk SSSR **155**, 54–56 (1964). Eng. Transl. *Soviet Math. Dokl.* **5** (1964), 361–363. http://mi.mathnet.ru/eng/dan29255

Chapter 4
Effective Unique Ergodicity and Weak Mixing of Translation Flows

Giovanni Forni

Abstract This text is an introduction to the author's cohomological approach, based on Hodge theory, to (effective) unique ergodicity and weak mixing of translation flows. Compared to previous expositions, it emphasizes the analogy between the two problems by introducing a point of view on weak mixing based on the appropriate twisted cohomology. In particular, a new cohomological proof, based on Hodge theory for the twisted cohomology, of Veech's classical criterion for weak mixing is presented here for the first time. The exposition also aims to give a general introduction to the ergodic theory of translation flows and related systems and includes references to background material and related developments in the theory, as well as several exercises, open problems and conjectures.

4.1 Introduction

Translation surfaces and translation flows were introduced in the study of the dynamics of billiards in rational polygons and interval exchange transformations (IET's). Later the renormalization theory of translation flows and IET's provided a fundamental tool for the study of the dynamics of smooth locally Hamiltonian (area-preserving) flows on higher genus surfaces with singularity of saddle type.

In these notes, after reviewing basic definitions and results in the ergodic theory of translation flows, we outline a cohomological approach, based on Hodge theory, to the basic effective ergodic properties of translation flows: polynomial unique ergodicity and polynomial weak mixing.

We emphasize how this point of view allows for a largely unified treatment to the theory of unique ergodicity and weak mixing of translation flows (and related dynamical systems). Methods of Hodge theory were introduced in the subject in the seminal papers of M. Kontsevich [1] and M. Kontsevich and A. Zorich [2],

G. Forni (✉)
Department of Mathematics, University of Maryland, College Park, MD, USA
e-mail: gforni@umd.edu

developed in the work of the author on effective unique ergodicity starting in [3] (and in a different direction by M. Möller [4]), and perfected in the more recent work of S. Filip [5–7].

The power of the Hodge theoretic approach consists in formulas and results about the Lyapunov exponents of a linear cocycle (the so-called Kontsevich–Zorich cocycle) which on the one hand encodes the topological (homological) behavior of trajectories of translation flows, and on the other hand is related to the tangent cocycle of the Teichmüller flow on the moduli space of Abelian differentials. Results on the Lyapunov structure of the Kontsevich–Zorich cocycle are therefore fundamental for both the ergodic theory of translation flows and for the theory of the Teichmüller flow and of the related $SL(2, \mathbb{R})$ action (see for instance [8, 9]).

Similar methods of Hodge theory (for the twisted cohomology) were first introduced in the author's work on effective weak mixing [10], and have provided a new point of view on earlier work on weak mixing (in particular a new cohomological proof of a fundamental weak mixing criterion of Veech [11]), as well as formulas and estimates on the Lyapunov exponents of a "twisted cohomology cocycle" which lead to bounds on the speed of weak mixing of typical translation flows. A similar linear cocycle was introduced in the work of A. Bufetov and B. Solomyak [12–15] on Hölder properties of spectral measures which motivated our own, although the Bufetov-Solomyak cocycle differs as it is not cohomological and the analysis of its Lyapunov spectrum is based on a different method (Erdös-Kahane argument), not related to Hodge theory.

These notes are thus based mostly on [3, 16, 17] (and also [18, 19]) for unique ergodicity and on [10, 20] for the weak mixing property of translation flows. The cohomological proof of the Veech criterion for weak mixing was so far unpublished.

There are many excellent surveys on interval exchange transformations, translation surfaces and flows, Veech surfaces, polygonal billiards, dynamics on moduli space, which can complement these notes, for instance (the list is not exhaustive): [17, 21–26]. However, with the exception of [17] (which does not cover weak mixing) none of the above sources covers applications of Hodge theory to the ergodic theory of translation flows.

The notes are an expanded version of lectures given at the CIME school *Modern Aspects of Dynamical Systems* in Cetraro, Italy, in August 2021.

I am very grateful to the organizers Claudio Bonanno, Alfonso Sorrentino and Corinna Ulcigrai for the opportunity to lecture there.

4.2 Background: Definitions and Results

4.2.1 Polygonal Billiards

A billiard in an euclidean (planar) polygon P is a dynamical systems defined by the motion of a point mass on the billiard table which undergoes "elastic" collisions at the boundary edges (according to the reflection law of geometric optics: the angle of incidence is equal to the angle of reflection). We refer to the survey of H. Masur and S. Tabachnikov [22] for an introduction to billiards in polygons and translation surfaces and flows.

We note that the phase space of the polygonal billiard in P, restricted to an energy surface, can be taken to be $P \times S^1$ since the dynamics on all energy surfaces are isomorphic and the velocity can be taken to be of unit norm, hence an element of $S^1 = \{v \in \mathbb{R}^2 | \|v\| = 1\}$. We also note that the dynamics of a polygonal billiard is not defined (for all times) for all trajectories that end up in corners, hence it is defined almost everywhere (on the complement of the union of a countable set of lines in the phase space).

Polygonal billiards in right triangles are famously related to the motion of two point masses on an interval with elastic collisions between the masses and at the endpoints.

Exercise 4.1 *Prove that the Hamiltonian system given by point masses m_1 and m_2 on the interval $[0, 1]$ with elastic collisions between the masses and between the masses and the endpoints of the interval is isomorphic to the Hamiltonian system given by a right triangle billiard with an angle $\tan^{-1} \sqrt{m_1/m_2}$.*

The *unfolding construction* of Zemlyakov and Katok [27] uncovers a fundamental dichotomy in the dynamics of polygonal billiards between the so-called *rational billiards* and *non-rational billiards*. The geometric idea of the construction is to continue a trajectory as a straight line beyond the reflection at an edge by reflecting the table with respect to that edge. It replaces therefore the motion with reflections on the billiard table by a straight line flow on a surface endowed with a flat (indeed a translation) structure. The dichotomy appears since this (minimal) unfolding surface can be a closed (finite genus) orientable surface (rational case) or a non-closed, infinite surface (non-rational case). Indeed, the unfolding surface is given by the glueing (by translations) along boundary edges of a number of copies of the given table P equal to the cardinality of the group G_P generated by reflections with respect to lines (through the origin) parallel to edges of P.

Definition 4.1 The table P is called rational if and only if the group G_P has finite order. If P is simply connected the condition that G_P has finite order is equivalent to the condition that all angles of P are rational multiples of π.

If P is rational, the billiard flow in P is equivalent to the straight line flow on a *translation surface*, that is, on a closed orientable surface M_P with a well-defined horizontal (and vertical) direction (away from finitely many conical points). The straight line flow on a translation surface has a first integral: the angle between the unit tangent vector and the horizontal direction. It follows that the phase space $P \times S^1$ is foliated by invariant surfaces (level surfaces of the prime integral), which are isomorphic to M_P (as translation surfaces).

In other terms, the angle of the unit tangent vector with the horizontal on the original table, although not invariant, since the trajectory undergoes reflections at the boundary edges, is invariant modulo the action of G_P on S^1. If G_P is finite (rational case), then the angle defines a prime integral with values in the interval S^1/G_P, hence the phase space of the billiard flow is foliated by invariant surfaces. Billiards in rational polygons are a main example of *pseudo-integrable* Hamiltonian systems in the terminology of P. J. Richens and M. V. Berry [28].

An alternative construction going back to R. Fox and R. Kershner [29] consists in forming the double S_P of the table P (given by two copies of P glued along corresponding edges), which is a flat sphere with finitely many conical singularities with angle which are twice the angle of P.

If all angles are rational multiple of π, it is possible to consider a branched cover of S_P with suitable branching orders at the cone points to get a higher genus surface endowed with a flat metric with trivial holonomy (that is, with well defined horizontal and vertical directions) and, in particular, with cone singularities of total angles which are *integer multiples of* 2π.

We conclude this brief introduction to billiards in polygons with the formula for the genus of the unfolding surface of a rational polygonal billiard.

Proposition 4.2 (see for instance [22]) *Let G_P be as above the group generated by all reflections with respect to the edges of the rational polygon P. Let $N_P := \#G_P/2$ (since G_P is a dihedral group, it has even order). Let $\{\pi m_i/n_i | i = 1, \ldots, \sigma\}$ denote the (rational) angles of P. Then the genus g_P of M_P is given by the formula:*

$$g_P = 1 + \frac{N_P}{2}(\sigma - 2 - \sum_{i=1}^{\sigma} \frac{1}{n_i}).$$

Exercise 4.3 *Prove the above formula and find all the completely integrable polygons (characterized by $g_P = 1$).*

4.2.2 Moduli Space of Translation Surfaces

The unfolding construction motivated the introduction of the notion of a translation surface.

Definition 4.2 The following definitions of a **translation surface** are equivalent:

- A closed orientable surface endowed with a translation structure, that is, endowed with an equivalence class of atlases such that coordinate changes are given by translations of the plane;
- A closed orientable surface which can be obtained by glueing planar polygons along boundary edges such that all glueing maps are translations;
- A closed orientable surface endowed with a flat metric with finitely many cone singularities (of total angles integer multiples of 2π) and trivial holonomy (that is, such that the holonomy representation of the fundamental group of the complement of the cone points is trivial);

4 Effective Unique Ergodicity and Weak Mixing of Translation Flows

- A closed orientable surface endowed with a complex structure, hence a Riemann surface M, and a holomorphic differential h on M;
- A closed orientable surface endowed with a pair $\{X, Y\}$ of transverse vector fields defined on the complement of finitely many points $\{p_1, \ldots, p_\sigma\}$ such that $[X, Y] = 0$ and such that, for every $i \in \{1, \ldots, \sigma\}$ there exists $k_i \in \mathbb{N}$ and a complex coordinate z_i defined on a neighborhood U_i of p_i such that

$$X = \operatorname{Re}(z_i^{-k_i} \frac{\partial}{\partial z_i}) \quad \text{and} \quad Y = -\operatorname{Im}(z_i^{-k_i} \frac{\partial}{\partial z_i}) \quad \text{on } U_i.$$

The vector fields X and Y are called respectively the horizontal and vertical vector fields of the translation surface.

Exercise 4.4 *Prove the equivalence of the above definitions.*

Linear flows on translation surfaces are called *translation flows*:

Definition 4.3 A **translation flow** on a translation surface M is the flow (defined almost everywhere, outside the union of countable many lines) generated by a vector field V which is parallel with respect to the flat metric. All translation flows are generated by linear combinations with constant coefficients of the horizontal and vertical vector fields of the translation surface.

We are interested in the ergodic theory of the *typical* translation flow. A precise notion of a typical translation flow can be immediately given on every given translation surface since linear flows for normalized generators are naturally parametrized by their angle with the horizontal direction and the set of angles is naturally endowed with the (normalized) Lebesgue measure. However, most ergodic properties have been proved, at least initially, for *typical* translation surfaces. This notion requires the introduction of finitely many parameters on the space of all translation surfaces and of natural measures on such a space, thereby providing motivation for the introduction of a *Teichmüller space* and a *moduli space* of translation surfaces.

Definition 4.4 The moduli space $\mathcal{H}_g$ of translation surfaces of genus $g \geq 1$ is the quotient of the space of all translation structures on a given smooth closed orientable surface S of genus g under the action of the group $\operatorname{Diff}^\infty(S)$ of all smooth diffeomorphisms of the surface S. The Teichmüller space $\hat{\mathcal{H}}_g$ of translation surfaces of genus $g \geq 1$ is the quotient of the space of all translation structures on a given smooth closed orientable surface S of genus g under the action of the subgroup $\operatorname{Diff}^\infty_0(S)$ of all smooth diffeomorphisms isotopic to the identity. Consequently, the moduli space $\mathcal{H}_g$ can be naturally identified to the quotient $\hat{\mathcal{H}}_g / \Gamma_g$ of the Teichmüller space $\hat{\mathcal{H}}_g$ under the action of the mapping class group $\Gamma_g := \operatorname{Diff}^\infty(S) / \operatorname{Diff}^\infty_0(S)$.

Taking into account the equivalent definition of a translation surface as the data of a holomorphic differential on a Riemann surface, it can be proved that the Teichmüller and moduli spaces of translation surfaces can be respectively identified with finite dimensional vector bundles (of dimension g) over the classical Teichmüller and moduli spaces of Riemann surfaces. This connection allows for methods of Riemann surface theory to be extended and applied to the study of translation surfaces. In particular, the various extensions and refinements of the Deligne-Mumford compactification of the moduli space of Riemann surfaces have played a crucial role in the theory of translation surfaces [30–32].

The Teichmüller and moduli spaces of translation surfaces are *stratified* according to the pattern of multiplicity of zeroes of Abelian differentials, or equivalently, of total cone angles. For every $\kappa = (\kappa_0, \ldots, \kappa_{\sigma-1}) \in \mathbb{N}^\sigma$, such that $\sum_{i=0}^{\sigma-1} \kappa_i = 2g-2$, the subsets of translation surfaces of genus $g \geq 1$ with a set of cone points of cardinality $\sigma \geq 1$ and multiplicities κ is invariant under the action of the diffeomorphism group, hence it descends to a subset $\hat{\mathcal{H}}_\kappa \subset \hat{\mathcal{H}}_g$ of the Teichmüller space and to a subset $\mathcal{H}_\kappa \subset \mathcal{H}_g$ of the moduli space, called a *stratum*.

We summarize below several fundamental structures on strata of Teichmüller and moduli spaces of translation surfaces:

- A function $\mathcal{A}_\kappa : \hat{\mathcal{H}}_\kappa \to \mathbb{R}^+$ which gives the total area of the flat metric of the translation surface; the function $\mathcal{A}_\kappa$ is invariant under the mapping class group hence is well-defined on the stratum $\mathcal{H}_\kappa$ of the moduli space;
- An atlas of affine charts given by the so-called period maps

$$\hat{\mathcal{H}}_\kappa \to H^1(M, \Sigma; \mathbb{C}),$$

with values in the relative cohomology (with complex coefficients) of the surface M relative to a set of $\Sigma = \{p_0, \ldots, p_{\sigma-1}\}$ of cone points; the period maps are locally defined on the stratum $\mathcal{H}_\kappa$ of the moduli space since the mapping class group acts properly discontinuously on $\hat{\mathcal{H}}_\kappa$;
- A measure class defined on $\mathcal{H}_\kappa$, and on $\hat{\mathcal{H}}_\kappa$, by the pull-back of the Lebesgue measure class on the euclidean space $H^1(M, \Sigma, \mathbb{C})$ via the local coordinates given by the period map;
- A natural action of the Lie group $GL(2, \mathbb{R})$ on the stratum $\hat{\mathcal{H}}_\kappa$, which descends to an action on the stratum $\mathcal{H}_\kappa$ since the action of $GL(2, \mathbb{R})$ and of the mapping class group Γ_g commute.

The period maps can be defined as the (locally defined) maps into the relative cohomology with complex coefficients

$$(M, h) \to [h] \in H^1(M, \Sigma, \mathbb{C})$$

or more explicitly as follows: let $\{a_1, b_1, \ldots, a_g, b_g, \gamma_1, \ldots, \gamma_{\sigma-1}\}$ denote a standard basis of the relative homology $H_1(M, \Sigma, \mathbb{Z})$ (recall that $\{a_1, b_1, \ldots, a_g, b_g\}$ is a canonical basis of the homology and $\{\gamma_1, \ldots, \gamma_{\sigma-1}\}$ is a set of relative cycles joining, for instance, p_0 to $p_1, \ldots, p_{\sigma-1}$, respectively). The period map can be

4 Effective Unique Ergodicity and Weak Mixing of Translation Flows

defined as follows:

$$(M, h) \to (\int_{a_1} h, \int_{b_1} h, \ldots, \int_{a_g} h, \int_{b_g} h, \int_{\gamma_1} h, \ldots, \int_{\gamma_{\sigma-1}} h) \in \mathbb{C}^{2g+\sigma-1}.$$

The action of the group $GL(2, \mathbb{R})$ on $\mathcal{H}_\kappa$ (hence on $\mathcal{H}_g$) can be defined as follows: on translation structures given as the data of a translation atlas, the group $GL(2, \mathbb{R})$ acts by post-composition of all charts (which are maps with values in $\mathbb{R}^2$) by a given element of the group; on translation structures as the data of planar polygons glued along the edges by translations, the group $GL(2, \mathbb{R})$ acts by the natural action on $GL(2, \mathbb{R})$ on planar polygons; on translation structures given as the data (M, h) of a Riemann surface M and a holomorphic differential h, the action can be described as follows: for all $A \in GL(2, \mathbb{R})$,

$$A \cdot (M, h) = (M_A, h_A) \quad \text{with} \quad \begin{pmatrix} \text{Re}(h_A) \\ \text{Im}(h_A) \end{pmatrix} = (A^t)^{-1} \begin{pmatrix} \text{Re}(h) \\ \text{Im}(h) \end{pmatrix}$$

and M_A the unique Riemann surface such that h_A is holomorphic on M_A; on translation structures given as the data (X, Y) of a pair of infinitesimally commuting vector fields, the action can be described as follows: for all $A = (a_{ij}) \in GL(2, \mathbb{R})$,

$$A \cdot (X, Y) = (X_A, Y_A) \quad \text{with} \quad \begin{pmatrix} X_A \\ Y_A \end{pmatrix} = A \begin{pmatrix} X \\ Y \end{pmatrix} = \begin{pmatrix} a_{11}X + a_{12}Y \\ a_{21}X + a_{22}Y \end{pmatrix}. \quad (4.1)$$

Since dilations of the translation structure do not affect the dynamical properties of its translation flows, it is natural to restrict consideration to the subset of translation surfaces of unit total area:

$$\hat{\mathcal{H}}_\kappa^{(1)} = \mathcal{A}_\kappa^{-1}\{1\} = \hat{\mathcal{H}}_\kappa / \mathbb{R}^+ \quad \text{and} \quad \mathcal{H}_\kappa^{(1)} := \hat{\mathcal{H}}_\kappa^{(1)} / \Gamma_g.$$

Exercise 4.5 *Prove that by Riemann bilinear relations the subset $\hat{\mathcal{H}}_\kappa^{(1)} \subset \hat{\mathcal{H}}_\kappa$ is given locally by a quadratic equation with respect to period coordinates.*

It can be verified that the area function $\mathcal{A}_\kappa$ is invariant under the action of the group $SL(2, \mathbb{R}) < GL(2, \mathbb{R})$ of matrices of determinant equal to 1, hence the action of $GL(2, \mathbb{R})$ on $\hat{\mathcal{H}}_\kappa$ introduced above induces an action of $SL(2, \mathbb{R})$ on $\hat{\mathcal{H}}_\kappa^{(1)}$, that descends to an action of $SL(2, \mathbb{R})$ on $\mathcal{H}_\kappa^{(1)}$.

The sub-actions of the above action of $SL(2, \mathbb{R})$ on the union $\mathcal{H}_g^{(1)}$ of all strata $\mathcal{H}_\kappa^{(1)}$ given by the subgroups $g_\mathbb{R}$ and $h_\mathbb{R}^\pm$, defined (for $t \in \mathbb{R}$) as

$$g_t = \begin{pmatrix} e^t & 0 \\ 0 & e^{-t} \end{pmatrix} \quad \text{and} \quad h_t^+ = \begin{pmatrix} 1 & t \\ 0 & 1 \end{pmatrix}, \quad h_t^- = \begin{pmatrix} 1 & 0 \\ t & 1 \end{pmatrix}, \quad (4.2)$$

are known as the *Teichmüller geodesic flow* $g_\mathbb{R}$ (introduced by H. Masur [33]), and the unstable and stable *Teichmüller horocycle flows* $h_\mathbb{R}^\pm$ (introduced in [34]).

A fundamental theorem of H. Masur [33] and W. Veech [35] establishes for every stratum the existence of a probability absolutely continuous measure in the Lebesgue measure class.

Theorem 4.6 ([33, 35]) *For every stratum $\mathcal{H}_\kappa$, there exists a unique $\mathrm{GL}(2,\mathbb{R})$-invariant measure μ_κ on $\mathcal{H}_\kappa$, which belongs to the Lebesgue measure class and projects to a probability $\mathrm{SL}(2,\mathbb{R})$-invariant measure $\mu_\kappa^{(1)}$ on $\mathcal{H}_\kappa^{(1)}$.*

We can now introduce several precise definitions of *typical* (in measure sense) for translation surfaces and flows.

Definition 4.5 Let μ denote any $\mathrm{SL}(2,\mathbb{R})$-invariant probability measure on $\mathcal{H}_\kappa^{(1)}$. A translation surface in $\mathcal{H}_\kappa^{(1)}$ is called μ**-typical** if it is chosen randomly with respect to μ. It is called **Masur–Veech typical** if it is μ-typical for $\mu = \mu_\kappa^{(1)}$, the Masur–Veech measure. A translation flow is called μ-typical if it is the horizontal (or vertical) flow of a μ-typical translation surface and Masur–Veech typical if $\mu = \mu_\kappa^{(1)}$ is the Masur–Veech measure. Finally, a translation flow is called **directionally typical** for a given translation surface with horizontal/vertical vector fields (X, Y), if it is chosen randomly, with respect to the Lebesgue measure on the circle S^1, among the flows generated by vector fields in the one parameter family

$$\{V_\theta := \cos\theta \cdot X + \sin\theta \cdot Y, \quad \theta \in S^1\}.$$

A well-known difficulty in the study of the dynamics of rational polygonal billiards is that they are never Masur–Veech typical (for genus $g \geq 2$), hence the wealth of results for Masur–Veech typical translation surfaces are not directly relevant for the dynamics of rational billiards.

The strategy to go beyond results that can be proved *for all translation surfaces*, for directionally typical translation flows (which of course are relevant for rational billiards), is to consider a natural $\mathrm{SL}(2,\mathbb{R})$-invariant measure μ supported on the $\mathrm{SL}(2,\mathbb{R})$-orbit closures of a given translation surface, and to prove results for μ-typical translation surfaces, hoping that they can be extended to the initial surface.

This program presents a priori several very serious difficulties, beginning with the fact that since $\mathrm{SL}(2,\mathbb{R})$ is not an amenable group, there is no guarantee that an $\mathrm{SL}(2,\mathbb{R})$-invariant measure supported on any $\mathrm{SL}(2,\mathbb{R})$-orbit closure even exists. The celebrated theorems of Eskin and Mirzakhani [9] and Eskin, Mirzakhani and Mohammadi [36] made the above strategy possible.

Theorem 4.7 ([9, 36]) *For any $(M, h) \in \mathcal{H}_\kappa$, its $\mathrm{GL}(2,\mathbb{R})$-orbit closure $\mathcal{M} := \overline{\mathrm{GL}(2,\mathbb{R})(M, h)}$ is locally in period coordinates an affine manifold. In addition, every ergodic probability measure invariant for the (amenable) maximal parabolic subgroup P of upper triangular matrices is $\mathrm{SL}(2,\mathbb{R})$-invariant, hence every $\mathrm{GL}(2,\mathbb{R})$-orbit closure supports a $\mathrm{GL}(2,\mathbb{R})$-invariant ergodic measure. Conversely, every ergodic $\mathrm{SL}(2,\mathbb{R})$-invariant probability measure on $\mathcal{H}_\kappa$ is supported on a codimension 1 submanifold of an orbit closure (given by translation surfaces of a given total area) and belongs to the Lebesgue measure class on it.*

S. Filip [5, 6] later refined the above result and proved several algebraic properties (real multiplication and torsion) characterizing (together with a dimensional constraint) orbit closures, which imply that they are always algebraic varieties defined over $\overline{\mathbb{Q}}$.

The above mentioned results of Eskin, Mirzakhani and Mohammadi, and Filip, have established the foundations for a program of classification of orbit closures. However, from the point of view of questions concerning the dynamical properties of arbitrary translation surfaces, and of rational polygonal billiards in particular, the above theorem do not establish that the initial translation surface is *typical* in its orbit closure (with respect to the associated Lebesgue measure class).

In fact, for the purpose of understanding the dynamics of the directionally typical translation flow, for all translation surfaces, it may be necessary to understand the limits of orbit segments of the circle subgroup or of the unipotent subgroups of $SL(2, \mathbb{R})$ as they are pushed under the action of the diagonal subgroup.

Recent results (see [37, 38]) on the dynamics of the unipotent subgroup (the Teichmüller horocycle flow) have confirmed that it has a complicated dynamical behavior, although they do not address in general the question of limits of geodesic push-forwards of horocycle arcs or invariant measures (see [39] for a partial result).

4.2.3 Ergodic Properties of Translation Flows: Unique Ergodicity

A celebrated theorem proved independently by H. Masur and W. Veech states that, in the terminology introduced above:

Theorem 4.8 ([33, 35]) *For all strata of translation surfaces, the Masur–Veech typical translation flow is uniquely ergodic.*

Note that, as remarked above, this theorem does not apply to rational billiards.

An equivalent statement for IET's had been conjectured by M. Keane for IET's and was then known as the *Keane conjecture*. Keane had investigated conditions for the minimality of IET's [40] and had first conjectured that minimality implies unique ergodicity. Counterexamples to this first version (see for instance [41]) prompted him to formulate a revised form of the conjecture.

Equivalent results on the minimality of translation flows had been proved independently by Zemlyakov and Katok [27], and it appears their paper was not immediately accessible to mathematicians in the West.

We recall that a dynamical system is uniquely ergodic if the cone of all invariant measures is one-dimensional, a property that, for minimal continuous dynamical systems, is equivalent to the uniform convergence of ergodic averages to the mean, for all continuous functions. This characterization holds for IET's and translation flows (despite the fact that they are not continuous dynamical systems) for all points with well defined semi-orbits.

The above mentioned Masur–Veech theorem was later extended by Kerckhoff, Masur and Smillie to all translation surfaces.

Theorem 4.9 ([42]) *For all translation surfaces, the directionally typical translation flow is uniquely ergodic. In particular, for any rational billiard table and for Lebesgue almost all directions, the billiard flow is uniquely ergodic on the corresponding invariant surface.*

An effective version of the above unique ergodicity result was given by Y. Vorobets [43], who proved bounds in mean which however are not polynomial (power-law) in time, but have the significant advantage to come with explicit estimates on the constants in terms of the genus of the surface. Indeed, the main result of [43] is a quantitative condition for the ergodicity (on the 3-dimensional phase space) of a non-rational polygonal billiard flow in terms of its rational approximations.

The polynomial unique ergodicity of directionally typical translation flows on all translation surfaces was proved in [16].

Theorem 4.10 *For each stratum $\mathcal{H}_\kappa^{(1)}$ of translation surfaces there exists a constant $\alpha_\kappa > 0$ such that, for all translation surfaces $(M, h) \in \mathcal{H}_\kappa^{(1)}$ and for the directionally typical translation flow $\phi_\mathbb{R}^{V_\theta}$ on (M, h) there exists a constant $K_h(\theta) > 0$ such that the following holds. For all functions $f \in H^1(M)$ (the Sobolev space of functions with square integrable first weak derivative) of zero average, and for all $(x, T) \in M \times \mathbb{R}^+$ such that x has infinite forward orbit, we have*

$$\left| \frac{1}{T} \int_0^T f \circ \phi_t^{V_\theta}(x) dt \right| \leq K_h(\theta) T^{-\alpha_\kappa}.$$

We outline below a cohomological proof of the Masur–Veech unique ergodicity theorem which, thanks to the results of J. Athreya [44], can be strengthened to a proof of the above theorem on effective unique ergodicity.

4.2.4 Ergodic Properties of Translation Flows: Weak Mixing

We recall the definition of the weak mixing property. It is a standard result of ergodic theory that weak mixing can be defined in several equivalent ways. We consider below the equivalent formulations which are more relevant for our purposes.

Definition 4.6 A flow $\phi_\mathbb{R}$ on a probability space (M, μ) is **weakly mixing** if it satisfies any of the following equivalent properties:

- The flow $\phi_\mathbb{R}$ has no non-constant (square-integrable) eigenfunctions, that is, there exists no non-zero function $u \in L^2(M, \mu)$ of zero average with the property that for some $\lambda \in \mathbb{R}$

$$u \circ \phi_t = e^{2\pi i \lambda t} u, \quad \text{for all } t \in \mathbb{R}.$$

4 Effective Unique Ergodicity and Weak Mixing of Translation Flows

- The spectral measures of the flow $\phi_\mathbb{R}$ for all functions $f \in L^2(M, \mu)$ of zero average, defined as the Fourier transforms of the self-correlation functions $\langle f \circ \phi_t, f \rangle$ for $t \in \mathbb{R}$, are continuous, that is, have no atoms.
- The Cesaro averages of correlations with respect to the flow $\phi_\mathbb{R}$ of all pairs of square integrable functions of zero average converge to zero: for all $f, g \in L^2(M, \mu)$ of zero average, we have

$$\frac{1}{T} \int_0^T |\langle f \circ \phi_t, g \rangle| dt \;\to\; 0.$$

- For all $c \in \mathbb{R}$, the product $\phi_\mathbb{R} \times R_\mathbb{R}^c$ of the flow $\phi_\mathbb{R}$ with the linear flow $R_\mathbb{R}^c$ on $\mathbb{T}$, defined as

$$R_t^c(y) = y + ct \;\; \mathrm{mod.} \; \mathbb{Z}, \qquad \text{for all } (y, t) \in \mathbb{T} \times \mathbb{R},$$

is ergodic on the product space $(M \times \mathbb{T}, \mu \times \mathrm{Leb}_\mathbb{T})$.

Exercise 4.11 *Prove the equivalence of all the properties listed in Definition 4.6.*

We recall that weak mixing is a G_δ dense property in the space of measure preserving transformations [45], while mixing is not [46]. For IET's and translation flows, A. Katok [47] proved that they are never mixing.

The weak mixing property of Masur–Veech typical translation flows on surfaces of higher genus was first proved in [20].

Theorem 4.12 *For all strata of translation surfaces of genus $g \geq 2$, the Masur–Veech typical translation flow is weakly mixing.*

We note that on the related problem of weak mixing of interval exchange transformations which are not rotations, Katok and Stepin [48] proved the weak mixing property for IET's on 3 intervals, and Veech [11] generalized the result to IET's of rotation type on any number of intervals by introducing an important criterion for weak mixing which will be explained below.

We also note that, as pointed out above, Theorem 4.12 does not imply any weak mixing property for rational polygonal billiards. The argument can be generalized to prove the weak mixing property for μ-typical translation flows for all $SL(2, \mathbb{R})$-invariant measures supported on orbit closure of *rank* at least 2 (in the sense of A. Wright [49]). In the rank one case, directionally typical weak mixing was proved by Avila and Delecroix [50] for translation surfaces, which are not torus covers, on closed $SL(2, \mathbb{R})$ orbits (non-arithmetic Veech surfaces). This class includes billiards in regular polygons with at least 5 edges. A generalization to all rank one orbifolds has been announced by Aulicino, Avila and Delecroix.

Finally, it is well known that weak mixing cannot be directionally typical for translation surfaces which factor over a circle $\mathbb{T} := \mathbb{R}/\mathbb{Z}$ (that is, that have a completely periodic directional foliation with commensurable cylinders), since non-constant eigenfunctions for the linear flow on $\mathbb{T}$ give by pull-back non-constant eigenfunctions for all ergodic directional flow on the translation surface. The presence of exceptional surfaces makes the problem of characterizing surfaces with directionally typical weak mixing more challenging. The most natural and simplest conjecture can be stated as follows:

Conjecture 4.13 *A proof of the conjecture has been announced in 2024 by F. Arana Herrera, J. Chaika and the author. For all translation surface which do not factor over the circle (and in particular are not torus covers), the directionally typical translation flow is weakly mixing.*

We now turn to the notion of polynomial weak mixing. The study of effective weak mixing (in particular of the Hölder property of spectral measures) for substitution systems and translation flows was initiated by A. Bufetov and B. Solomyak in a series of papers [12–15].

Definition 4.7 A smooth flow $\phi_\mathbb{R}$ on a probability space (M, μ) is **polynomially weakly mixing** for functions in a Banach space $W \subset L^2(M, \mu)$ if it satisfies any of the following (roughly equivalent) properties:

- The spectral measures of the flow $\phi_\mathbb{R}$ for all functions $f \in W$ of zero average satisfy a Hölder property, in the following sense. There exists $\alpha > 0$ such that, for all $\lambda \in \mathbb{R}$ there exists a constant $C(\lambda) > 0$ such that for any $f \in W$ the spectral measure σ_f of $f \in W$ satisfies the bound

$$\sigma_f(\lambda - r, \lambda + r) \leq C(\lambda) \|f\|_W \cdot r^\alpha, \quad \text{for all } r > 0.$$

- The Cesaro averages of correlations with respect to the flow $\phi_\mathbb{R}$ of all pairs of functions of zero average in W converge to zero polynomially: there exist $\alpha' > 0$ and $C > 0$ such that for all $f, g \in W$ of zero average, we have

$$\frac{1}{T} \int_0^T |\langle f \circ \phi_t, g \rangle| dt \leq C \|f\|_W \|g\|_W \cdot T^{-\alpha'}.$$

- The twisted ergodic integrals of functions in W decay polynomially in mean: there exists $\alpha'' > 0$, and, for all $\lambda \in \mathbb{R}$, there exists a constant $C(\lambda) > 0$ such that, for all $f \in W$ and all $T > 0$, we have

$$\|\frac{1}{T} \int_0^T e^{2\pi i \lambda t} f \circ \phi_t dt\|_{L^2(M,\mu)} \leq C(\lambda) \|f\|_W \cdot T^{-\alpha''}.$$

Exercise 4.14 *Discuss the equivalence of all the properties listed in Definition 4.7.*

The polynomial weak mixing of Masur–Veech typical translation flows in all higher genus strata was proved in [10] and, by a somewhat different approach, by A. Bufetov and B. Solomyak [15]. We will outline below our cohomological approach to (effective) weak mixing based on (twisted) Hodge theory.

Theorem 4.15 *For all strata of translation surfaces of genus $g \geq 2$, the Masur–Veech typical translation flow is polynomially weakly mixing.*

Once more, the argument can be generalized to prove the polynomial weak mixing property for μ-typical translation flows for all $SL(2, \mathbb{R})$-invariant measures supported on orbit closures of *rank* at least 2. The rank one case, including the case of non-arithmetic Veech surfaces, is open.

Problem 4.16 *Establish polynomial weak mixing for directionally typical translation flows on non-arithmetic Veech translation surfaces.*

All of the above results are concerned with translation flows and with rational billiard flows We conclude this section with a digression on general polygonal billiards, whose ergodic theory is poorly understood.

Conjecture 4.17 *For almost all (simply connected) polygonal tables the billiard flow is ergodic and weakly mixing (in the 3-dimensional phase space).*

It was proved by S. Kerckhoff et al. [42] that there exists a G_δ dense set of polygons with ergodic billiard flow. Recently J. Chaika and the author [51] have proved that there exists a G_δ dense set of polygons with weakly mixing billiard flow. It is not known whether there exists a polygon with mixing billiard flow, although numerical evidence [52] suggests that it is the case for almost all acute triangles. However, while (unique) ergodicity and weak mixing are G_δ properties, dense in the space of measure preserving transformations [45], mixing is not [46]. The above-mentioned results on ergodicity and weak mixing of general polygons are proved by fast approximation based on the corresponding properties of rational billiards, an approach that cannot be applied to prove existence of mixing polygons.

The striking gap in our knowledge of the dynamics of billiards in general polygons when compared to rational billiards is related to the discovery (by Rauzy, Masur, Veech) of a renormalization dynamical system for the dynamics of Interval Exchange Transformations and Translation flows.

4.3 (Effective) Unique Ergodicity: A Cohomological Approach

In this section we review the cohomological approach, based on Hodge theory, to effective unique ergodicity expounded in a series of work including [3, 16, 17, 53].

Recently, it has been rediscovered by C. McMullen [54], who has often presented it without reference to the much earlier above-mentioned work in his subsequent papers (see for instance [55], page 210).

The cohomological approach has its roots in A. Katok's [56] proof of the finiteness of the cone of invariant measures for quasi-minimal flows on surfaces with saddle-like singularities.

4.3.1 Katok's Finiteness Theorem

In this section, we outline, following [56], the proof of the following fundamental result.

Theorem 4.18 ([56]) *The cone of invariant measures for any minimal translation flow on a surface M of genus $g \geq 1$ has dimension at most g.*

This result sharpens an earlier finiteness result by Oseledets (who proved for IET's an upper bound equivalent to $2g + \#\Sigma - 1$ for the dimension of the cone of invariant measures of a translation flows on a surface of genus g with singularity set Σ).

In the case $M = \mathbb{T}^2$ ($g = 1$), Katok's upper bound already implies that all (linear) minimal flows are uniquely ergodic.

The first step of Katok's argument consists in introducing the (flux) cohomology class of an invariant probability measure. Let $\phi_{\mathbb{R}}^V$ denote a flow with generator V on M and let μ an invariant probability measure for $\phi_{\mathbb{R}}^V$.

Definition 4.8 The flux current $\iota_V \mu$ of the invariant probability measure μ for the flow $\phi_{\mathbb{R}}^V$ is the closed current of dimension 1 (and degree 1, since M has dimension 2) defined as follows:

$$(\iota_V \mu)(\alpha) = \int_M \iota_V \alpha \, d\mu, \quad \text{for every 1-form } \alpha \in \Omega^1(M, \mathbb{C}).$$

(The symbol ι_V denote the contraction operator with the vector field V on differential forms, which can be extended to currents by duality).

The (flux) cohomology class $F(\mu) \in H^1(M, \mathbb{R})$ of the invariant probability measure μ is the (de Rham) cohomology class of the flux form of μ, that is,

$$F(\mu) := [\iota_V \mu] \in H^1(M, \mathbb{R}).$$

Note: We adopt here the point of view of L. Schwartz and G. de Rham that a current of dimension $k \in \mathbb{N}$ is a continuous linear functional on the Fréchet space of smooth k-forms, and that every closed current of dimension k has a well-defined k-cohomology class, defined as the de Rham cohomology class of an appropriate de Rham regularization (that is, of a smooth closed k-form).

In the above definition, it remains to prove that the current $\iota_V \mu$ is indeed closed. This property follows from the invariance of μ with respect to $\phi_{\mathbb{R}}^V$. In fact, we have to prove that the boundary $b(\iota_V \mu)$ or, equivalently, its differential $d(\iota_V \mu)$ vanishes.

For every $f \in \Omega^0(M, \mathbb{C})$, we can compute

$$d(\iota_V \mu)(f) = (\iota_V \mu)(df) = \int_M \iota_V df\, d\mu$$

$$= \int_M Vf\, d\mu = \frac{d}{dt}\left(\int_M f \circ \phi_t^V\, d\mu\right)\bigg|_{t=0} = 0.$$

We have introduced the language of currents since it is essential in proving effective results. However, following [56] it is possible to bypass the flux current and directly define the flux cohomology class as an element of the dual $H_1(M, \mathbb{R})^*$. It is sufficient to define the flux class $F(\mu)$ on a basis of the homology $H_1(M, \mathbb{Z})$ and then extend it by linearity. As a basis of the homology, we can choose a canonical basis $\{a_1, b_1, \ldots, a_g, b_g\}$ of cycles. It is therefore sufficient to define the flux through a loop and prove that it only depends on the homology class of the loop.

Indeed, the contraction $\iota_V \mu$ of the invariant measure μ can be interpreted as a *transverse invariant measure* for the orbit foliation of $\phi_\mathbb{R}^V$ in the following sense. For every rectifiable arc $I \subset M$, we define its (transverse) measure $(\iota_V \mu)(I)$ as

$$(\iota_V \mu)(I) := \lim_{t \to 0} \frac{1}{t}\mu\left(\cup_{s \in [0,t]}\phi_t^V(I)\right)$$

It can be proved that the limit exists and that, in addition, the one-dimensional transverse measure $\iota_V \mu$ is invariant for the orbit foliation: for every pair of transverse arcs $[p, q]$ and $[p', q']$ with endpoints p, q and p', q' respectively, such that $p' \in \phi_{\mathbb{R}_+}^V(p)$ and $q' \in \phi_{\mathbb{R}_+}^V(q)$ we have

$$(\iota_V \mu)([p, q]) = (\iota_V \mu)([p', q']).$$

The above property follows from the fact that by Jordan curve theorem the union of the transverse intervals $[p, q]$, $[p', q']$ and of the orbits segments joining p to p' and q to q' bounds a (simply connected) domain $D(p, q, p', q')$ such that, for all $t > 0$, by the invariance of the measure μ with respect to $\phi_\mathbb{R}^V$, we have

$$\mu\left(\cup_{s \in [0,t]}\phi_t^V([p', q']) \setminus \cup_{s \in [0,t]}\phi_t^V([p, q])\right)$$
$$= \mu(\phi_t^V(D(p, q, p', q'))) - \mu(D(p, q, p', q')) = 0.$$

Once the notion of a cohomology class of an invariant measure is established, the argument proceeds by the following steps:

- if the flow $\phi_\mathbb{R}^V$ is minimal, then the flux class map $F : C_V \to H^1(M, \mathbb{R})$ is injective from the cone C_V of its invariant measures into the cohomology;

- the image of the map $F : C_V \to H^1(M, \mathbb{R})$ is contained in a Lagrangian subspace for the natural symplectic structure on $H^1(M, \mathbb{R})$ (hence it has dimension $\leq g$).

The natural symplectic structure on $H^1(M, \mathbb{R})$ encodes a fundamental symmetry in the structure of deviations of ergodic averages of translation flows and related dynamical systems.

We now outline the proof of the above two steps:

Lemma 4.19 *If $\phi_{\mathbb{R}}^V$ is minimal, then $F : C_V \to H^1(M, \mathbb{R})$ is injective.*

Proof Let us assume that for $\mu_1, \mu_2 \in C_V$ we have $[\iota_V(\mu_1)] = [\iota_V(\mu_2)] \in H^1(M, \mathbb{R})$. By the de Rham theorem there exists a current U of dimension 2 (and degree 0) such that

$$dU = \iota_V(\mu_1) - \iota_V(\mu_2).$$

From the above identity, since $\iota_V^2 = 0$, it follows immediately that

$$VU = \iota_V(dU) = \iota_V^2(\mu_1) - \iota_V^2(\mu_2) = 0,$$

hence U is invariant with respect to $\phi_{\mathbb{R}}^V$. In addition, for any transverse interval $I \subset M$ and for any $p, q \in I$ we have, by integrating along the subsegment $[p, q] \subset I$ with endpoints p, q,

$$U(p) - U(q) = \int_{[p,q]} \iota_V(\mu_1 - \mu_2).$$

Since by minimality of the flow, the current $\iota_V(\mu_1 - \mu_2)$ is a continuous transverse measure (it has no atoms), the current U is a continuous function. Since it is invariant, U is a constant function, again by the minimality of the flow. It follows that $\iota_V(\mu_1) = \iota_V(\mu_2)$ and, since the contraction operator $\iota_V : \Omega^1(M, \mathbb{R}) \to \Omega^0(M, \mathbb{R})$ is surjective, it follows that $\mu_1 = \mu_2$, thereby concluding the argument. □

For every $(p, T) \in M \times \mathbb{R}^+$, let $\gamma_T^V(p)$ denote the (oriented) orbit segment (defined for almost all $p \in M$ for all $T > 0$)

$$\gamma_T^V(p) = \bigcup_{0 \leq t \leq T} \{\phi_t^V(p)\},.$$

For any pair of points $p, q \in M$, let $I(p, q) \subset M$ denote an (oriented) arc of uniformly bounded length (with respect to a fixed Riemannian metric) joining p to q, union of a subsegment of a transverse arc $I \subset M$ with orbit segments joining p and q to I. Let then

$$\overline{\gamma}_T^V(p) := \gamma_T^V(p) \cup I(\phi_T^V(p), p).$$

4 Effective Unique Ergodicity and Weak Mixing of Translation Flows

Lemma 4.20 *For any pair of ergodic invariant measures μ_1 and μ_2 for $\phi_\mathbb{R}^V$, we have that the intersection of their flux forms vanishes*

$$F(\mu_1) \wedge F(\mu_2) = 0.$$

Proof The argument is based on Poincaré duality $P : H^1(M, \mathbb{R}) \to H_1(M, \mathbb{R})$ and Birkhoff ergodic theorem. By the ergodic theorem, there exist p_1, p_2 in M, with distinct orbits, such that the following identity of currents holds:

$$\lim_{T \to +\infty} \frac{1}{T} \overline{\gamma}_T^V(p_1) = \iota_V(\mu_1) \quad \text{and} \quad \lim_{T \to +\infty} \frac{1}{T} \overline{\gamma}_T^V(p_2) = \iota_V(\mu_2).$$

The above identity concerns currents of dimension 1, hence it can be checked on smooth 1-forms. For any ergodic measure μ and μ-almost all $p \in M$, for every smooth 1-form α, by the ergodic theorem we have

$$\lim_{T \to +\infty} \frac{1}{T} \langle \overline{\gamma}_T^V(p), \alpha \rangle = \lim_{T \to +\infty} \frac{1}{T} \int_0^T \iota_V \alpha(\phi_t^V(p)) dt = \int_M \iota_V \alpha d\mu = F(\mu)(\alpha).$$

The cup product $F(\mu_1) \wedge F(\mu_2)$ is therefore given as the limit

$$F(\mu_1) \wedge F(\mu_2) = \lim_{T \to +\infty} \frac{1}{T^2} \cdot ([\overline{\gamma}_T^V(p_1)] \cap [\overline{\gamma}_T^V(p_2)]).$$

By Poincaré duality, the above cup product between cohomology classes is equal to the algebraic intersection number (in homology) between their Poincaré duals, which in turn is bounded by the total number of intersections between the loops $\overline{\gamma}_T^V(p_1)$ and $\overline{\gamma}_T^V(p_1)$:

$$[\overline{\gamma}_T^V(p_1)] \cap [\overline{\gamma}_T^V(p_2)] = P[\overline{\gamma}_T^V(p_1)] \cap P[\overline{\gamma}_T^V(p_2)] \le \#\left(\overline{\gamma}_T^V(p_1) \cap \overline{\gamma}_T^V(p_2)\right).$$

Finally, we have

$$\#\left(\overline{\gamma}_T^V(p_1) \cap \overline{\gamma}_T^V(p_2)\right) = \#\left(\gamma_T^V(p_1) \cap \gamma_T^V(p_2)\right) + \#(\gamma_T^V(p_1) \cap I(\phi_T^V(p_2), p_2))$$

$$+ \#\left(I(\phi_T^V(p_1), p_1) \cap \gamma_T^V(p_2)\right) + \#\left(I(\phi_T^V(p_1), p_1) \cap I(\phi_T^V(p_2), p_2)\right),$$

so that since $\gamma_T^V(p_1) \cap \gamma_T^V(p_2) = \emptyset$, as they are distinct orbits of a flow, and since the arcs $I(\phi_T^V(p_1), p_1)$ have uniformly bounded length, we conclude that there exists a constant $C > 1$ such that, for all $T > 1$

$$\#\left(\overline{\gamma}_T^V(p_1) \cap \overline{\gamma}_T^V(p_2)\right) \le CT,$$

hence

$$F(\mu_1) \wedge F(\mu_2) = \lim_{T\to+\infty} \frac{1}{T^2} \cdot ([\overline{\gamma}_T^V(p_1)] \cap [\overline{\gamma}_T^V(p_2)])$$
$$\leq \lim_{T\to+\infty} \frac{1}{T^2} \cdot (\#\left(\overline{\gamma}_T^V(p_1)] \cap [\overline{\gamma}_T^V(p_2)\right)) = 0.$$

□

Proof of Theorem 4.18 By Lemma 4.19, the map $F : C_V \to H^1(M, \mathbb{R})$ is injective and, since all probability invariant measures are convex combinations of ergodic measures, by Lemma 4.20 it has image contained in a Lagrangian subspace of the symplectic space $H^1(M, \mathbb{R})$, endowed with the cup product structure. Since all Lagrangian subspaces have dimension equal to half the dimension of the space, and $H^1(M, \mathbb{R})$ has dimension $2g$, the bound on the dimension of C_V follows. □

4.3.2 Self-Similar Translation Flows

A self-similar translation flow is a translation flow rescaled by a (linear) *pseudo-Anosov* map, according to the definition below. Pseudo-Anosov maps were introduce by W. Thurston in his work on diffeomorphisms of surfaces. Those which stabilize translation flows are of a special type, since their invariant foliations are orientable. In this section we outline a renormalization approach to the effective unique ergodicity of self-similar translation flows. Since the (effective, polynomial) unique ergodicity of the unstable foliation of a hyperbolic (Anosov) diffeomorphism is closely related to the (effective, exponential) mixing of the diffeomorphism, the argument outlined below is the core of a cohomological approach to exponential mixing for a class of pseudo-Anosov diffeomorphisms (see [57]).

Definition 4.9 A translation flow $\phi_\mathbb{R}^X$ is self-similar if there exists a translation structure (M, h) with horizontal and vertical vector fields (X, Y), a homeomorphism $\psi : M \to M$ and a real number $\lambda > 1$ such that

$$\psi_*(X) = \lambda X \quad \text{and} \quad \psi_*(Y) = \lambda^{-1} Y.$$

The real number λ is called the dilation factor of the pseudo-Anosov map ψ.

We note that the homeomorphism $\psi : M \to M$ which appears in the above definition is a pseudo-Anosov map with orientable invariant foliations given by the orbit foliations of the translation flows $\phi_\mathbb{R}^X$ and $\phi_\mathbb{R}^Y$ and it is a diffeomorphism on the complement of the set $\Sigma_h = \{h = 0\}$ of conical singularities of the translation surface. Let $\psi^* : H^1(M, \mathbb{R}) \to H^1(M, \mathbb{R})$ denote the linear homomorphism induced by the pseudo-Anosov map ψ on the first cohomology of the surface. It follows from the definition that

$$\psi^*(\operatorname{Re}(h)) = \lambda^{-1} \operatorname{Re}(h) \quad \text{and} \quad \psi^*(\operatorname{Im}(h)) = \lambda \operatorname{Im}(h).$$

Since $\operatorname{Re}(h)$ and $\operatorname{Im}(h)$ are closed 1-form, by the de Rham theorem they have well-defined cohomology classes $[\operatorname{Re}(h)]$ and $[\operatorname{Im}(h)] \in H^1(M, \mathbb{R})$. It follows that the spectrum of the linear map ψ^* contains $\{\lambda, \lambda^{-1}\}$. Since $\lambda > 1$ equals the maximal dilation of ψ at all points in $M \setminus \Sigma_h$, it follows that

$$\lambda = \rho(\psi^*), \quad \text{the spectral radius of the linear map } \psi^*.$$

It can be proved, as a consequence of the Perron-Frobenius theorem, that λ and λ^{-1} are simple eigenvalues of the linear map ψ^*. Another proof, based on Hodge theory, which holds in greater generality, is presented below (see Exercise 4.32).

We now explain how to derive a proof of unique ergodicity of self-similar translation flows based on the ideas of the previous section.

Theorem 4.21 *All self-similar translation flows are uniquely ergodic.*

Proof We establish first that all self-similar translation flows are minimal. Let (X, Y) and ψ be as in the above Definition 4.9. If the translation flow $\phi_\mathbb{R}^X$ is not minimal, then it has a saddle connection γ, as the surface can be decomposed as the union of finitely many cylindrical component (foliated by periodic orbits) and minimal components. In particular, there exists a saddle connection of minimal length. However, by the action of ψ^{-1}, since ψ and ψ^{-1} stabilize the orbit foliation of $\phi_\mathbb{R}^X$ and ψ^{-1} scales the time variable along the orbits by a factor λ^{-1}, we can find a sequence of saddle connections of $\phi_\mathbb{R}^X$ with lengths converging to zero, which is a contradiction. □

Unique ergodicity then follows from the injectivity Lemma 4.19 and from the following claim:

Claim 4.22 *For every ergodic probability measure μ for the flow $\phi_\mathbb{R}^X$ on M, we have*

$$\psi^*(F(\mu)) = \lambda F(\mu).$$

By the above claim, since $\lambda > 1$ is a simple eigenvalue and $\psi^*[\operatorname{Im}(h)] = \lambda[\operatorname{Im}(h)]$, it follows that, for every ergodic measure μ of $\phi_\mathbb{R}^X$ on M, the flux cohomology class $F(\mu) \in \mathbb{R}[\operatorname{Im}(h)] \subset H^1(M, \mathbb{R})$, and, by the injectivity of the map $F : C_X \to H^1(M, \mathbb{R})$, the cone C_X of invariant measures for the flow $\phi_\mathbb{R}^X$ is one-dimensional, hence $\phi_\mathbb{R}^X$ is uniquely ergodic.

Proof of Claim 4.22 The Claim follows from Birkhoff ergodic theorem. Let μ be any ergodic probability measure. Let us denote $\gamma_T^X(p)$ the orbit of $\phi_\mathbb{R}^X$ starting at p and ending at $\phi_T^X(p)$, as in the previous section. By the ergodic theorem, for

μ-almost all $p \in M$ we have (as currents of dimension 1)

$$F(\mu) = \lim_{T \to +\infty} \frac{1}{T} \gamma_T^X(p) = \lim_{T \to +\infty} \frac{1}{T} \gamma_T^X(\psi(p)),$$

hence, by taking into account that $\psi_*(X) = \lambda X$, we can compute

$$\psi^*(F(\mu)) = \lim_{T \to +\infty} \frac{1}{T} \psi^*(\gamma_T^X(p)) = \lim_{T \to +\infty} \frac{1}{T} \gamma_{\lambda T}^X(\psi(p))$$

$$= \lambda \lim_{T \to +\infty} \frac{1}{\lambda T} \gamma_{\lambda T}^X(\psi(p)) = \lambda F(\mu).$$

□

The proof of Claim 4.22 completes the proof of Theorem 4.21.

The above argument already exhibits the structure of the cohomological argument for unique ergodicity, which derives it from the simplicity of the top eigenvalue of a linear dynamical system (in this case the action of the pseudo-Anosov map on cohomology).

We now turn to the problem of proving effective unique ergodicity for self-similar translation flows. The key point is that, not only $\psi^* : H^1(M, \mathbb{R}) \to H^1(M, \mathbb{R})$ has simple top eigenvalue, but it also has a *spectral gap*, that is, the rest of the spectrum is contained within a disk of radius strictly less than the spectral radius (in the case of homomorphisms of finite dimensional vector spaces, simplicity implies the existence of a spectral gap).

Theorem 4.23 *There exist $C > 0$ and $\alpha > 0$ such that, for all functions $f \in H^1(M)$ (the Sobolev space of square integrable functions with square integrable weak first derivative), of zero average, for all $(p, T) \in M \times \mathbb{R}^+$ (such that p has infinite forward orbit) we have*

$$\left| \int_0^T f \circ \phi_t^X(p) dt \right| \le C \|f\|_{H^1(M)} \cdot T^{1-\alpha}.$$

Proof *The first step* in the argument is to write ergodic integrals in terms of the one-dimensional currents given by orbit segments: since by definition $\iota_X(\operatorname{Re}(h)) = 1$, we have, for all $(p, \mathbb{R}^+)$ and for all $f \in H^1(M)$,

$$\int_0^T f \circ \phi_t^X(p) dt = \langle \gamma_T^X(p), f \operatorname{Re}(h) \rangle. \tag{4.3}$$

It is therefore enough to prove bounds on the currents $\gamma_T^X(p)$ with respect to the dual Sobolev norm on the space $W^{-1}(M, h)$ of currents (defined as the dual Banach space of the space $W^1(M, h)$ of 1-forms with coefficients, with respect to the frame $\{\operatorname{Re}(h), \operatorname{Im}(h)\}$, in the Sobolev space $H^1(M)$). We note that, by the *Sobolev trace theorem*, the integration currents along smooth 1-dimensional submanifolds of a surface belong to the dual Sobolev space $W^{-1}(M, h)$.

The second step is to generate orbit segments, and the associate currents, for large $T > 0$, by iterated applications of the pseudo-Anosov map $\psi : M \to M$ on uniformly bounded orbit segments. In fact, we have the following identity.

For any $T > 1$, let $n_T = [\log T / \log \lambda]$ and let $\tau = T/\lambda^{n_T} \in [1, \lambda)$. We then have

$$\gamma_T^X(p) = \psi_*^{n_T}(\gamma_\tau^X(\psi^{-n_T}(p))) . \tag{4.4}$$

Bounds (and even asymptotics) for the currents $\gamma_T^X(p)$ as a function of T for $T > 1$ can then in principle be derived from the knowledge of the spectrum of the operator $\psi_* : W^{-1}(M, h) \to W^{-1}(M, h)$. However, this approach does not work since the space $W^{-1}(M, h)$ is too large and bounds on its spectral radius are too weak to derive interesting bounds on ergodic integrals.

The third step is based on the remark that although the currents $\gamma_T^X(p)$ are never closed (for minimal translation flows), they are always at uniformly bounded distance from closed 1-currents, given by the loops

$$\overline{\gamma}_T^X(p) = \gamma_T^X(p) \cup I(\phi_T^X(p), p) .$$

In fact, the integration currents along rectifiable loops are closed and there exists a constant such that

$$\|\gamma_T^X(p) - \overline{\gamma}_T^X(p)\|_{W^{-1}(M,h)} = \|I(\phi_T^X(p), p)\|_{W^{-1}(M,h)} \leq C \operatorname{diam}(M, h) .$$

Let $K^{-1}(M, h) \subset W^{-1}(M, h)$ denote the closed subspace defined as

$$K^{-1}(M, h) = \{\gamma \in W^{-1}(M, h) | \langle \gamma, \operatorname{Re}(h) \rangle = 0\} .$$

We note that, since $\psi^*(\operatorname{Re}(h)) = \lambda^{-1} \operatorname{Re}(h)$, the subspace $K^{-1}(M, h)$ is ψ_*-invariant.

Let then $\mathcal{Z}^{-1}(M, h) \subset W^{-1}(M, h)$ denote the subspace of closed 1-currents. For all $k \in \mathbb{N}$ we write

$$\psi_*^k(\gamma_\tau^X(f^{-n_T}(p)) + (\lambda^k \tau) \operatorname{Im}(h) = z_k + r_k , \tag{4.5}$$

with $z_k \in \mathcal{Z}^{-1}(M, h) \cap K^{-1}(M, h)$ and $r_k \in \mathcal{Z}^{-1}(M, h)^\perp \cap K^{-1}(M, h)$ such that

$$\|r_k\|_{W^{-1}(M,h)} \leq C \operatorname{diam}(M, h) . \tag{4.6}$$

By definition, for all $k \in \mathbb{N}$ we have

$$z_{k+1} + r_{k+1} = \psi_*(z_k + r_k) = \psi_*(z_k) + \psi_*(r_k) ,$$

and by orthogonal projection there exists $r'_k \in \mathcal{Z}^{-1}(M,h) \cap K^{-1}(M,h)$ such that

$$z_{k+1} = \psi_*(z_k) + r'_k. \tag{4.7}$$

Since the current $r'_k \in \mathcal{Z}^{-1}(M,h)^\perp \cap K^{-1}(M,h)$ is the orthogonal projection of the current $\psi_*(r_k)$, by the bound in formula (4.6) there exists a constant $C' > 0$ such that

$$\|r'_k\|_{W^{-1}(M,h)} \le \|\psi_*(r_k)\|_{W^{-1}(M,h)} \le C' \operatorname{diam}(M,h). \tag{4.8}$$

Finally let $\rho_\psi \in [1, \lambda)$ denote the spectral radius of ψ_* on $\mathcal{Z}^{-1}(M,h)$. By the recursive formula (4.7), we have, for all $k \in \mathbb{N}$,

$$\|z_{k+1}\|_{W^{-1}(M,h)} \le \rho_\psi \|z_k\|_{W^{-1}(M,h)} + C' \operatorname{diam}(M,h).$$

which by comparison leads to the estimate

$$\|z_k\|_{W^{-1}(M,h)} \le \rho_\psi^k \left(\|z_0\|_{W^{-1}(M,h)} + C' \operatorname{diam}(M,h) \sum_{j=0}^{k-1} \rho_\psi^{-(j+1)} \right).$$

For $k = n_T$, by formulas (4.4) and (4.5), we finally conclude that there exists a constant $C'' > 0$ such that, for all $(p, T) \in M \times \mathbb{R}^+$

$$\|\gamma_T^X(p) + T \operatorname{Im}(h)\|_{W^{-1}(M,h)} = \|z_{n_T} + r_{n_T}\|_{W^{-1}(M,h)}$$
$$\le \rho_\psi^{n_T} \left(\|z_0\|_{W^{-1}(M,h)} + C' \operatorname{diam}(M,h) n_T \right)$$
$$\le C'' T^{\frac{\log \rho_\psi}{\log \lambda}} \left(\lambda + \operatorname{diam}(M,h) \frac{\log T}{\log \lambda} \right),$$

which completes the argument (since $1 - \log \rho_\psi / \log \lambda > 0$) up to the *fourth and last step*: the proof of the following claim on the spectrum of ψ_* on the space $\mathcal{Z}^{-1}(M,h)$ of closed 1-currents. □

Claim 4.24 *The spectrum of ψ_* on the space $\mathcal{Z}^{-1}(M,h)$ of closed 1-currents outside of the unit circle equals the spectrum of the finite dimensional homomorphism ψ^* on $H^1(M, \mathbb{R})$. In fact, the restriction of ψ_* to the space $\mathcal{E}^{-1}(M,h)$ of exact 1-currents is isometric, hence its spectrum is contained in the unit circle.*

Proof of Claim 4.24 By the de Rham theorem (for currents), there is an isomorphism

$$H^1(M, \mathbb{R}) \approx \mathcal{Z}^{-1}(M,h)/\mathcal{E}^{-1}(M,h).$$

4 Effective Unique Ergodicity and Weak Mixing of Translation Flows 183

Since $\mathcal{E}^{-1}(M,h)$ is a closed ψ_*-invariant space, it is enough to prove that $\psi_*|\mathcal{E}^{-1}(M,h)$ is isometric with respect to an equivalent norm. By definition for any exact current γ of dimension 1 there exists a unique current U_γ of dimension 2 such that $\langle U_\gamma, \text{area}_h \rangle = 0$ and

$$\gamma = dU_\gamma.$$

Since the exterior derivative complex is elliptic with a gain of one derivative, the current U_γ is given by integration against a square integrable function.

By the de Rham theorem the exterior derivative operator $d : \Omega^1(M) \to \Omega^2(M)$ is onto the closed subspace $\Omega_0^2(M)$ of 2-forms with vanishing total integral. For every smooth 1-form α we can define

$$\langle U_\gamma, d\alpha \rangle := \langle \gamma, \alpha \rangle,$$

and the definition is well-posed since whenever $d\alpha = d\beta$, then $d(\alpha - \beta) = 0$ and since γ is exact, by definition

$$\langle \gamma, \alpha - \beta \rangle = 0.$$

Thus U_γ is well-defined on the subspace $\Omega_0^2(M)$ and can be extended to a bounded linear functional on $\Omega^2(M)$ by the condition $\langle U_\gamma, \text{area}_h \rangle = 0$.

In addition, by the open mapping theorem, there exists a constant $C > 0$ such that, for any smooth function f of zero average on M, there exists a smooth 1 form α_f such that $d\alpha_f = f \, \text{area}_h$, with

$$\|\alpha_f\|_{W^1(M,h)} \leq C\|f\|_{L^2(M,h)}.$$

It follows that

$$|\langle U_\gamma, f \rangle| = |\langle \gamma, \alpha_f \rangle| \leq \|\gamma\|_{W^{-1}(M,h)} \|\alpha_f\|_{W^1(M,h)} \leq C\|\gamma\|_{W^{-1}(M,h)} \|f\|_{L^2(M,h)},$$

which implies that $U_\gamma \in L^2(M,h)$ and that

$$\|U_\gamma\|_{L^2(M,h)} \leq C\|\gamma\|_{W^{-1}(M,h)}.$$

Conversely, for all smooth 1-form $\alpha \in \Omega^1(M)$, we have

$$|\langle \gamma, \alpha \rangle| = |\langle U_\gamma, d\alpha \rangle| \leq \|U_\gamma\|_{L^2(M,h)} \|\alpha\|_{W^1(M,h)},$$

which implies the converse estimate

$$\|\gamma\|_{W^{-1}(M,h)} \leq \|U_\gamma\|_{L^2(M,h)}.$$

We have thus proved that the norm

$$\|\|\gamma\|\| := \|U_\gamma\|_{L^2(M,h)}, \quad \text{for all } \gamma \in \mathcal{E}^{-1}(M,h),$$

is an equivalent norm on the Banach space $\mathcal{E}^{-1}(M,h)$ (endowed with the norm induced from $W^{-1}(M,h)$).

Finally, since the action of a diffeomorphism commutes with the exterior derivative and the pseudo-Anosov map is area-preserving on (M,h), hence unitary on $L^2(M,h)$, we have

$$\|\|\psi_*(\gamma)\|\| = \|U_{\psi_*(\gamma)}\|_{L^2(M,h)} = \|\psi^*(U_\gamma)\|_{L^2(M,h)} = \|\|\gamma\|\|,$$

which completes the proof that ψ_* is isometric on $\mathcal{E}^{-1}(M,h)$ with respect to an equivalent norm. □

The proof of Claim 4.24 completes the proof of Theorem 4.23.

4.3.3 The Kontsevich–Zorich Cocycle

The typical translation flow is not self-similar, and in fact the results of the previous section apply only to a countable set of translation flows (those which are self-similar). To extend the results to typical translation flows, it is necessary to introduce a renormalization cocycle over the Teichmüller flow (or a related dynamical system such as the Rauzy–Veech induction and the Veech "zippered rectangles" flow [35]) on a cohomology vector bundle over the moduli space of translation surfaces, which generalizes the action of a pseudo-Anosov map on the cohomology vector space.

Definition 4.10 ([1, 2]) The **complex Hodge bundle** $H^1_\kappa(M, \mathbb{C})$ over a stratum $\mathcal{H}^{(1)}_\kappa$ of the moduli space of translation surfaces is the quotient (orbifold) vector bundle $H^1_\kappa(M, \mathbb{R})$ of the product bundle $\hat{\mathcal{H}}^{(1)}_\kappa \times H^1(M, \mathbb{R})$ under the product action of the mapping class group Γ_g by pull-back:

$$H^1_\kappa(M, \mathbb{C}) := \left(\hat{\mathcal{H}}^{(1)}_\kappa \times H^1(M, \mathbb{C})\right) / \Gamma_g.$$

The **real Hodge bundle** $H^1_\kappa(M, \mathbb{R})$ is the real part of the complex Hodge bundle $H^1_\kappa(M, \mathbb{C})$, that is, the quotient of the trivial bundle $\hat{\mathcal{H}}^{(1)}_\kappa \times H^1(M, \mathbb{R})$.

The **Kontsevich–Zorich cocycle** $g^{KZ}_\mathbb{R}$ on $H^1_\kappa(M, \mathbb{C})$ is the projection to the complex Hodge bundle under the action of the mapping class group Γ_g of the product cocycle (over the Teichmüller geodesic flow $g_\mathbb{R}$)

$$g_\mathbb{R} \times \mathrm{Id} : \hat{\mathcal{H}}^{(1)}_\kappa \times H^1(M, \mathbb{C}) \to \hat{\mathcal{H}}^{(1)}_\kappa \times H^1(M, \mathbb{C}).$$

The Kontsevich–Zorich cocycle on the complex Hodge bundle $H^1_\kappa(M, \mathbb{C})$ has a well-defined restriction to the real Hodge bundle $H^1_\kappa(M, \mathbb{R})$.

Remark 4.1 The trivial bundle $\hat{\mathcal{H}}^{(1)}_\kappa \times H^1(M, \mathbb{C})$ is well-defined over the Teichmüller space of Riemann surfaces, which is a space of equivalence classes of surfaces with respect to the action of the group $\text{Diff}^+_0(M)$ of diffeomorphisms isotopic to the identity, since the action of $\text{Diff}^+_0(M)$ on $H^1(M, \mathbb{C})$ is trivial.

The bundle $H^1_\kappa(M, \mathbb{C})$ is therefore the pull-back of a bundle well-defined over the moduli space of Riemann surfaces (forgetting the translation surface structure). The trivial connection on $\hat{\mathcal{H}}^{(1)}_\kappa \times H^1(M, \mathbb{C})$ projects to a connection on $H^1_\kappa(M, \mathbb{R})$ called the *Gauss-Manin connection*.

The Kontsevich–Zorich cocycle is given by the parallel transport of cohomology classes along orbits of the Teichmüller flow with respect to the Gauss-Manin connection.

Exercise 4.25 Let $(M, h) \in \mathcal{H}^{(1)}_\kappa$ such that there exists a pseudo-Anosov map $\psi : (M, h) \to (M, h)$ (as in Definition 4.9) of dilation factor $\lambda > 1$. Derive from the definition of the Teichmüller flow $g_\mathbb{R}$ (see formulas (4.1) and (4.2) in Sect. 4.2.2) that (M, h) is a periodic point of $g_\mathbb{R}$ on $\mathcal{H}^{(1)}_\kappa$ of period $T = \log \lambda$. Prove that all periodic orbits of the Teichmüller flow on $\mathcal{H}^{(1)}_\kappa$ are of this type. Prove that the time T-map of the Kontsevich–Zorich cocycle is given by the formula

$$g^{KZ}_T((M, h), c) = ((M, h), (\psi^{-1})^*(c)), \quad \text{for all } c \in H^1(M, \mathbb{R}).$$

The Kontsevich–Zorich cocycle generalizes the action of a pseudo-Anosov map on cohomology, and provides a framework to extend the results of Sect. 4.3.2 to typical translation flows. In terms of the action of elements of the mapping class group on the cohomology, the returns of an orbit of the Teichmüller geodesic flow to a fundamental domain of a stratum of the moduli space of translation surfaces encode a sequence of (pseudo-Anosov) maps whose compositions replace the iteration of a single map, as in the self-similar case. A natural class of translation flows for which it is possible to prove (effective) unique ergodicity by these methods is therefore the class of horizontal flows of translation surfaces which are forward recurrent under the Teichmüller flow (with strictly positive frequency).

As the Kontsevich–Zorich cocycle generalizes the action of a single diffeomorphism on the cohomology vector space, the notion of a *Lyapunov spectrum* generalizes the notion of the spectrum of a homomorphism (strictly speaking the Lyapunov exponents generalize the logarithms of the moduli of the eigenvalues). Since the symplectic structure given by the cup product on the cohomology $H^1(M, \mathbb{R})$ is invariant under the action of diffeomorphisms, the Kontsevich-Zorich cocycle is a *symplectic cocycle*, hence its Lyapunov spectrum, with respect to any probability $g_\mathbb{R}$-invariant measure μ on a stratum $\mathcal{H}^{(1)}_\kappa$, is symmetric:

$$\lambda^\mu_1 \geq \lambda^\mu_2 \geq \cdots \geq \lambda^\mu_g \geq -\lambda^\mu_g \geq \cdots \geq -\lambda^\mu_2 \geq -\lambda^\mu_1.$$

The *Kontsevich–Zorich conjectures* [1, 2] stated that, for $\mu = \mu_\kappa^{(1)}$ the Masur-Veech measure on a stratum of the moduli space,

$$\lambda_1^\mu = 1 > \lambda_2^\mu > \cdots > \lambda_g^\mu > 0, \qquad (4.9)$$

and that these exponents are also the deviation exponents of ergodic averages of *Masur–Veech typical* translation flows (as well as Masur–Veech typical locally Hamiltonian flows with non-degenerate saddle singularities) in the sense that, there exist linear functionals $D_2, \ldots, D_g$ on $C^\infty(M)$ (invariant distributions) such that, for all functions $f \in C^\infty(M)$ and for all $\epsilon > 0$,

$$\int_0^T f(\phi_t^X(x))dt = T \int_M f \, \text{area}_h + \sum_{i=2}^g D_i(f) T^{\lambda_i + o(1)} + O(T^\epsilon). \qquad (4.10)$$

The identity $\lambda_1^\mu = 1$ is elementary (as it is explained below), the *spectral gap* property $\lambda_2^\mu < 1$ can be derived from Veech's proof of non-uniform hyperbolicity of the Masur–Veech measures, and will be proved below (following [3]) for all probability ergodic measures. In fact, it follows from the existence of period coordinates that the *non-negative* half of the symmetric Lyapunov spectrum of the tangent cocycle $Tg_\mathbb{R}$ of the Teichmüller flow are

$$2 \geq 1 + \lambda_2^\mu \geq \cdots \geq 1 + \lambda_g^\mu \geq 1 = \cdots = 1 \geq 1 - \lambda_g^\mu \geq \cdots \geq 1 - \lambda_2^\mu \geq 0$$

(with the exponent 1 appearing with multiplicity at least equal to the number of cone points of surfaces in the stratum minus one), hence $\lambda_2^\mu < 1$ if and only if all the non-trivial Lyapunov exponents of the tangent cocycle $Tg_\mathbb{R}$ are non-zero.

The non-vanishing of the Kontsevich–Zorich exponents, that is, the strict inequality $\lambda_g^\mu > 0$ (for $\mu = \mu_\kappa^{(1)}$), as well as the asymptotic of ergodic averages (4.10) for translation flows, were proved by the author in [3] (see also [58]).

The *simplicity* of the Kontsevich–Zorich spectrum, that is, all the strict inequalities in formula (4.9), was later proved by A. Avila and M. Viana [59].

The Kontsevich–Zorich conjecture on the deviation of ergodic averages for locally Hamiltonian flows has been completely proved only recently [60], and generalized to flows with degenerate (canonical) saddle like singularities in [61].

By the Oseledets theorem, for almost all translation surfaces (M, h) (with respect to any $g_\mathbb{R}$-ergodic measure μ on $\mathcal{H}_\kappa^{(1)}$) there exists a splitting of the cohomology

$$H^1(M, \mathbb{R}) = E(\lambda_1^\mu) \oplus E(\lambda_2^\mu) \cdots \oplus E(\lambda_s^\mu) \oplus E(-\lambda_s^\mu) \oplus \cdots \oplus E(-\lambda_2^\mu) \oplus E(-\lambda_1^\mu)$$

into subspaces corresponding to the distinct non-negative Lyapunov exponents

$$\lambda_1^\mu > \lambda_2^\mu > \cdots > \lambda_s^\mu > -\lambda_s^\mu > \cdots > -\lambda_1^\mu,$$

such that for all $i \in \{1, \ldots, s\}$ and for all $c \in E(\pm \lambda_i)$,

$$\lim_{t \to \pm \infty} \frac{1}{t} \log \|g_t^{KZ}(c)\| = \lambda_i^\mu.$$

In particular, if $\lambda_1^\mu = 1$ is simple and $\lambda_2^\mu < \lambda_1^\mu$, then for $c \in H^1(M, \mathbb{R})$ such that $c \wedge E(-\lambda_1^\mu) = 0$ we have

$$\lim_{t \to \pm \infty} \frac{1}{t} \log \|g_t^{KZ}(c)\| < \lambda_1^\mu = 1.$$

In the following we will prove a similar, but stronger, upper bound for all translation surfaces (M, h) with the property that the Teichmüller geodesic ray $g_\mathbb{R}(M, h)$ returns with positive frequency to any given compact subset of the moduli space (of Riemann surfaces).

4.3.4 The Hodge Norm and its First Variation

A powerful tool to analyze the Kontsevich–Zorich cocycle (hence the tangent cocycle of the geodesic flow) is given by the Hodge norm on the cohomology bundle (see for instance [1–3, 8, 9, 54, 62, 63]).

Definition 4.11 Let M be a Riemann surface. The **Hodge norm on the complex cohomology** $H^1(M, \mathbb{C})$ is defined as follows. For every $c \in H^1(M, \mathbb{C})$ and for every complex closed 1-form α on M such such that $c = [\alpha]$ we set

$$\|c\|_{H^1(M,\mathbb{C})} := \|\alpha\|_M = \left(\frac{\iota}{2} \int_M \alpha \wedge \overline{\alpha} \right)^{1/2}.$$

The **Hodge norm on the real cohomology** $H^1(M, \mathbb{R})$ can be defined via the identification (Hodge representation theorem for Riemann surfaces) of the real cohomology $H^1(M, \mathbb{R})$ with the subspace of holomorphic 1-forms $H^{1,0}(M) \subset H^1(M, \mathbb{C})$.

In other terms, for every $c \in H^1(M, \mathbb{R})$ and for every holomorphic 1-form α on M such that $c = \text{Re}[\alpha]$ we set

$$\|c\|_{H^1(M,\mathbb{R})} := \|\alpha\|_M = \left(\frac{\iota}{2} \int_M \alpha \wedge \overline{\alpha} \right)^{1/2}.$$

Exercise 4.26 Prove that the Hodge norm is well-defined, that is, the definition is independent of the 1-form representing the (de Rham) cohomology class, and that it gives an hermitian norm on $H^1(M, \mathbb{C})$ and an euclidean norm on $H^1(M, \mathbb{R})$.

Exercise 4.27 *Prove that the Hodge norm is equivariant under the action of the group* $\text{Diff}^+(S)$ *of diffeomorphisms of the underlying smooth surface* S, *in the sense that, for every* $\phi \in \text{Diff}^+(S)$ *and for every* $c \in H^1(M, \mathbb{C})$, *we have*

$$\|\phi^*(c)\|_{H^1(\phi_*(M),\mathbb{C})} = \|c\|_{H^1(M,\mathbb{C})}.$$

The Riemann surface $\phi_*(M)$ *is defined by pre-composition of the charts of an atlas of the Riemann surface* M *with the diffeomorphism* ϕ *on* S.

Prove that, since it is equivariant, the Hodge norm induces a norm on the Hodge bundle $H^1_\kappa(M, \mathbb{C})$, *and on* $H^1_\kappa(M, \mathbb{R})$.

Notation: We will adopt below the following notation for the Hodge norm of a cohomology class $c \in H^1(M, \mathbb{R})$ at a translation surface $(M, h) \in \mathcal{H}_\kappa^{(1)}$:

$$\|(c, M, h)\| = \|c\|_{H^1(M,\mathbb{R})}.$$

The first variation of the Hodge norm along a Teichmüller geodesic is given by the following formula (see [3], Lemma 2.1', [17], Theorem 29, or [19], Lemma 2.5):

Lemma 4.28 *Let* $(M, h) \in \mathcal{H}_\kappa^{(1)}$. *For any* $c \in H^1(M, \mathbb{R})$, *let* $\alpha \in H^{1,0}(M)$ *denote the unique holomorphic differential such that* $c = [\text{Re}(\alpha)]$. *We have*

$$\frac{d}{dt}\|g_t^{KZ}(M, h, c)\|^2\bigg|_{t=0} = 2\,\text{Re}\, B_{(M,h)}(c) := 2\,\text{Re}\left(\int_M \left(\frac{\alpha}{h}\right)^2 \text{area}_h\right).$$

Proof Since the Hodge norm is equivariant under the action of the mapping class group Γ_g we prove the identity for the product cocycle on $\mathcal{H}_\kappa^{(1)} \times H^1(M, \mathbb{R})$.

For all $t \in \mathbb{R}$, let $g_t(M, h) = (M_t, h_t)$ and let $\alpha_t = f_t h_t \in H^{1,0}(M_t)$ denote the unique 1-form, holomorphic on the Riemann surface M_t, such that $c = [\text{Re}(\alpha_t)]$.

By definition of the Hodge norm we have

$$\frac{d}{dt}\|g_t^{KZ}(M, h, c)\|^2 = \frac{d}{dt}\left(\frac{\imath}{2}\int_M \alpha_t \wedge \overline{\alpha_t}\right) = -\text{Re}\int_M \alpha_t \wedge \overline{\frac{d\alpha_t}{dt}}.$$

Since $[\text{Re}(\alpha_t)]$ is constant, it follows that the 1-form

$$\text{Re}(\frac{d\alpha_t}{dt}) \quad \text{is an exact 1-form.}$$

Since α_t is a holomorphic form, hence closed, and $\text{Re}(\frac{d\alpha_t}{dt})$ is exact, we have

$$\int_M \alpha_t \wedge \overline{\frac{d\alpha_t}{dt}} = \int_M \alpha_t \wedge \left(\overline{\frac{d\alpha_t}{dt}} - 2\,\text{Re}(\frac{d\alpha_t}{dt})\right) = -\int_M \alpha_t \wedge \frac{d\alpha_t}{dt}.$$

4 Effective Unique Ergodicity and Weak Mixing of Translation Flows

By definition of the Teichmüller flow, we have $h_t = e^{-t} \operatorname{Re}(h) + \imath e^t \operatorname{Im}(h)$, hence

$$\frac{dh_t}{dt} = -e^{-t} \operatorname{Re}(h) + \imath e^t \operatorname{Im}(h) = -\overline{h_t}, \tag{4.11}$$

and, as a consequence,

$$\frac{d\alpha_t}{dt} = \frac{df_t}{dt} h_t + f_t \frac{dh_t}{dt} = \frac{df_t}{dt} h_t - f_t \overline{h_t},$$

hence, by taking into account that h_t is holomorphic,

$$\int_M \alpha_t \wedge \frac{d\alpha_t}{dt} = \int_M \alpha_t \wedge \frac{d\overline{\alpha_t}}{dt} = \int_M f_t h_t \wedge (\frac{df_t}{dt} h_t - f_t \overline{h_t}) = -\int_M f_t^2 \, h_t \wedge \overline{h_t}.$$

We conclude that, since $h_t \wedge \overline{h_t} = -2\imath \, \text{area}_h$, for all $t \in \mathbb{R}$,

$$\frac{d}{dt} \|g_t^{KZ}(M, h, c)\|^2 = -\operatorname{Im}\left(\int_M f_t^2 \, h_t \wedge \overline{h_t}\right) = 2\operatorname{Re}\left(\int_M f_t^2 \, \text{area}_h\right),$$

as stated (since by definition $f_t = \alpha_t / h_t$, for all $t \in \mathbb{R}$). □

Remark 4.2 For every $(M, h) \in \mathcal{H}_\kappa^{(1)}$, the bilinear form

$$B_{(M,h)}(\alpha, \beta) = \frac{\imath}{2} \int_M \frac{\alpha}{h} \frac{\beta}{h} h \wedge \bar{h} = \frac{\imath}{2} \int_M (\alpha\beta) \frac{\bar{h}}{h}, \quad \text{for all } \alpha, \beta \in H^{1,0}(M)$$

can be interpreted as the *second fundamental form* of the holomorphic connection with respect to the (flat) Gauss-Manin connection (see [19], section 2.3).

We prove a crucial estimate on the second fundamental form:

Lemma 4.29 *For every $(M, h) \in \mathcal{H}_\kappa^{(1)}$ and for all $c \in H^1(M, \mathbb{R})$, we have*

$$|B_{(M,h)}(c)| \le \|(M, h, c)\|^2.$$

Let then $\mathcal{T}(M, h) := \mathbb{R}[\operatorname{Re}(h)] \oplus \mathbb{R}[\operatorname{Im}(h)] \subset H^1(M, \mathbb{R})$ denote the so-called *tautological plane* and let $\mathcal{T}(M, h)^\perp$ denote its symplectic orthogonal:

$$\mathcal{T}(M, h)^\perp = \{c \in H^1(M, \mathbb{R}) | c \wedge [\operatorname{Re}(h)] = c \wedge [\operatorname{Im}(h)] = 0\}.$$

We have the following estimate:

$$\Lambda(M, h) = \max_{c \in \mathcal{T}(M,h)^\perp \setminus \{0\}} \frac{|B_{(M,h)}(c)|}{\|(M, h, c)\|^2} < 1. \tag{4.12}$$

Proof Let $c = \text{Re}[\alpha]$ with $\alpha \in H^{1,0}(M)$ and let $f = \alpha/h$. A straightforward estimate based on the Schwarz inequality gives

$$|B_{(M,h)}(c)| = \left|\int_M f^2 \, \text{area}_h\right| = |\langle f, \bar{f}\rangle_{L^2(M,\,\text{area}_h)}| \leq \|f\|^2_{L^2(M,\text{area}_h)}$$
$$= \|(M, h, c)\|^2.$$

The first bound in the statement is therefore proved.

In addition, by the Cauchy-Schwarz inequality, equality holds if and only if $\bar{f} \in \mathbb{C}f$. Since the differential $\alpha = fh \in H^{1,0}(M)$, so that the function f is meromorphic (and not equal to zero), if equality holds then f is meromorphic and ant-meromorphic (as $\bar{f}$ is also meromorphic), that is, if and only if f is a non-zero constant function. However, for all cohomology classes $c \in \mathcal{T}(M, h)^\perp$ f is not a non-zero constant function since

$$\int_M f \, \text{area}_h = \imath \int_M \text{Re}(fh) \wedge \bar{h} = 0,$$

hence the above Cauchy-Schwarz inequality is strict. The proof of the second inequality is therefore complete. □

The (tautological) $g_\mathbb{R}^{KZ}$-invariant expanding and contracting line bundles $\mathcal{T}^\pm$, defined as

$$\mathcal{T}^+(M, h) = \mathbb{R}[\text{Re}(h)] \quad \text{and} \quad \mathcal{T}^-(M, h) = \mathbb{R}[\text{Im}(h)],$$

have well defined Lyapunov exponents equal to ± 1 since

$$g_t^{KZ}([\text{Re}(h)]) = e^t[\text{Re}(g_t(h))] \quad \text{and} \quad g_t^{KZ}([\text{Im}(h)]) = e^{-t}[\text{Im}(g_t(h))].$$

Exercise 4.30 *Prove that for any $g_\mathbb{R}$-ergodic invariant measure μ on $\mathcal{H}_\kappa^{(1)}$ the top Kontsevich–Zorich exponent $\lambda_1^\mu = 1$.*

The following lemma is the key "spectral gap" result for the Kontsevich–Zorich cocycle:

Lemma 4.31 *For all $(M, h) \in \mathcal{H}_\kappa^{(1)}$, for all $c \in \mathcal{T}(M, h)^\perp$ and for all $t \in \mathbb{R}$, we have*

$$\|g_t^{KZ}(M, h, c)\| \leq \|(M, h, c)\| \exp\left(\int_0^t \Lambda(g_s(M, h))ds\right).$$

In particular, if the forward Teichmüller orbit $g_{\mathbb{R}}(M, h)$ visits a compact set $K \subset \mathcal{H}_\kappa^{(1)}$ with positive frequency, in the sense that

$$f_K := \liminf_{t \to +\infty} \frac{1}{t} \operatorname{Leb}(\{t \geq 0 | g_t(M, h) \in K\}) > 0, \tag{4.13}$$

then, for all $c \in \mathcal{T}(M, h)^\perp$,

$$-1 < \liminf_{t \to +\infty} \frac{1}{t} \log \|g_t^{KZ}(M, h, c)\| \leq \limsup_{t \to +\infty} \frac{1}{t} \log \|g_t^{KZ}(M, h, c)\| < 1.$$

Proof From Lemma 4.28, for all $(M, h) \in \mathcal{H}_\kappa^{(1)}$, all $c \in H^1(M, \mathbb{R})$ and for all $t \in \mathbb{R}$, we have

$$\frac{d}{dt} \log \|g_t^{KZ}(M, h, c)\| = \frac{\operatorname{Re} B_{g_t(M,h)}(c)}{\|g_t^{KZ}(M, h, c)\|}.$$

By the above formula and by the definition of the function Λ in formula (4.12), it follows that for all $(M, h) \in \mathcal{H}_\kappa^{(1)}$ and for all $c \in \mathcal{T}(M, h)^\perp$,

$$\frac{d}{dt} \log \|g_t^{KZ}(M, h, c)\| \leq \Lambda(g_t(M, h)),$$

hence the estimate in the statement follows by integration.

It can be proved that the function Λ is continuous on $\mathcal{H}_\kappa^{(1)}$ (as the maximum of a continuous functions on the Hodge bundle over the unit sphere in the fiber). Since $\Lambda < 1$ everywhere, it follows that for any compact set $K \subset \mathcal{H}_\kappa^{(1)}$,

$$\Lambda_{\max} := \sup_{(M,h) \in K} \Lambda(M, h) = \max_{(M,h) \in K} \Lambda(M, h) < 1.$$

Under the assumption that the visit frequency $f_K > 0$ (see formula (4.13)) we can estimate

$$\limsup_{t \to \infty} \frac{1}{t} \int_0^t \Lambda(g_s(M, h)) ds \leq 1 - (1 - \Lambda_{\max}) f_K < 1,$$

from which the estimate in the statement follows immediately. □

Exercise 4.32 *Derive from Lemma 4.31 a proof (announced in Sect. 4.3.2) that the action on cohomology of any pseudo-Anosov map with orientable invariant foliations has a simple top eigenvalue.*

In fact, the Hodge theory approach easily gives a more general result:

Exercise 4.33 *Derive from Lemma 4.31 a proof that for any ergodic $g_\mathbb{R}$ invariant measure the top Lyapunov exponent is simple and we have the spectral gap*

$$\lambda_2^\mu < \lambda_1^\mu = 1.$$

Exercise 4.34 *Derive from Lemma 4.31 a proof that any ergodic $g_\mathbb{R}$ invariant measure is non-uniformly hyperbolic, in the sense that all Lyapunov exponents of the tangent cocycle $Tg_\mathbb{R}$ are non-zero, except the (trivial) zero exponent corresponding to the flow direction.*

Remark 4.3 Lower bounds on Kontsevich–Zorich exponents of $SL(2, \mathbb{R})$-invariant measures can be derived from second variation formulas for the Hodge norm [3], in particular from a formula for the (hyperbolic) Laplacian of the Hodge norm along $SL(2, \mathbb{R})$ orbits (Teichmüller disks). However, the examples of the so-called *Eierlegende Wollmilchsau* [18] and *Platypus* [64, 65] made clear that there are no general lower bounds, as the second exponent can be equal to zero Lower bounds for the sum of the first k Kontsevich–Zorich exponents in terms of the degrees of holomorphic k-bundles were conjectured by Yu [69], proved by Eskin et al. [70] and later refined by Costantini [71]. A criterion for the positivity of Kontsevich–Zorich exponents was given in [58], improving upon [3].

4.3.5 Typical (Effective) Unique Ergodicity

In this section we outline the cohomological proof of (effective) unique ergodicity. The arguments generalize those given in the self-similar case in Sect. 4.3.2, however we adopt a different, more geometric, perspective, which replaces the analysis of the action of the mapping class group on cohomology with control of the flat geometry given by the renormalization dynamics (Teichmüller flow) on the moduli space.

We begin with a cohomological proof of a version of a fundamental unique ergodicity criterion first proved by H. Masur (see [33], Prop. 6.2):

Theorem 4.35 (Masur's Criterion) *Assume that the forward Teichmüller orbit $g_{\mathbb{R}^+}(M, h)$ is recurrent to a compact set $K \subset \mathcal{H}_g^{(1)}$ of the moduli space of unit area Abelian differentials. Then the horizontal flow $\phi_\mathbb{R}^X$ of the translation surface (M, h) is uniquely ergodic.*

Proof Let μ be any probability $\phi_\mathbb{R}^X$-invariant measure on M and let $F(\mu) \in H^1(M, \mathbb{R})$ denote its flux class. As in the proof of Claim 4.22, by the ergodic theorem the flux class is given, for μ-almost all $p \in M$, by the formula

$$F(\mu) = \lim_{T \to +\infty} \frac{1}{T} \gamma_T^X(p).$$

For all $t > 0$, the segment $\gamma_T^X(p)$ has length $e^{-t}T$ on the translation surface $g_t(M, h)$, hence for any compact set K there exists a constant $C_K > 0$ such that, whenever $g_t(M, h) \in K$, the norm of the current $\gamma_T^X(p)$ in the dual Sobolev space $W^{-1}(g_t(M, h))$ of the (flat) translation surface $g_t(M, h)$ satisfies (for $T > e^t$) the bound

$$\|\gamma_T(p)\|_{W^{-1}(g_t(M,h))} \leq C_K e^{-t} T.$$

Since over K all norms on the cohomology bundle are equivalent, we derive the following bound for the Hodge norm of the flux class: there exists a constant $C'_K > 0$ such that, for all $t > 0$, we have

$$\|g_t^{KZ}(M, h, F(\mu))\| \leq C'_K e^{-t}.$$

Under the recurrence hypothesis, the above estimate implies that $F(\mu) \in \mathcal{T}(M, h)$. In fact, if $F(\mu)$ has a component $F_0(\mu) \neq 0$ in $\mathcal{T}(M, h)^\perp$, by Lemma 4.31 there exists a constant $C''_K > 0$ such that

$$\|(M, h, F_0(\mu))\| \leq \|g_t^{KZ}(M, h, F_0(\mu))\| \exp\left(\int_0^t \Lambda(g_s(M, h))ds\right) \quad (4.14)$$
$$\leq C''_K \exp\left(-t + \int_0^t \Lambda(g_s(M, h))ds\right)$$

The assumption that $F_0(\mu) \neq 0$ hen implies that

$$\limsup_{t \to +\infty} \left(t - \int_0^t \Lambda(g_s(M, h))ds \right) < +\infty,$$

in contradiction with the assumption that the forward orbit $g_{\mathbb{R}^+}(M, h)$ is recurrent to the compact set K. In fact, since for any compact set K' such that $K \subset K'$ we have $\max_{K'} \Lambda < 1$, and the set K' can be chose so that, since the trajectory is recurrent to K, the total Lebesgue measure of the time intervals that it spends in K' is infinite.

We have thus proved that $F(\mu) \in \mathcal{T}(M, h)$, and again by the estimate in formula (4.14), it follows that $F(\mu) \in \mathbb{R}[\text{Im}(h)]$ (since $[\text{Im}(h)]$ has Lyapunov exponent equal to -1). Thus by the injectivity Lemma 4.19 the cone of invariant measures is one-dimensional and $\phi_\mathbb{R}^X$ is uniquely ergodic. □

The above Masur's criterion implies, by the Poincaré recurrence theorem, that all $g_\mathbb{R}$-invariant probability measures are supported on the set of translation surfaces with uniquely ergodic horizontal and vertical flows.

It implies in particular Theorem 4.8, which states that the Masur–Veech typical translation flow is uniquely ergodic (Keane conjecture).

The stronger unique ergodicity result given in Theorem 4.8, which states that unique ergodicity is directionally typical for any given translation surface, follows from Masur's criterion and from the statement first proved in [42] that forward

Teichmüller geodesics are recurrent for almost all translation surfaces in *every* $SO(2, \mathbb{R})$-orbit on the moduli space.

An effective (polynomial) unique ergodicity theorem (Theorem 4.10) was later derived in [16], from Lemma 4.31 and from the effective version of the recurrence result of [42] proved by J. Athreya in his Ph. D. thesis [44].

In fact, the spectral gap result implies the following conditional effective unique ergodicity result.

Theorem 4.36 *Assume that the forward Teichmüller orbit* $g_{\mathbb{R}^+}(M, h)$ *visits a compact set* $K \subset \mathcal{H}_\kappa^{(1)}$ *with positive frequency, that is,*

$$f_K := \liminf_{t \to +\infty} \frac{1}{t} \operatorname{Leb}(\{t \geq 0 | g_t(M, h) \in K\}) > 0 \,.$$

Then there exist constants $C(M, h) > 0$ *and* $\alpha > 0$ *such that, for all functions* $f \in H^1(M)$ *of zero average and for all* $(p, T) \in M \times \mathbb{R}^+$, *such that* p *has an infinite forward orbit under* $\phi_\mathbb{R}^X$, *we have*

$$\left| \frac{1}{T} \int_0^T f \circ \phi_t^X(p) dt \right| \leq C(M, h) \|f\|_{H^1(M)} T^{-\alpha} \,.$$

The proof of Theorem 4.36 is analogous to that of Theorem 4.23 in Sect. 4.3.2. It is based on the proof of a spectral gap result for a distributional "transfer cocycle", which is derived from Lemma 4.31 by de Rham theorem.

We give below the precise definition of the Sobolev bundle of currents and of the transfer cocycle.

Let (M, h) be a translation surface with horizontal and vertical vector fields (X, Y). Let $W^1(M, h)$ denote the Sobolev space of 1-forms on M endowed with the Sobolev norm induced by the flat metric of (M, h). The Sobolev norm of the space $W^1(M, h)$ is defined as follows: for every smooth 1-form α on M

$$\|\alpha\|_{W^1(M,h)} := \left(\|\iota_X \alpha\|_{L^2(M, \operatorname{area}_h)}^2 + \|\iota_Y \alpha\|_{L^2(M, \operatorname{area}_h)}^2 \right)^{1/2} \,.$$

Let $W^{-1}(M, h)$ denote the dual Sobolev space of currents of dimension 1 (and degree 1) on M endowed with the Sobolev norm induced by the flat metric of (M, h).

Definition 4.12 The **Sobolev bundle of 1-currents** W_κ^{-1} over a stratum $\mathcal{H}_\kappa^{(1)}$ is the quotient of the Sobolev (Hilbert) bundle

$$\bigcup_{(M,h) \in \hat{\mathcal{H}}_\kappa^{(1)}} \{(M, h)\} \times W^{-1}(M, h)$$

under the action of the mapping class group Γ_g on $\mathcal{H}_\kappa^{(1)}$ and, for every $(M, h) \in \mathcal{H}_\kappa^{(1)}$, on the Sobolev space $W^{-1}(M, h)$ of currents by push-forward. The bundle

## 4 Effective Unique Ergodicity and Weak Mixing of Translation Flows

W_κ^{-1} is a Hilbert bundle with norm

$$\|(M, h, \gamma)\|_{-1} := \|\gamma\|_{W^{-1}(M,h)}, \quad \text{for all } (M, h) \in \mathcal{H}_\kappa^{(1)} \text{ and } \gamma \in W^{-1}(M, h).$$

The **transfer cocycle** $g_t^{(1)}$ on W_κ^{-1} is defined as the quotient of the trivial cocycle

$$g_t \times \mathrm{Id} : (M, h) \times W^{-1}(M, h) \to g_t(M, h) \times W^{-1}(g_t(M, h)),$$

under the action of the mapping class group.

We now state the spectral gap result for the transfer cocycle (analogous of Claim 4.24) in Sect. 4.3.2.

Lemma 4.37 *The sub-bundle $\mathcal{Z}_\kappa^{-1} \subset W_\kappa^{-1}$ of closed 1-currents is invariant under the transfer cocycle $g_\mathbb{R}^{(1)}$ and the restriction of $g_\mathbb{R}^{(1)}$ to $\mathcal{Z}_\kappa^{-1}$ has a spectral gap, in the following sense. If the forward Teichmüller orbit $g_\mathbb{R}(M, h)$ visits a compact set $K \subset \mathcal{H}_\kappa^{(1)}$ with positive frequency, in the sense that*

$$f_K := \liminf_{t \to +\infty} \frac{1}{t} \mathrm{Leb}(\{t \geq 0 | g_t(M, h) \in K\}) > 0, \tag{4.15}$$

then, there exist constants $C > 0$ and $\alpha > 0$ such that, for all $\gamma \in \mathcal{Z}_\kappa^{-1}(M, h)$, such that $\langle \gamma, \mathrm{Re}(h) \rangle = \langle \gamma, \mathrm{Im}(h) \rangle = 0$, and for all $t > 0$,

$$C^{-1} e^{-(1-\alpha)t} \|(M, h, \gamma)\|_{-1} \leq \|g_t^{(1)}(M, h, \gamma)\|_{-1} \leq C e^{(1-\alpha)t} \|(M, h, \gamma)\|_{-1}.$$

Proof The argument is analogous to the one given in the proof of Claim 4.24 and will be generalized in the proof of the effective Veech criterion (see Theorem 4.53) in the part on effective weak mixing. We briefly summarize the main steps here.

It is based on the de Rham theorem for currents, on the spectral gap Lemma 4.31 for the cohomology bundle and on a direct proof that the (invariant) sub-bundle of exact currents carries the single Lyapunov exponent 0 (with infinite multiplicity).

The latter statement in turn follows from the basic fact that exact currents in $W^{-1}(M, h)$ are exterior derivatives of square-integrable functions, and that the norm on $L^2(M, h)$ is invariant under the Teichmüller flow, in the sense that

$$\|u\|_{L^2(g_t(M,h))} = \|u\|_{L^2(M,h)}, \quad \text{for all } u \in L^2(M, h) \text{ and } t \in \mathbb{R},$$

since the area form area_h is invariant under the Teichmüller flow. □

Proof of Theorem 4.36 We outline the argument which, as announced, is similar to the proof of Theorem 4.23. However, the point of view is different and geometric, as we prove below bounds on a given orbit segment under the Teichmüller deformation of the flat metric and of the related Sobolev norms, while in the proof of the self-similar case (Theorem 4.23) we proved bounds on the push-forwards of a given orbit segment under the action of the iterates of a pseudo-Anosov map.

We again write ergodic integrals of the horizontal translation flow in terms of one-dimensional currents $\gamma_T^X(p) \in W^{-1}(M, h)$, as in formula (4.3). It is therefore enough to prove bounds on the currents $\gamma_T^X(p)$ with respect to the dual Sobolev norm on the space $W^{-1}(M, h)$.

We consider a sequence (t_n) of return times of the forward Teichmüller orbit $g_{\mathbb{R}^+}$ to K with positive frequency, that is, such that

$$\lim_{n \to +\infty} \frac{t_n}{n} = f > 0. \qquad (4.16)$$

Let $\gamma_{T_n}^X(p)$ be the first forward return orbit of the horizontal flow to a vertical interval centered at $p \in M$ of unit length on $g_{t_n}(M, h)$. We note that a vertical interval of unit length on $g_{t_n}(M, p)$ has length e^{-t_n} on (M, h) and that $\gamma_{T_n}^X(p)$ is the also the first forward return orbit $\gamma_{e^{-t_n}T_n}^{X_{t_n}}(p)$ for the horizontal flow $\phi_{\mathbb{R}}^{X_{t_n}}$ of $g_{t_n}(M, h)$. Since $g_{t_n}(M, h) \in K$, there exists a constant $C_K > 0$ such that

$$e^{t_n}/C_K \leq T_n(p) \leq C_K e^{t_n}.$$

By a standard decomposition argument (a generalization of the so-called Ostrowski expansion of an irrational number), by the condition in formula (4.16), it is then possible to reduce estimates on ergodic integrals for arbitrary times to estimates for the sequence of times $(T_n(p))$ defined above. We then estimate the Sobolev norm of the currents $\gamma_{T_n}^X(p) \in W^{-1}(M, h)$.

By its definition as a return orbit to a vertical interval, the horizontal orbit segment $\gamma_{T_n}^X(p)$ can be closed by the union $\bar{\gamma}_{T_n}^X(p) := \gamma_{T_n}^X(p) \cup I_n$ with a vertical segment I_n of uniformly bounded length on $g_{t_n}(M, h)$ connecting its endpoints. Since $g_{t_n}(M, h) \in K$, by definition the loop $\bar{\gamma}_{T_n}^X(p)$ has bounded length on $g_{t_n}(M, h)$ and there exists a constant $C'_K > 0$ such that

$$\|g_{t_n}^{(1)}(M, h, \bar{\gamma}_{T_n}^X(p)) + T_n \operatorname{Im}(h)\|_{-1} \leq C'_K, \quad \text{for all } n \in \mathbb{N}.$$

Since $\bar{\gamma}_{T_n}^X(p)) \in \mathcal{Z}^{-1}(g_{t_n}(M, h))$ is a closed current, by the spectral gap Lemma 4.37 for the transfer cocycle, there exist $C''_K > 0$ and $\alpha > 0$ such that we have

$$\|(M, h, \bar{\gamma}_{T_n}^X(p)) + T_n \operatorname{Im}(h)\|_{-1} \leq C''_K e^{(1-\alpha)t_n},$$

Since the *vertical* interval I_n has length at most e^{-t_n} on (M, h) (by construction it has at most unit length on $g_{t_n}(M, h)$), we conclude that

$$\|\gamma_{T_n}^X(p) + T_n \operatorname{Im}(h)\|_{W^{-1}(M,h)} \leq C''_K T_n^{1-\alpha},$$

which completes the proof of the lemma (up to the Ostrowski-type decomposition result, for which we refer to [3], Lemma 9.4). □

Finally, the polynomial unique ergodicity Theorem 4.10 for directionally typical translation flows follows from Lemma 4.36 and from the result of J. Athreya's [44]. In fact, Athreya's results imply that the positive frequency condition in formula (4.13) holds for Lebesgue almost all directions on *every* translation surface (that is, for the horizontal translation flow of $(M, e^{\iota\theta} h)$, for *every* translation surface (M, h) and for Lebesgue almost all $\theta \in \mathbb{T}$).

4.4 (Effective) Weak Mixing: A Twisted Cohomology Approach

The cohomological approach to weak mixing of translation flows is based on a notion of twisted cohomology, which arises naturally if we want to attach cohomology classes to eigenfunctions, generalizing the flux cohomology class of an invariant function (or an invariant measure). It emphasizes a close analogy between (effective) unique ergodicity and (effective) weak mixing, viewed as the (effective) unique ergodicity of twisted flows (the products with linear flows on the circle). The relevant twisted cohomology has a well-known Hodge decomposition and a Hodge norm, and it is possible to compute once again the first variation of the Hodge norm, and derive a spectral gap result for the corresponding twisted cocycle. Bounds on twisted integrals of translation flows can then be derived from the spectral gap of the twisted cocycle, in analogy with the (untwisted) case of ergodic averages.

4.4.1 Twisted Cohomology

We associate to every eigenfunction of a translation flow a twisted cohomology class.

Let (X, Y) denote the horizontal and vertical flows of a translation surface (M, h). Let $u \in L^2(M, \text{area}_h)$ be an eigenfunction of the horizontal translation flow $\phi_{\mathbb{R}}^X$ of eigenvalue $-2\pi \iota \lambda$ (with $\lambda \in \mathbb{R}$):

$$u \circ \phi_t^X = e^{-2\pi \iota \lambda t} u, \quad \text{for all } t \in \mathbb{R}, \quad \text{or} \quad Xu + 2\pi \iota \lambda u = 0.$$

The 1-form $u \, \text{Im}(h)$ is closed for the twisted differential

$$d_{h,\lambda} = d + 2\pi \iota \lambda \, \text{Re}(h) \wedge,$$

in fact, since h is d-closed,

$$\begin{aligned} d_{h,\lambda}(u\operatorname{Im}(h)) &= du \wedge \operatorname{Im}(h) + 2\pi\imath\lambda u \operatorname{Re}(h) \wedge \operatorname{Im}(h) \\ &= (Xu \cdot \operatorname{Re}(h) + Yu \cdot \operatorname{Im}(h)) \wedge \operatorname{Im}(h) + 2\pi\imath\lambda u \cdot \operatorname{Re}(h) \wedge \operatorname{Im}(h) \\ &= (Xu + 2\pi i\lambda u)\operatorname{area}_h = 0. \end{aligned}$$

In the above calculation, all derivatives are taken in the weak L^2 sense.

Exercise 4.38

(a) Prove that in the case $\lambda = 0$, square integrable functions with zero cohomology class are distributional derivatives of continuous invariant functions, hence they vanish identically if the horizontal translation flow is minimal.
(b) Prove also that, for $\lambda \neq 0$, square integrable functions with zero twisted cohomology class are distributional derivatives of continuous invariant functions.
(c) Prove that non-trivial continuous eigenfunctions for the horizontal flow exist if and only if the translation surface is a (translation) cover of the circle $\mathbb{T} = \mathbb{R}/\mathbb{Z}$, if and only if it has a completely periodic directional foliation, with cylinders of commensurable heights.

We generalize the above definition of twisted differential to arbitrary closed 1-forms and we introduce the corresponding twisted cohomology.

Definition 4.13 For any closed 1-form η, the **twisted cohomology** $H^1_\eta(M, \mathbb{C})$ is defined as the cohomology of the **twisted differential**

$$d_\eta = d + 2\pi\imath\eta\wedge,$$

that is, the cohomology of the differential complex $(\Omega^*(M), d_\eta)$ of differential forms, which is defined as as the quotient

$$H^1_\eta(M, \mathbb{C}) := \frac{\operatorname{Ker}(d_\eta : \Omega^1(M, \mathbb{C}) \to \Omega^2(M, \mathbb{C}))}{\operatorname{Im}(d_\eta : \Omega^0(M, \mathbb{C}) \to \Omega^1(M, \mathbb{C}))}.$$

We note that the above formula gives a well-defined space since for all smooth differential form $\alpha \in \Omega^*(M, \mathbb{C})$ we have, as η is a closed 1-form,

$$d_\eta^2 \alpha = (d + 2\pi\imath\eta\wedge) \circ (d + 2\pi\imath\eta\wedge)\alpha = d^2\alpha + 2\pi\imath\eta \wedge d\alpha - 2\pi i\eta \wedge d\alpha = 0.$$

It follows that $(\Omega^*(M), d_\eta)$ is a well-defined differential complex and the twisted cohomology $H^1_\eta(M, \mathbb{C})$ is well-defined.

The twisted differential $d_{h,\lambda}$ defined above is a special case of the twisted differential d_η in Definition 4.13 when the closed 1-form $\eta = \lambda \operatorname{Re}(h)$.

4 Effective Unique Ergodicity and Weak Mixing of Translation Flows

Lemma 4.39 (See [10], Lemma 4.2) *For every closed 1-form η on M, the twisted cohomology $H_\eta^1(M, \mathbb{C})$ only depends, up to a unitary (gauge) transformation, on the cohomology class $[\eta] \in H^1(M, \mathbb{R})/H^1(M, \mathbb{Z})$.*

Proof If $[\eta_1] = [\eta_2] \in H^1(M, \mathbb{R})/H^1(M, \mathbb{Z})$, then there exists $f : M \to \mathbb{T}$ such that $\eta_1 - \eta_2 = df$, and, given any point $p \in M$ we can define

$$f(x) = \int_p^x \eta_1 - \eta_2, \quad \text{for all } x \in M.$$

The above definition is well-posed since the value of the integral does not depend modulo $\mathbb{Z}$ on the path of integration joining the points p and $x \in M$.

We then define the "gauge transformation" as

$$U_f = e^{2\pi \imath f} : H_{\eta_1}^1(M, \mathbb{C}) \to H_{\eta_2}^1(M, \mathbb{C}).$$

In fact, since $f : M \to \mathbb{T}$ the function $e^{2\pi \imath f}$ is well-defined on M with modulus equal to 1 and, for all 1-forms $\alpha \in \Omega^1(M, \mathbb{C})$ we have

$$\begin{aligned}
d_{\eta_2}(U_f \alpha) &= d(e^{2\pi \imath f} \alpha) + 2\pi \imath\, e^{2\pi \imath f} \eta_2 \wedge \alpha \\
&= e^{2\pi \imath f}(d\alpha + 2\pi \imath df \wedge \alpha + 2\pi \imath\, \eta_2 \wedge \alpha) \\
&= e^{2\pi \imath f}(d\alpha + 2\pi \imath (\eta_1 - \eta_2) \wedge \alpha + 2\pi \imath\, \eta_2 \wedge \alpha) = U_f(d_{\eta_1} \alpha).
\end{aligned}$$

Thus the transformation U_f intertwines the twisted differentials d_{η_1} and d_{η_2} and therefore maps $H_{\eta_1}^1(M, \mathbb{C})$ onto $H_{\eta_2}^1(M, \mathbb{C})$. □

It follows in particular that for $[\eta] \in H^1(M, \mathbb{Z})$ we have that $H_\eta^1(M, \mathbb{C}) \equiv H^1(M, \mathbb{C})$, hence it has dimension $2g$.

For $[\eta] \notin H^1(M, \mathbb{Z})$, it can be proved that the twisted cohomology $H_\eta^1(M, \mathbb{C})$ has dimension $2g - 2$ (prove as an Exercise or see [10], Lemma 4.3).

This dimension loss can be accounted for since $[\eta] \in H^1(M, \mathbb{Z})$ if and only if $H_\eta^0(M, \mathbb{C})$ is non-trivial, and in fact it is one-dimensional (prove as an exercise or see [10], Lemma 4.1).

4.4.2 The Twisted Cocycle

In Definition 4.10 we have introduced the real Hodge bundle $H_\kappa^1(M, \mathbb{R})$ over a stratum $\mathcal{H}_\kappa^{(1)}$ of the moduli space of translation surfaces. We introduce below the toral Hodge bundle and the twisted cohomology bundle, on which the key renormalization cocycles for weak mixing are defined.

Definition 4.14 The **toral Hodge bundle** $H^1_\kappa(M, \mathbb{T})$ over $\mathcal{H}^{(1)}_\kappa$ is the quotient

$$H^1_\kappa(M, \mathbb{T}) := H^1_\kappa(M, \mathbb{R})/H^1_\kappa(M, \mathbb{Z})$$

of the real Hodge bundle over the sub-bundle with fibers given by the integral cohomology. The **twisted cohomology bundle** $\mathcal{T}^1_\kappa(M, \mathbb{C})$ is the bundle over $H^1_\kappa(M, \mathbb{T})$ with fibers given by the twisted cohomology spaces, defined as the quotient of the bundle

$$\hat{\mathcal{T}}^1_\kappa(M, C) = \{(h, \eta, c) | (h, \eta) \in \hat{\mathcal{H}}^{(1)}_\kappa \times H^1(M, \mathbb{R}) \text{ and } c \in H^1_\eta(M, \mathbb{C})\}/\Gamma_g$$

with respect to the action of the group $H^1(M, \mathbb{Z})$ on $H^1(M, \mathbb{R})$ by translations and on $H^1_\eta(M, \mathbb{C})$ by unitary transformations.

Parallel transport yields cocycles over the Teichmüller flow $g_\mathbb{R}$ and, more generally, over the $SL(2, \mathbb{R})$ action on the moduli spaces of translation surfaces.

Definition 4.15 The **twisted cocycle** (denoted as $g^\#_\mathbb{R}$) is the lift to the twisted cohomology bundle of the toral Kontsevich–Zorich cocycle (the projection of the Kontsevich–Zorich cocycle onto the toral Hodge bundle $H^1_\kappa(M, \mathbb{T})$) by parallel transport: for all $[(h, \eta, c)] \in \mathcal{T}^1_\kappa(M, \mathbb{C})$ and for all $t \in \mathbb{R}$, we define

$$g^\#_t[(M, h, \eta, c)] = [(g_t(M, h), \eta, c)].$$

In the above formulas the symbol $[(h, \eta, c)]$ denotes the equivalence class of the triple (h, η, c) such that $(h, \eta) \in \hat{\mathcal{H}}^{(1)} \times H^1(M, \mathbb{R})$ and $c \in H^1_\eta(M, \mathbb{C})$ with respect to the action of the mapping class group Γ_g by pull-back and to the action of the group $H^1(M, \mathbb{Z})$ on the space $\{(\eta, c) | \eta \in H^1(M, \mathbb{R}) \text{ and } c \in H^1_\eta(M, \mathbb{C})\}$.

4.4.3 The Twisted Hodge Norm and its First Variation

The twisted cohomology, hence the twisted cohomology bundle, can be endowed with a *Hodge norm* and a variational formula can be computed.

Any (closed) real 1-form η on a translation surface (M, h) (in fact, on any Riemann surface) has a Hodge decomposition into holomorphic and anti-holomorphic part, that is, $\eta = \eta^{1,0} + \eta^{0,1}$, hence we have the decomposition

$$d_\eta = d^{1,0}_\eta + d^{0,1}_\eta = d^{1,0} + 2\pi\iota\eta^{1,0} + d^{0,1} + 2\pi\iota\eta^{0,1}.$$

Let $\mathcal{H}_\eta^{1,0}(M, \mathbb{C})$ and $\mathcal{H}_\eta^{0,1}(M, \mathbb{C})$ denote the space of twisted holomorphic, respectively anti-holomorphic, differentials:

$$\mathcal{H}_\eta^{1,0}(M, \mathbb{C}) := \{\alpha \in \Omega^{1,0}(M, \mathbb{C}) | d_\eta^{1,0}\alpha = 0\},$$
$$\mathcal{H}_\eta^{0,1}(M, \mathbb{C}) := \{\alpha \in \Omega^{0,1}(M, \mathbb{C}) | d_\eta^{0,1}\alpha = 0\}.$$

Since η is a real form, and h is of type $(1, 0)$ (holomorphic) there exists a smooth function $f_\eta : M \to \mathbb{C}$ such that

$$\eta^{1,0} = f_\eta h \quad \text{and} \quad \eta^{0,1} = \overline{f_\eta h},$$

hence we have twisted Cauchy–Riemann operators

$$\begin{aligned} \partial_\eta^+ &= \partial^+ + 2\pi \imath \overline{f_\eta} = (X + \imath Y) + 2\pi \imath \overline{f_\eta}, \\ \partial_\eta^- &= \partial^- + 2\pi \imath f_\eta = (X - \imath Y) + 2\pi \imath f_\eta. \end{aligned} \quad (4.17)$$

Exercise 4.40 *Given a real 1-form η, prove that every twisted-holomorphic (resp. anti-holomorphic) differential α (for the 1-form η) can be written as $\alpha = m^+ h$ (resp. $\alpha = m^- \bar{h}$) with m^+ (resp. m^-) a twisted-meromorphic (resp. anti-meromorphic) function, that is, such that $\partial_\eta^+ m^+ = 0$ (resp. $\partial_\eta^- m^- = 0$).*

We note that, for any real closed 1-form $\eta \in \Omega^1(M, \mathbb{R})$ the twisted cohomology space $H_\eta^1(M, \mathbb{C})$ has no real subspace, hence for any real closed 1-form η we define the real twisted cohomology

$$H_\eta^1(M, \mathbb{R}) := \operatorname{Re}(H_\eta^1(M, \mathbb{R}) \oplus H_{-\eta}^1(M, \mathbb{R})).$$

It can be proved that any class $c \in H_\eta^1(M, \mathbb{R})$ has a twisted holomorphic representative, that is, there exists a 1-form $\alpha_\eta + \alpha_{-\eta} \in \mathcal{H}_\eta^{1,0} \oplus \mathcal{H}_{-\eta}^{1,0}$ such that

$$c = \operatorname{Re}([\alpha_\eta] + [\alpha_{-\eta}]) \in H_\eta^1(M, \mathbb{C}) \oplus H_{-\eta}^1(M, \mathbb{C}).$$

Definition 4.16 Let η be a real closed 1-form. The **twisted Hodge norm** of any twisted cohomology class $c \in H_\eta^1(M, \mathbb{C})$ is defined in terms of a d_η-closed representative α as follows:

$$\|c\|_{H_\eta^1(M,\mathbb{C})} := \left(\frac{\imath}{2} \int_M \alpha \wedge \bar{\alpha}\right)^{1/2}.$$

The **twisted Hodge norm** of a real twisted cohomology class $c \in H^1_\eta(M, \mathbb{R})$ is defined in terms of a twisted holomorphic representative $\alpha_\eta + \alpha_{-\eta}$ as

$$\|c\|_{H^1_\eta(M,\mathbb{R})} := \left(\|\alpha_\eta\|^2_{H^1_\eta(M,\mathbb{C})} + \|\alpha_{-\eta}\|^2_{H^1_{-\eta}(M,\mathbb{C})} \right)^{1/2}.$$

Notation For the Hodge norm on the real twisted cohomology bundle we adopt the notation

$$\|(M, h, \eta, c)\| = \|c\|_{H^1_\eta(M,\mathbb{R})}, \quad \text{for all } (h, \eta, c) \in \mathcal{T}^1_\kappa(M, \mathbb{C}).$$

The following variational formula, analogous to the first variational formula in Lemma 4.28, holds. The argument follows the original one in the proof of Lemma 5.2 of [10] with additional details.

Lemma 4.41 *For all* $(M, h, \eta) \in H^1_\kappa(M, \mathbb{R})$ *and* $c \in H^1_\eta(M, \mathbb{R})$, *we have*

$$\frac{d}{dt} \|g_t^\#(h, \eta, c)\|^2 \Big|_{t=0} = 4 \operatorname{Re} \left(\int_M \left(\frac{\alpha_\eta}{h} \right) \left(\frac{\alpha_{-\eta}}{h} \right) \operatorname{area}_h \right) := 2 \operatorname{Re} B_{(M,h,\eta)}(c).$$

Proof For $t \in \mathbb{R}$, let us denote $g_t^\#(M, h) = (M_t, h_t)$ and let

$$c = \operatorname{Re}[m_{\eta,t} h_t + m_{-\eta,t} h_t], \quad \text{with } m_{\pm\eta,t} h_t \in \mathcal{H}^{1,0}_{\pm\eta}(M_t, \mathbb{C}). \tag{4.18}$$

By the definition of the twisted cocycle and of the Hodge norm, we have

$$\|g_t^\#(M, h, \eta, c)\|^2 = \int_M (|m_{\eta,t}|^2 + |m_{-\eta,t}|^2) \operatorname{area}_h,$$

hence

$$\frac{d}{dt} \|g_t^\#(M, h, \eta, c)\|^2 = 2 \operatorname{Re} \left(\int_M \frac{dm_{\eta,t}}{dt} \cdot \overline{m_{\eta,t}} + \frac{dm_{-\eta,t}}{dt} \cdot \overline{m_{-\eta,t}} \right) \operatorname{area}_h \right). \tag{4.19}$$

By differentiating in formula (4.18), since the cohomology class $c \in H^1_\eta(M, \mathbb{C}) \oplus H^1_{-\eta}(M, \mathbb{C})$ does not depend on $t \in \mathbb{R}$, since $dh_t/dt = -\bar{h}_t$ by the definition of the Teichmüller flow (see formula (4.11)) and by taking into account that $d_{-\eta} = \bar{d}_\eta$, there exists a unique smooth function $w_{\eta,t}$ of zero average such that

$$\operatorname{Re} \left(\frac{dm_{\eta,t}}{dt} h_t - m_{\eta,t} \bar{h}_t + \frac{dm_{-\eta,t}}{dt} h_t - m_{-\eta,t} \bar{h}_t - d_\eta w_{\eta,t} \right) \equiv 0. \tag{4.20}$$

Now we observe that the 1-forms $\frac{dm_{\pm\eta,t}}{dt} h_t$ are $d^{1,0}_{\pm\eta,t}$-closed, and the 1-forms $m_{\pm\eta,t}\bar{h}_t$ are $d^{0,1}_{\pm\eta,t}$-closed, hence there exist $\alpha^{1,0}_{\pm\eta,t} \in \mathcal{H}^{1,0}_\eta(M_t, \mathbb{C})$ and $\alpha^{0,1}_{\pm\eta,t} \in \mathcal{H}^{0,1}_\eta(M_t, \mathbb{C})$, and smooth functions $u_{\pm\eta,t}, v_{\pm\eta,t}$ such that

$$\frac{dm_{\pm\eta,t}}{dt} h_t = \alpha^{1,0}_{\pm\eta,t} + d^{1,0}_{\pm\eta,t} u_{\pm\eta,t} \quad \text{and} \quad m_{\pm\eta,t}\bar{h}_t = \alpha^{0,1}_{\pm\eta,t} + d^{0,1}_{\pm\eta,t} v_{\pm\eta,t}.$$

The functions $u_{\pm\eta,t}, v_{\pm\eta,t}$ can be chosen so that

$$u_{\pm\eta,t} \in \ker(d^{1,0}_{\pm\eta,t})^\perp = \ker(\partial^-_{\pm\eta,t})^\perp \quad \text{and} \quad v_{\pm\eta,t} \in \ker(d^{0,1}_{\pm\eta,t})^\perp = \ker(\partial^+_{\pm\eta,t})^\perp.$$

In addition, since $m_{\pm\eta,t} h_t$ are $d_{\pm\eta}$-closed, and the twisted differential does not depend on $t \in \mathbb{R}$, it follows that their derivatives are $d_{\pm\eta}$-closed, so that

$$d_{\pm\eta}\left(\frac{dm_{\pm\eta,t}}{dt} h_t - m_{\pm\eta,t}\bar{h}_t\right) = d_{\pm\eta}(\alpha^{1,0}_{\pm\eta,t} - \alpha^{0,1}_{\pm\eta,t} + d^{1,0}_{\pm\eta,t} u_{\pm\eta,t} - d^{0,1}_{\pm\eta,t} v_{\pm\eta,t})$$
$$= d_{\pm\eta}(d^{1,0}_{\pm\eta,t} u_{\pm\eta,t} - d^{0,1}_{\pm\eta,t} v_{\pm\eta,t}) = 0,$$

from which we claim that we can derive that $u_{\pm\eta,t} + v_{\pm\eta,t} \equiv 0$, hence

$$d^{1,0}_{\pm\eta,t} u_{\pm\eta,t} - d^{0,1}_{\pm\eta,t} v_{\pm\eta,t} = d_{\pm\eta} u_{\pm\eta,t}.$$

In fact, we prove below our claim that for all (M, h), for all real harmonic form η on M and for functions $w \in H^1(M)$ such that $w \in \left((\ker(\partial^+_{\pm\eta}) \cap \ker(\partial^-_{\pm\eta}))\right)^\perp$, we have

$$d^{1,0}_{\pm\eta} d^{0,1}_{\pm\eta} w = 0 \Longrightarrow w = 0,$$

The identity $d^{1,0}_{\pm\eta} d^{0,1}_{\pm\eta} w = 0$ can be written in terms of the twisted Cauchy-Riemann operators (since the function f_η in formula (4.17) is meromorphic) as

$$\partial^+_{\pm\eta} \partial^-_{\pm\eta} w = \partial^+_{\pm\eta} \partial^-_\eta w = \partial^+ \partial^- w + 2\pi \iota(f_{\pm\eta} \partial^+ + \overline{f_{\pm\eta}} \partial^-) w - 4\pi |f_{\pm\eta}|^2 w = 0.$$

Since the adjoint $(\partial^\pm_{\pm\eta})^* = -\partial^\mp_{\pm\eta}$ as densely defined closed operators on $L^2(M, \text{area}_h)$ with domain the Sobolev space $H^1(M)$, it follows that

$$\|\partial^\pm_{\pm\eta} w\|^2 = -\langle \partial^\mp_{\pm\eta} \partial^\pm_{\pm\eta} w, w \rangle = 0,$$

hence $w \in \ker(\partial^+_{\pm\eta}) \cap \ker(\partial^-_{\pm\eta})$ and, by the assumption that it also belongs to its orthogonal, we conclude that $w = 0$, as claimed.

From formula (4.20) we then conclude that $\operatorname{Re}(d_{\pm\eta}(u_{\eta,t}+u_{-\eta,t})-d_{\pm\eta}w_{\pm\eta,t}) \equiv 0$ and

$$\operatorname{Re}\left(\alpha^{1,0}_{\eta,t} + \alpha^{1,0}_{-\eta,t} - \alpha^{0,1}_{\eta,t} - \alpha^{0,1}_{-\eta,t}\right) \equiv 0$$

and from the above equation we derive

$$\alpha^{1,0}_{\pm\eta,t} = \overline{\alpha^{0,1}_{\mp\eta,t}}.$$

By taking into account that the cup $d^{1,0}_{\pm\eta,t}C^\infty(M)$ is Hodge orthogonal to $\mathcal{H}^{1,0}_{\pm\eta}(M_t,\mathbb{C})$ and $d^{0,1}_{\pm\eta,t}C^\infty(M)$ is Hodge orthogonal to $\mathcal{H}^{0,1}_{\pm\eta}(M_t,\mathbb{C})$ follows that

$$\operatorname{Re}\int_M \frac{dm_{\pm\eta,t}}{dt}\overline{m_{\pm\eta,t}} \cdot \operatorname{area}_h = -\frac{1}{2}\operatorname{Im}\int_M (\alpha^{1,0}_{\pm\eta,t} + d^{1,0}_{\eta,t}u_{\pm\eta}) \wedge \overline{m_{\pm\eta,t}h_t}$$

$$= -\frac{1}{2}\operatorname{Im}\int_M (\alpha^{1,0}_{\pm\eta,t}) \wedge \overline{m_{\pm\eta,t}h_t} = \frac{1}{2}\operatorname{Im}\int_M (\alpha^{0,1}_{\mp\eta,t}) \wedge m_{\pm\eta,t}h_t$$

$$= \frac{1}{2}\operatorname{Im}\int_M (m_{\mp\eta,t}\bar{h}_t - d^{0,1}_{\eta,t}v_{\pm\eta}) \wedge m_{\pm\eta,t}h_t = \operatorname{Re}\int_M (m_{\mp\eta,t}m_{\pm\eta,t})\operatorname{area}_h.$$

By formula (4.19) we conclude that

$$\frac{d}{dt}\|g^\#_t(M,h,\eta,c)\|^2 = 4\operatorname{Re}\int_M (m_{\eta,t}m_{-\eta,t})\operatorname{area}_h,$$

as stated, hence the argument is completed. □

We now derive from the above variation formula a "spectral gap" result for the twisted cocycle. Strictly speaking, since the twisted cocycle does not have "tautological" exponents (the ± 1 exponent in the untwisted case) we will prove below an upper bound on the top exponent of the twisted cocycle.

For all $(M,h) \in \mathcal{H}^{(1)}_\kappa$, all $\eta \in H^1(M,\mathbb{R})$, we let

$$\Lambda^\#(M,h,\eta) := \max_{c \in H^1_\eta(M,\mathbb{R})\setminus\{0\}} \frac{|B_{(M,h,\eta)}(c)|}{\|(M,h,\eta,c)\|^2}. \tag{4.21}$$

From Lemma 4.41 we immediately derive

Lemma 4.42 *For all $(M,h,\eta) \in H^1_\kappa(M,\mathbb{T})$, for all $c \in H^1_\eta(M,\mathbb{R})$ and for all $t \in \mathbb{R}$, we have*

$$\|g^\#_t(M,h,\eta,c)\| \leq \|(M,h,\eta,c)\|\exp\left(\int_0^t \Lambda^\#(g^{KZ}_s(M,h,\eta))ds\right).$$

We prove below a key statement about the function $\Lambda^{\#}$ analogous to Lemma 4.29 in Sect. 4.3.4. The main difference is that while for the function of Lemma 4.29 we have $\Lambda < 1$ on compact sets, the identity $\Lambda^{\#} = 1$ is verified on compact set on integral cohomology classes.

Lemma 4.43 ([10], Lemma 5.4) *The function $\Lambda^{\#}$ is well-defined on the toral Hodge bundle $H^1_\kappa(M, \mathbb{T})$, it satisfies the inequality $0 \leq \Lambda^{\#} \leq 1$ everywhere and*

$$\Lambda^{\#}(M, h, \eta) = 1 \Longrightarrow [\eta] \in H^1(M, \mathbb{Z}).$$

Proof The bound follows from the definition and from the Cauchy-Schwarz inequality. The condition $\Lambda^{\#}(M, h, \eta) = 1$ is equivalent to equality in the Cauchy-Schwarz inequality. It implies that there exist functions $m_{\pm\eta} \in \ker(\partial^+_{\pm\eta})$ such that $m_\eta \in \mathbb{C}\overline{m_{-\eta}}$, hence

$$m_\eta \in \ker(\partial^+_\eta) \cap \ker(\partial^-_\eta).$$

It follows that

$$dm_\eta + 2\pi \imath m_\eta \eta = d_\eta(m_\eta) = (\partial^-_\eta m_\eta)h + (\partial^+_\eta m_\eta)\bar{h} = 0,$$

hence (see Exercise 4.44) we have $|m_\eta| \in \mathbb{C}$ and, since $m_\eta \neq 0$ (as $c \neq 0$),

$$\eta = \frac{1}{2\pi \imath} \frac{dm_\eta}{m_\eta} \in H^1(M, \mathbb{Z}). \tag{4.22}$$

We have thus proved that $\Lambda^{\#}(M, h, \eta) = 1$ implies $[\eta] \in H^1(M, \mathbb{Z})$ as stated. □

Exercise 4.44 *Prove that the identity $du + 2\pi \imath u \cdot \eta = 0$ for a function $u \in L^2(M, \text{area}_h)$ implies that $u \in C^\infty(M)$, as well as $|u|^2 \in \mathbb{C}$, and then that, if $u \neq 0$,*

$$\eta = \frac{1}{2\pi \imath} \frac{du}{u} \Longrightarrow \eta \in H^1(M, \mathbb{Z}).$$

4.4.4 The Veech Criterion Revisited

We prove below a version of Veech's criterion [11] for weak mixing of translation flows and state a corresponding effective version. The argument is based on the first variation Lemma 4.41 for the Hodge norm on the twisted cohomology bundle and has not appeared before. It is analogous to the cohomological Hodge theory proof of Masur's criterion (see Theorem 4.35).

Let $g_{\mathbb{R}}^{KZ} : H_\kappa^1(M, \mathbb{T}) \to H_\kappa^1(M, \mathbb{T})$ denote the toral Kontsevich–Zorich cocycle over the stratum $\mathcal{H}_\kappa^{(1)}$ of the moduli space of translation surfaces.

Theorem 4.45 *Let $\lambda \in \mathbb{R} \setminus \{0\}$ be such that $-2\pi\imath\lambda$ is an eigenvalue for the horizontal translation flow $\phi_{\mathbb{R}}^X$ of a translation surface $(M, h) \in \mathcal{H}_\kappa^{(1)}$, with no continuous eigenfunction. Then we have the inclusion*

$$\lim_{t \to +\infty} g_t^{KZ}(M, h, [\lambda \operatorname{Re}(h)]) \subset [H_\kappa^1(M, \mathbb{Z})] \subset H_\kappa^1(M, \mathbb{T}),$$

that is, every limit of the orbit $g_{\mathbb{R}}^{KZ}(M, h, [\lambda \operatorname{Re}(h)])$ belongs to the set of integer points of the real Hodge bundle over the moduli space of translation surfaces.

The first step in the argument is given by the following lemma. For any $\lambda \in \mathbb{C}$, let $d_{h,\lambda}$ denote the twisted differential for the closed 1-form $\eta := \lambda \operatorname{Re}(h)$, that is,

$$d_{h,\lambda} := d + 2\pi\imath\lambda \operatorname{Re}(h) \wedge$$

and let $H_{h,\lambda}^1(M, \mathbb{C})$ denote the corresponding twisted cohomology.

Lemma 4.46 *Let $u \in L^2(M, \operatorname{area}_h)$ be an eigenfunction for the horizontal flow with eigenvalue $-2\pi\imath\lambda$. Then $u \operatorname{Im}(h)$ is a $d_{h,\lambda}$-closed 1-form. In addition, if the twisted class $[u \operatorname{Im}(h)]_{h,\lambda} = 0$, then $u = YU$ is the vertical derivative of a continuous eigenfunction $U \in H^1(M)$ of eigenvalue $-2\pi\imath\lambda$.*

Proof Let $\{X, Y\}$ denote the generators of the horizontal and vertical flows, so that, for all $f \in C^\infty(M)$,

$$df = Xf \operatorname{Re}(h) + Yf \operatorname{Im}(h).$$

By assumption we have (in the weak sense)

$$(X + 2\pi\imath\lambda)u = 0$$

As we have mentioned at the beginning of Sect. 4.4.1, by a calculation (in the weak sense) we derive

$$d_{h,\lambda}(u \operatorname{Im}(h)) = d(u \operatorname{Im}(h)) + 2\pi\imath\lambda u \operatorname{Re}(h) \wedge \operatorname{Im}(h) = (Xu + 2\pi\imath\lambda u)\operatorname{area}_h = 0.$$

Let us now assume that the twisted cohomology class $[u \operatorname{Im}(h)]_{h,\lambda} = 0 \in H_{h,\lambda}^1(M, \mathbb{C})$. There exists a function $U \in L^2(M, \operatorname{area}_h)$ such that

$$dU + 2\pi\imath\lambda U \operatorname{Re}(h) = u \operatorname{Im}(h).$$

Since the exterior derivative is an elliptic operator, it follows that $U \in H^1(M)$ and by contraction we have

$$XU + 2\pi \imath \lambda U = 0 \quad \text{and} \quad YU = u.$$

It follows that U is also smooth in the horizontal direction, hence it is continuous (in fact, it is Hölder) by the Sobolev trace theorem along vertical trajectories. □

Exercise 4.47 *Prove that every function $U \in H^1(M)$ such that $(X + 2\pi \imath \lambda)U = 0$ is of Hölder class $1/2$ on the complement of the set $\Sigma_h := \{h = 0\}$.*

Let $\partial_{h,\lambda}^{\pm}$ denote the twisted Cauchy-Riemann operators:

$$\partial_{h,\lambda}^{\pm} = \partial_h^{\pm} + 2\pi \imath \lambda = (X \pm \imath Y) + 2\pi \imath \lambda .$$

By Forni [10], Prop. 3.2, the maximal closed extension of the operator $\partial_{h,\lambda}^{\pm}$ is the operator $-(\partial_{h,\lambda}^{\mp})^*$ and there exist orthogonal decompositions

$$L^2(M, \text{area}_h) = \text{Ran}(\partial_{h,\lambda}^{\pm}) \oplus \text{Ker}\,((\partial_{h,\lambda}^{\pm})^*) = \text{Ran}(\partial_{h,\lambda}^{\pm}) \oplus \text{Ker}\,(\partial_{h,\lambda}^{\mp}).$$

For every $u \in L^2(M, \text{area}_h)$ there exist functions $v^{\pm} \in \text{Ker}\,(\partial_{h,\lambda}^{\mp})^{\perp} \cap H^1(M)$ and functions $m^{\pm} \in \text{Ker}\,((\partial_{h,\lambda}^{\pm})^*) = \text{Ker}\,(\partial_{h,\lambda}^{\mp}) \subset L^2(M, \text{area}_h)$ such that we have a decomposition

$$u = \partial_{h,\lambda}^{+} v^{+} + m^{-} = \partial_{h,\lambda}^{-} v^{-} + m^{+} . \tag{4.23}$$

Lemma 4.48 *Let $u \in L^2(M, \text{area}_h)$ be an eigenfunction for the horizontal flow with eigenvalue $-2\pi \imath \lambda$. Let us assume that $\lambda \, \text{Re}(h) \notin H^1(M, \mathbb{Z})$. Then we have the following identity between twisted cohomology classes:*

$$[u \, \text{Im}(h)]_{h,\lambda} = \frac{1}{2\imath}[m^{+}h - m^{-}\bar{h}]_{h,\lambda} \in H^1_{h,\lambda}(M, \mathbb{C}) .$$

Proof Since $m^{\pm} \in \text{Ker}\,(\partial_{h,\lambda}^{\mp})$, it follows that the 1-forms $m^{+}h$ and $m^{-}\bar{h}$ are twisted closed. In fact,

$$d(m^{+}h) + (2\pi \imath \lambda m^{+}) \, \text{Re}(h) \wedge h = \frac{1}{2}(\partial_h^{+}m^{+} + 2\pi \imath \lambda m^{+})\bar{h} \wedge h = 0,$$

$$d(m^{-}\bar{h}) + (2\pi \imath \lambda m^{-}) \, \text{Re}(h) \wedge \bar{h} = \frac{1}{2}(\partial_h^{-}m^{-} + 2\pi \imath \lambda m^{-})h \wedge \bar{h} = 0$$

We have the identity

$$u \operatorname{Im}(h) = u \frac{h - \bar{h}}{2\imath} = \frac{(\partial_{h,\lambda}^- v^-)h - (\partial_{h,\lambda}^+ v^+)\bar{h}}{2\imath} + \frac{m^+ h - m^- \bar{h}}{2\imath}.$$

It follows that the 1-form $(\partial_{h,\lambda}^+ v^+)h - (\partial_{h,\lambda}^- v^-)\bar{h}$ is $d_{h,\lambda}$-closed, which implies that the 1-forms $\partial_{h,\lambda}^+(v^+ + v^-)h$ is $d_{h,\lambda}$-closed, hence $\partial_{h,\lambda}^- \partial_{h,\lambda}^+ (v^+ + v^-) = 0$, or

$$\partial_{h,\lambda}^-(v^+ + v^-) = \partial_{h,\lambda}^+(v^+ + v^-) = 0.$$

Thus the function $v^+ + v^- \in \operatorname{Ker}(\partial_{h,\lambda}^+) \cap \operatorname{Ker}(\partial_{h,\lambda}^+)$ and by assumption

$$v^+ + v^- \in \left(\operatorname{Ker}(\partial_{h,\lambda}^+) \cap \operatorname{Ker}(\partial_{h,\lambda}^+) \right)^{\perp}$$

hence $v^+ + v^- = 0$, and we have

$$u \operatorname{Im}(h) = \frac{1}{2\imath} d_{h,\lambda} v^+ + \frac{m^+ h - m^- \bar{h}}{2\imath},$$

as stated. The argument is therefore complete. □

We then consider the orbit of the twisted cohomology class $[u \operatorname{Im}(h)]_{h,\lambda}$ under the twisted cocycle and prove that it decays exponentially (with respect to the Hodge norm) in forward time.

Lemma 4.49 *Let $u \in L^2(M, \operatorname{area}_h)$ be an eigenfunction for the horizontal flow with eigenvalue $-2\pi\imath\lambda$. For all $t \in \mathbb{R}$ we have*

$$\|g_t^{\#}(M, h, [\lambda \operatorname{Re}(h)], [u \operatorname{Im}(h)])\| \leq \|u\|_{L^2(M, \operatorname{area}_h)} e^{-t}.$$

Proof Let $(M_t, h_t) = g_t(M, h)$. We recall that, by definition of the Teichmüller flow, we have, for all $t \in \mathbb{R}$,

$$h_t = e^{-t} \operatorname{Re}(h) + \imath e^t \operatorname{Im}(h),$$

and by the definition of the cocycle we also have

$$g_t^{\#}(M, h, [\lambda \operatorname{Re}(h)], [u \operatorname{Im}(h)]) = (M_t, h_t, [e^t \lambda \operatorname{Re}(h_t)], [e^{-t} u \operatorname{Im}(h_t)])$$
$$= (M_t, h_t, [\lambda_t \operatorname{Re}(h_t)], [u_t \operatorname{Im}(h_t)]).$$

Let us now write the decomposition of the function u in formula (4.23) for $(M_t, h_t, \lambda_t \operatorname{Re}(h_t))$: there exist functions $m_t^\pm \in \ker(\partial_{h_t,\lambda_t}^\pm)$ and $v_t^\pm \in H^1(M)$ such that

$$u_t = \partial_{h_t,\lambda_t}^+ v_t^+ + m_t^- = \partial_{h_t,\lambda_t}^- v_t^- + m_t^+.$$

By Lemma 4.48 we have that, for all $t \in \mathbb{R}$,

$$[u_t \operatorname{Im}(h_t)]_{h_t,\lambda_t} = \frac{1}{2t}[m_t^+ h_t - m_t^- \bar{h}_t]_{h_t,\lambda_t},$$

hence by the definition of the Hodge norm and by orthogonality

$$\|g_t^\#(M, h, [\lambda \operatorname{Re}(h)], [u \operatorname{Im}(h)])\| = \frac{1}{2}(\|m_t^+\|_{L^2(M,\operatorname{area}_h)}^2 + \|m_t^-\|_{L^2(M,\operatorname{area}_h)}^2)^{1/2}$$

$$\leq \|u_t\|_{L^2(M,\operatorname{area}_h)} = e^{-t}\|u\|_{L^2(M,\operatorname{area}_h)}.$$

□

Let $\Lambda^\# : H_\kappa^1(M, \mathbb{T}) \to [0, 1]$ denote the function defined in formula (4.21).

Lemma 4.50 *Let $\lambda \in \mathbb{R}\setminus\{0\}$ be such that $-2\pi\iota\lambda$ is an eigenvalue for the horizontal flow of the translation surface $(M, h) \in \mathcal{H}(\kappa)$ which has no continuous eigenfunction. There exists a constant $C_{h,\lambda} > 0$ such that, for all $t > 0$, we have*

$$\int_0^t \Lambda^\#(g_s^{KZ}(M, h, [\lambda \operatorname{Re}(h)])ds \geq t - C_{h,\lambda}, \quad \text{for all } t \in \mathbb{R}^+.$$

Proof Let $u \in L^2(M, \operatorname{area}_h)$ denote an eigenfunction of the horizontal flow with eigenvalue $-2\pi\iota\lambda$. By Lemma 4.42, we have

$$\|(M, h, [\lambda \operatorname{Re}(h)], [u \operatorname{Im}(h)])\| \leq \|g_t^{KZ}(h, [\lambda \operatorname{Re}(h)], [u \operatorname{Im}(h)])\|$$

$$\times \exp\left(\int_0^t \Lambda^\#(g_{-s}^{KZ} g_t^{KZ}(h, [\lambda \operatorname{Re}(h)], [u \operatorname{Im}(h)]))ds\right),$$

hence by Lemma 4.49 we have

$$\|(M, h, [\lambda \operatorname{Re}(h)], [u \operatorname{Im}(h)])\| \leq \|u\|_{L^2(M,\operatorname{area}_h)}$$

$$\times \exp\left(-t + \int_0^t \Lambda^\#(g_t(h, [\lambda \operatorname{Re}(h)], [u \operatorname{Im}(h)]))ds\right)$$

Since by assumption $-2\pi\iota\lambda$ has no continuous eigenfunction, the twisted cohomology class $[u \operatorname{Im}(h)]_{h,\lambda} \neq 0$. The statement follows after taking logarithms in the above inequality. □

Proof of Theorem 4.45 We argue by contraposition. Suppose there exists a sequence $(t_n) \subset \mathbb{R}^+$ of positive times such that

$$g_{t_n}^{KZ}(M, h, [\lambda \operatorname{Re}(h)]) \to (M_0, h_0, [\eta_0]) \notin H^1(M, \mathbb{Z}).$$

Since by Lemma 4.43 the strict inequality $\Lambda^\# < 1$ holds on $H_\kappa(M, \mathbb{R}) \setminus H_\kappa(M, \mathbb{Z})$, and since by Forni [10], Lemma 5.3, the function $\Lambda^\#$ is continuous, there exist $\Lambda_0^\# < 1$ and a neighborhood $\mathcal{U}_0$ of $(M_0, h_0, [\eta_0])$ such that $\Lambda^\# | \mathcal{U}_0 \leq \Lambda_0^\#$.

Since the forward orbit $g_{\mathbb{R}^+}(M, h, [\lambda \operatorname{Re}(h)])$ spends an infinite time in $\mathcal{U}_0$, it follows that for any constant $C > 0$ there exists $t_C > 0$ such that for all $t \geq t_C$ we have

$$\int_0^t \Lambda^\#(g_t^{KZ}(h, [\lambda \operatorname{Re}(h)], [u \operatorname{Im}(h)])) ds \leq t - C,$$

thereby contradicting Lemma 4.50. The criterion is therefore proved. □

4.4.5 Weak Mixing of Typical Translation Flows

The proof of weak mixing for typical translation flows (and interval exchange transformations) is based on the Veech criterion for weak mixing [11], established in Sect. 4.4.4. In this section we outline the main steps of the proof for translation flows following [20]. The case of interval exchange transformations is harder and the proof required a different strategy.

The Veech criterion suggests the introduction of the notion of a subspace of the real Hodge bundle which contains all translation surfaces with non-weakly mixing horizontal flow:

Definition 4.17 The **weak stable space** $W^s_{(M,h)} \subset H^1(M, \mathbb{R})$ at a translation surface $(M, h) \subset \mathcal{H}_\kappa^{(1)}$ is the set defined as follows. Let $\mathcal{K}$ be a (countable) exhaustion of $\mathcal{H}_\kappa^{(1)}$ by compact subset. We then define

$$W^s_{(M,h)} := \cap_{K \in \mathcal{K}} \{c \in H^1(M, \mathbb{R}) | g_t^{KZ}(M, h, c) \to H_\kappa^1(M, \mathbb{Z}) \text{ as } g_t(M, h) \in K\}.$$

In other terms, after projection onto the toral Hodge bundle $H_\kappa^1(M, \mathbb{T})$, the weak stable space consists of all cohomology classes which converge to the zero section of $H_\kappa^1(M, \mathbb{T})$ under the Kontsevich–Zorich cocycle $g_{\mathbb{R}}^{KZ}$ as the orbit of the Teichmüller flow returns to a compact set.

We list below the essential properties of the weak stable space.

4 Effective Unique Ergodicity and Weak Mixing of Translation Flows

For almost all (M, h) with respect to any $g_\mathbb{R}$-invariant probability ergodic measure on $\mathcal{H}_\kappa^{(1)}$ we have:

- $\mathcal{W}_{(M,h)}^s$ depends only on the forward $g_\mathbb{R}$ trajectory of (M, h), hence it only depends on $[\text{Im}(h)] \in H^1(M, \Sigma_h, \mathbb{R})$, which determines a stable manifold of the Teichmüller flow $g_\mathbb{R}$;
- $\mathcal{W}_{(M,h)}^s$ is saturated by translates of the stable Oseledets subspaces of the Kontsevich–Zorich cocycle and equals the union of all integer translates in case the forward $g_\mathbb{R}$-orbit of (M, h) is relatively compact in $\mathcal{H}_\kappa^{(1)}$.

We note that in general we do not expect the weak stable space to be contained in the union of integer translates of the Oseledets central stable space since in the condition which defines the weak stable space we make no assumption on the behavior of the trajectory of the cohomology class during excursions of the forward Teichmüller trajectory $g_{\mathbb{R}^+}(M, h)$ outside a compact subset of the moduli space. It is therefore possible that the cohomology class converges to the set $H_\kappa^1(M, \mathbb{Z})$ without converging to any given element of it (that is, by jumping during excursions from one integer point to another).

The key lemma on the weak stable space (which we state below without proof) gives a bound on its dimension.

Lemma 4.51 (See [20] Th. A.1) *For μ-almost all (Oseledets regular) translation surface $(M, h) \in \mathcal{H}_\kappa^{(1)}$, with respect to any $g_\mathbb{R}$ ergodic probability measure, the Hausdorff codimension H-codim $(\mathcal{W}_{(M,h)})$ of the weak stable space is greater or equal than the dimension of the unstable Oseledets space $E_{(M,h)}^+ \subset H^1(M, \mathbb{R})$, that is, it satisfies the bound*

$$\text{H-codim }(\mathcal{W}_{(M,h)}^s) \geq \dim(E_{(M,h)}^+) = \sum_{\lambda_i^\mu > 0} \dim(E_{(M,h)}(\lambda_i^\mu)).$$

From the Veech criterion and from above dimension bound, we derive the following

Theorem 4.52 (Linear Elimination) *Let μ be any $g_\mathbb{R}$-invariant ergodic measure with a product structure with respect to the invariant foliations of the Teichmüller flow and with unstable dimension d_μ^u (in the absolute cohomology, that is, after the projection $H^1(M, \Sigma, \mathbb{C}) \to H^1(M, \mathbb{C})$ in period coordinates) satisfying the bound*

$$d_\mu^u > 1 + \text{codim}(E_\mu^+) = 1 + 2g - \sum_{\lambda_i^\mu > 0} \dim(E_{(M,h)}(\lambda_i^\mu)),$$

then, for μ-almost all $(M, h) \in \mathcal{H}_\kappa^{(1)}$, the horizontal translation flow $\phi_\mathbb{R}^X$ is weakly mixing. In particular, the horizontal translation flow $\phi_\mathbb{R}^X$ is weakly mixing for the Masur–Veech typical translation surface in every stratum of the moduli space of translation surfaces of genus $g \geq 2$.

Proof By the Veech criterion [11] (see also Theorem 4.45) the horizontal flow is weakly mixing under the condition that

$$\mathbb{R}[\mathrm{Re}(h)] \not\subset \mathcal{W}^s_{(M,h)} \iff [\mathrm{Re}(h)] \notin \mathbb{R} \cdot \mathcal{W}^s_{(M,h)}.$$

For all (M, h) which is an Oseledets regular point for the Kontsevich–Zorich cocycle $g^{KZ}_\mathbb{R}$, the set $\mathbb{R} \cdot \mathcal{W}^s_{(M,h)}$ depends only on $[\mathrm{Im}(h)]$ (which in turn determines the stable manifold of the Teichmüller flow). In addition we have the dimension bound

$$\text{H-dim}\,(\mathbb{R} \cdot \mathcal{W}^s_{(M,h)}) \le 1 + \text{H-dim}\,(\mathcal{W}^s_{(M,h)}) \le 1 + \text{codim}(E^+_\mu) < d^u_\mu.$$

It follows that, since μ has a product structure $\mu = \mu_s \otimes \mu_u$, for μ_s-almost all $[\mathrm{Im}(h)] \in H^1(M, \Sigma, \mathbb{R})$ the translation surface (M, h) is Oseledets regular for $g^{KZ}_\mathbb{R}$, and by the above inequality it follows that, for μ_u-almost all $[\mathrm{Re}(h)] \in H^1(M, \Sigma, \mathbb{R})$, we have that $\mathbb{R}[\mathrm{Re}(h)] \cap \mathcal{W}^s_{(M,h)} = \{0\}$, hence the horizontal flow is weakly mixing.

In the particular case of the Masur–Veech measure $\mu = \mu^{(1)}_\kappa$, it was proved in [3] (and later in [59]) that $\dim(E^+_\mu) = g$ (the genus). In addition, the Masur–Veech measure has a product structure with dimension $d^u_\mu = 2g$. Since for $g \ge 2$ we have

$$2g > 1 + 2g - g = g + 1,$$

it follows that the Masur–Veech typical translation flow is weakly mixing. The argument is complete. □

Remark 4.4 An analogous argument applies to canonical invariant measures supported on $\text{SL}(2, \mathbb{R})$-invariant orbifolds $\mathcal{M}$ of rank r at least 2. In fact, every canonical invariant measure μ has a product structure and the stable and unstable dimensions $d^u_\mu = d^s_\mu = 2r$. Since $\mathcal{M}$ is $g_\mathbb{R}$-invariant the analysis can be restricted to the invariant sub-bundle $H^1_\mathcal{M}(M, \mathbb{R}) \subset H^1_\kappa(M, \mathbb{R})$ given by the real part of the projection of the tangent bundle $T\mathcal{M} \subset H^1_\kappa(M, \Sigma, \mathbb{C})$ (the complex relative cohomology bundle) onto $H^1_\kappa(M, \mathbb{C})$ (the complex absolute cohomology bundle). It is known that the Kontsevich–Zorich exponents are all non-zero on $H^1_\mathcal{M}(M, \mathbb{R})$ (see [58] and [7], Cor. 1.3), which implies that $\dim(E^+_\mu) = r$. A condition analogous to that of Theorem 4.52 is therefore verified if and only if

$$2r = d^u_\mu > 1 + 2r - r = r + 1 \iff r \ge 2.$$

Remark 4.5 The case of $\text{SL}(2, \mathbb{R})$-invariant orbifolds of rank one is in general open. A solution of this problem has been announced in 2024 by F. Arana Herrera, J. Chaika and the author with the important exception of closed $\text{SL}(2, \mathbb{R})$-orbits (Veech translation surfaces). Avila and Delecroix [50] have investigated in depths this case and proved in particular directionally typical weak mixing on all non-arithmetic Veech surfaces. These include surfaces coming by the unfolding of billiards in regular polygons with at least 5 edges.

We conclude this section with a few words on the proof of weak mixing for Interval Exchange Transformations (IET's). The problem is essentially equivalent to establishing weak mixing for translation flows (M, h) with a *fixed* class $\text{Re}(h) \in H^1(M, \mathbb{R})$ (for almost all $\text{Im}(h) \in H^1(M, \mathbb{R})$). Indeed, an interval exchange transformation is equivalent to the (horizontal) translation flow obtained by suspension under a constant roof function. The roof function determines the class $\text{Re}(h)$. The linear elimination procedure outlined above therefore fails.

In fact, for many permutations (of rotation type) the constant roof function determines a class in the relative cohomology $H^1(M, \Sigma, \mathbb{R})$ which has a non-trivial component on the kernel of the forgetful map $H^1(M, \Sigma, \mathbb{R}) \to H^1(M, \mathbb{R})$ (with respect to the Oseledets decomposition). Since the extension of the Kontsevich–Zorich cocycle (or the Rauzy–Veech cocycle) to the kernel of the forgetful map is *isometric*, the Veech criterion immediately implies the weak mixing property for almost all IET's, since, by the isometric behavior of the cocycle, non-integer vector cannot converge to the integer lattice under the cocycle.

By this *isometric elimination* scheme, outlined above, Veech [11] established the weak mixing property for Lebesgue almost all IET's, for all permutations of rotation class, which are not rotations. Permutations of rotation class are those equivalent to rotations (cyclic) permutations under Rauzy operations. The particular case of IET's on 3-intervals (which are not rotations) was known earlier from the work of Katok and Stepin [48].

The completion of the proof of weak mixing for typical IET's (see [20, 66]) is based on a non-linear, probabilistic elimination scheme.

The *hyperbolic elimination* scheme analyzes the dynamics in $H^1(M, \mathbb{R})$ modulo $H^1(M, \mathbb{Z})$ of *line segments* approximately aligned with the top unstable Oseledets subspace contained in balls of given (sufficiently) small radius around integer vectors in the cohomology vector space. There are two main effects:

1. *Repulsion*: Under the hypothesis that the second Lyapunov exponent of the Kontsevich–Zorich cocycle $\lambda_2 > 0$ line segment are repelled away from the nearest integer point (under the action of the Kontsevich–Zorich cocycle);
2. *Spreading*: For sufficiently *long excursions* of the Teichmüller orbit, a line segment can get dilated and spread to several line segments near many different integer points (after intersection with balls centered at those points).

The strategy of the argument is to prove that the probability that a given segment remains inside a fixed neighborhood of the integer lattice converges to zero as time diverges: the probability of escaping elimination goes to zero with time. This conclusion follows from an estimate of the probability that the repulsion mechanism pushes a segment out of the given neighborhood against the probability that the segment "survives" by jumping near another lattice point during an excursion of the Teichmüller orbit. The latter is small since excursions of Teichmüller geodesics outside of the "thick part" of the moduli space are rare (see [44] or [67] although the original argument in [20] bypasses such stronger results on excursions).

The key underlying dynamical feature that makes the probabilistic argument possible are the mixing properties of the Rauzy–Veech induction (rather of the induced maps on compact sub-simplices) or of the Teichmüller flow, which are exponentially mixing dynamical systems (see [20, 67, 68]).

4.4.6 Effective Weak Mixing for Translation Flows

The quantitative theory of weak mixing for translation flows and Interval Exchange Transformations is based on an effective version of the Veech criterion, which, as the Veech criterion in Theorem 4.45, can be derived by methods of twisted Hodge theory.

Theorem 4.53 (Effective Veech Criterion) *Assume that for some compact set $K \subset \mathcal{H}_\kappa^{(1)}$ of positive measure and for some neighborhood $U \subset H_\kappa^1(M, \mathbb{T})$ of the zero section (or, equivalently, for some neighborhood of $H_\kappa^1(M, \mathbb{Z})$ in $H_\kappa^1(M, \mathbb{R})$), for a given $(M, h) \in \mathcal{H}_\kappa^{(1)}$ and $\lambda \in \mathbb{R} \setminus \{0\}$, we have*

$$f_{K,U}$$
$$:= \limsup_{t > 0} \frac{\text{Leb}(\{s \in [0, t] | g_t(M, h) \in K \text{ and } g_t^{KZ}(M, h, \lambda[\text{Re}(h)]) \in U\})}{\text{Leb}(\{s \in [0, t] | g_t(M, h) \in K\})} < 1.$$

Then there exist constants $\alpha > 0$ and $C(\lambda) > 0$ such that, for all weakly differentiable $f \in H^1(M)$ and for all $(p, T) \in M \times \mathbb{R}^+$ (such that p has an infinite forward horizontal orbit) we have

$$\left| \int_0^T e^{2\pi i \lambda t} f \circ \phi_t^X(p) dt \right| \leq C(\lambda) \|f\|_{H^1(M)} T^{1-\alpha}.$$

The proof of the effective Veech criterion in Theorem 4.53 is completely analogous to that of effective unique ergodicity in Theorem 4.23 (self-similar case) and Theorem 4.36 (typical case), in fact it is a "twisted" version of it. We outline this analogy and the resulting argument below.

Outline of the Proof of Theorem 4.53 *First Step*. For a fixed $\lambda \in \mathbb{R}$ (including in the untwisted case $\lambda = 0$), we write the twisted integrals in terms of 1-dimensional currents given by the orbit segments. We view orbit arcs $\gamma_T^X(p)$ of the horizontal flow $\phi_\mathbb{R}^X$ as 1-dimensional currents defined as

$$\gamma_T^X(p)(\alpha) := \int_0^T e^{2\pi i \lambda t} (\iota_X \alpha) \circ \phi_t^X(p) dt, \quad \text{for all smooth 1-form } \alpha \in W^1(M, h).$$

Since $\iota_X \operatorname{Re}(h) \equiv 1$, we have that

$$\gamma_{\lambda,T}^X(p)(f \operatorname{Re} h) = \int_0^T e^{2\pi i \lambda t} f \circ \phi_t^X(p) dt.$$

We note that the current $\gamma_{\lambda,T}^X(p)$, as current of integration along a codimension 1 submanifold belong to the space $W^{-1}(M,h)$ by the Sobolev trace theorem.

Second step. We argue that such currents are at bounded distance from the space of currents which are closed with respect to the twisted differential

$$d_{h,\lambda} = d + 2\pi i \lambda \operatorname{Re}(h) \wedge .$$

Indeed, we can compute the twisted differential of the current $\gamma_{\lambda,T}^X(p)$ as follows. Let $f \in C^\infty(M)$, then we have

$$|\gamma_{\lambda,T}^X(p)(d_{h,\lambda} f)| := \left| \int_0^T e^{2\pi i \lambda t} (\iota_X d_{h,\lambda} f) \circ \phi_t^X(p) dt \right|$$

$$= \left| \int_0^T e^{2\pi i \lambda t} (Xf + 2\pi i \lambda f) \circ \phi_t^X(p) dt \right|$$

$$= \left| \int_0^T \frac{d}{dt} \left(e^{2\pi i \lambda t} f \circ \phi_t^X(p) \right) dt \right|$$

$$= \left| e^{2\pi i \lambda T} f \circ \phi_T^X(p) - f(p) \right| \leq 2 \|f\|_{W^2(M,h)}.$$

Thus, $\|d_{h,\lambda} \gamma_{\lambda,T}^X(x)\|_{W^{-2}(M,h)} \leq 2$, for all $(p,T) \in M \times \mathbb{R}^+$, and since the operator $d_{h,\lambda}$ is elliptic, hence its restriction to the orthogonal of its kernel $\mathcal{Z}_\lambda^{-1}(M,h)$ has bounded inverse, it is possible to prove that there exists a constant $C(M) > 0$ (uniformly bounded on compact subsets of the moduli space) such that

$$\inf_{z \in \mathcal{Z}_\lambda^{-1}(M,h)} \|\gamma_{\lambda,T}^X(p) - z\|_{W^{-1}(M,h)} \leq C(M).$$

Since, as we have just proved, the current $\gamma_{\lambda,T}^X(p)$ is at uniformly bounded distance from twisted-closed currents, the argument is reduced to an estimate on the growth of the (Sobolev) norm of *twisted closed currents* under the restriction of the "transfer" cocycle on the bundle $W_\kappa^{-1}(M)$ of currents of Sobolev order -1 to the sub-bundle $\mathcal{Z}_{\kappa,\lambda}^{-1}(M)$ of twisted closed currents.

Third step. The proof of bounds on the growth of the (Sobolev) norm of order -1 of twisted closed currents can be reduced to the growth of the Hodge norm of twisted cohomology classes once we have proven that there exists an invariant norm on the sub-bundle $\mathcal{E}_{\kappa,\lambda}^{-1}(M)$ of twisted exact currents. The invariant norm is defined as follows. By definition, a current $\gamma \in W^{-1}(M,h)$ is twisted exact if $\gamma \in d_{h,\lambda}(L^2(M, \text{area}_h))$, hence we can define

$$\|\gamma\|_{h,\lambda} := \|U_{h,\lambda}\|_{L^2(M,\text{area}_h)} \quad \text{if } \gamma = d_{h,\lambda}U_{h,\lambda} \in \mathcal{E}_\lambda^{-1}(M,h).$$

Fourth step. At this point, since we have reduced bounds on the currents $\gamma_{\lambda,T}^X(p)$ (in Sobolev norm) to bounds on twisted cohomology classes (in Hodge norm), by Lemma 4.42 we can derive the following bound. We note that on compact sets of the moduli space, since the twisted cohomology is finite dimensional, the quotient norms induced by the Sobolev norms are equivalent to the Hodge norm.

Let $K \subset \mathcal{H}_\kappa^{(1)}$ be any given compact set. For any (M,h) there exists a constant $C_K(M,h) > 1$ such that, for any $T = e^t$ with $g_t(M,h) \in K$, for any $p \in M$ (with infinite horizontal forward orbit) we have

$$\|\gamma_{\lambda,T}^X(p)\|_{W^{-1}(M,h)} \leq C_K(M,h) \exp\left(\int_0^{\log T} \Lambda^\#(g_s^{KZ}(M,h,\lambda[\text{Re}(h)])ds\right) \tag{4.24}$$

Let K be a compact set of positive measure and (M,h) be such that

$$\liminf_{t>0} \frac{1}{t}\text{Leb}\{s \in [0,t] | g_s(M,h) \in K\} = \mu_K > 0.$$

Since by Lemma 4.43 there exists a constant $\Lambda_{\max}^\# < 1$ such that

$$\Lambda^\#(M,h,\eta) \leq \Lambda_{\max}^\# < 1, \text{ for all } (M,h) \in K \text{ and } (M,h,\eta) \notin U,$$

it follows from the bound in formula (4.24) that

$$\|\gamma_{\lambda,T}^X(p)\|_{W^{-1}(M,h)} \leq C_K(M,h) \exp\left(1 - \mu_K(1 - f_{K,U})(1 - \Lambda_{\max}^\#))\log T\right),$$

that is, for $\zeta := \mu_K(1 - f_U)(1 - \Lambda^\#) > 0$, we have

$$\|\gamma_{\lambda,T}^X(p)\|_{W^{-1}(M,h)} \leq C_K(M,h) T^{1-\zeta}. \tag{4.25}$$

Finally, let $(T_k) = (e^{t_k})$ with $t_0 = 0$ and (t_k) for $k \geq 1$ a sequence of return times of the forward orbit $g_{\mathbb{R}^+}(M,h)$ to the compact set $K \subset \mathcal{H}_\kappa^{(1)}$ such that

$$\lim_{k\to\infty} \frac{t_k}{k} = \mu_K > 0.$$

Fifth Step. The argument is completed by decomposing any horizontal orbit of length $T > 0$ into orbits arcs of lengths which belong to the sequence (T_k) (to which the bound (4.25) can be applied). In fact for any $T > 1$ we have the (Ostrowski's) decomposition

$$T = \sum_{k=0}^{N} m_k T_k$$

with $T_N \leq T \leq T_{N+1}$ and $m_k T_k \leq T_{k+1}$, for all $k < N$, hence by the bound in formula (4.25) we derive

$$\|\gamma_{\lambda,T}^X(p)\|_{W^{-1}(M,h)} \leq C_K(M,h) \sum_{k=1}^{N} m_k T_k^{1-\zeta} .$$

For every $\epsilon \in (0, \mu_K)$, there exists $k_\epsilon \in \mathbb{N}$ such that

$$(\mu_K - \epsilon)k \leq t_k \leq (\mu_K + \epsilon)k, \quad \text{for all } k \geq k_\epsilon ,$$

hence there exists a constant $C_{K,\epsilon} > 1$ such that

$$m_k \leq \frac{T_{k+1}}{T_k} = e^{t_{k+1} - t_k} \leq C_{K,\epsilon} e^{2\epsilon k} .$$

There exists therefore a constant $C_{K,\epsilon}(\zeta) > 1$ such that

$$\sum_{k=1}^{N} m_k T_k^{1-\zeta} \leq C_{K,\epsilon} \sum_{k=1}^{N} e^{[(1-\zeta)(\mu_K + \epsilon) + 2\epsilon]k}$$
$$\leq C_{K,\epsilon}(\zeta) e^{[(1-\zeta)(\mu_K + \epsilon) + 2\epsilon]N} \leq C_{K,\epsilon}(\zeta) T^{1-\alpha}$$

with

$$\alpha = 1 - (1-\zeta)\frac{\mu_K + \epsilon}{\mu_K - \epsilon} + 2\epsilon .$$

Since there exists $\epsilon > 0$ such that in the above formula $\alpha > 0$, the argument is complete (in fact, the bound in the statement holds for any exponent $\alpha < \zeta$). $\square$

The proof of (Masur–Veech) typical *effective* weak mixing of translation flows is completed by a linear approximation argument analogous to that outlined int the proof of typical weak mixing. Namely, motivated by the effective Veech criterion (Theorem 4.53), in analogy with Definition 4.17, we introduce the effective weak stable space:

Definition 4.18 The **effective weak stable space** $W^{eff}_{(M,h)} \subset H^1(M, \mathbb{R})$ at a translation surface $(M,h) \subset \mathcal{H}_\kappa^{(1)}$ is the set defined as follows.

Let $\mathcal{K}$ be a countable exhaustion of the moduli space $\mathcal{H}_\kappa^{(1)}$ by compact subsets and let $\mathcal{U}$ a countable basis of neighborhoods of $H_\kappa^1(M, \mathbb{Z}) \subset H_\kappa^1(M, \mathbb{R})$. We then define

$$\mathcal{W}_{(M,h)}^{eff} := \cap_{K \in \mathcal{K}} \cap_{U \in \mathcal{U}} \{c \in H^1(M, \mathbb{R}) |$$

$$\limsup_{t>0} \frac{\text{Leb}(\{s \in [0, t] | g_t(M, h) \in K \text{ and } g_t^{KZ}(M, h, \lambda[\text{Re}(h)]) \in U\})}{\text{Leb}(\{s \in [0, t] | g_t(M, h) \in K\})} = 1\}.$$

In other terms, in the definition of the effective stable space *the convergence takes place only in average* and excursions of the cohomology class outside of any given neighborhood of the integer lattice are possible provided their frequency converges to zero in the large time limit.

Despite this importance difference, all the basic properties of the stable space (in particular concerning its Hausdorff dimension) hold for the effective weak stable space, and the linear elimination procedure outlined in Sect. 4.4.5 can be carried out (see [10]), as well as the non-linear, probabilistic elimination (see [66]).

By the effective Veech criterion (Theorem 4.53) and by the linear elimination argument we can then derive the following:

Theorem 4.54 ([10]) *For each stratum $\mathcal{H}_\kappa^{(1)}$ of translation surfaces there exists a constant $\alpha_\kappa > 0$ such that, for the Masur–Veech typical translation surface $(M, h) \in \mathcal{H}_\kappa^{(1)}$ and for all $\lambda \in \mathbb{R}$ there exists a constant $K_h(\lambda) > 0$ such that the horizontal translation flow satisfies the following effective weak mixing estimate. For all functions $f \in H^1(M)$ (the Sobolev space of functions with square integrable first weak derivative) of zero average, and for all $(p, T) \in M \times \mathbb{R}^+$ such that p has infinite forward orbit, we have*

$$\left| \int_0^T e^{2\pi i \lambda t} f \circ \phi_t^X(p) dt \right| \leq K_h(\lambda) T^{1-\alpha_\kappa}.$$

The above theorem implies the last equivalent property in Definition 4.7 for the space $W^1(M)$, hence it concludes (see also Exercise 4.14) the outline of the proof of polynomial weak mixing for typical translation flows stated in Theorem 4.15. For the proof of effective weak mixing for IET's, based on the effective Veech criterion and the probabilistic elimination procedure, we refer the reader to [66].

References

1. Kontsevich, M.: Lyapunov exponents and Hodge theory. In: The Mathematical Beauty of Physics (Saclay, 1996). Advanced Series in Mathematical Physics, vol. 24, pp. 318–332. World Scientific Publishing, River Edge (1997)
2. Kontsevich, M., Zorich, A.: Lyapunov exponents and Hodge theory (1997). https://arxiv.org/pdf/hep-th/9701164.pdf

3. Forni, G.: Deviation of ergodic averages for area-preserving flows on surfaces of higher genus. Ann. Math. **155**(1), 1–103 (2002)
4. Möller, M.: Variations of hodge structures of a teichmüller curve. J. Am. Math. Soc. **19**(2), 327–344 (2006)
5. Filip, S.: Semisimplicity and rigidity of the Kontsevich-Zorich cocycle. Invent. Math. **205**(3), 617–670 (2016)
6. Filip, S.: Splitting mixed Hodge structures over affine invariant manifolds. Ann. Math. **183**(2), 681–713 (2017)
7. Filip, S.: Zero Lyapunov exponents and monodromy of the Kontsevich–Zorich cocycle. Duke Math. J. **166**(4), 657–706 (2017)
8. Athreya, J.S., Bufetov, A., Eskin, A., Mirzakhani, M.: Lattice point asymptotics and volume growth on Teichmüller space. Duke Math. J. **161**(6), 1055–1111 (2012)
9. Eskin, A., Mirzakhani, M.: Invariant and stationary measures for the SL(2, $\mathbb{R}$) action on moduli space. Publ. Math. Inst. Hautes Études Sci. **127**, 95–324 (2018)
10. Forni, G.: Twisted translation flows and effective weak mixing. J. Eur. Math. Soc. **24**, 4225–4276 (2022)
11. Veech, W.: The metric theory of interval exchange transformations. I. Generic spectral properties. Am. J. Math. **106**, 1331–1359 (1984)
12. Bufetov, A.I., Solomyak, B.: On the modulus of continuity for spectral measures in substitution dynamics. Adv. Math. **260**, 84–129 (2014)
13. Bufetov, A.I., Solomyak, B.: The Hölder property for the spectrum of translation flows in genus two. Isr. J. Math. **223**, 205–259 (2018)
14. Bufetov, A.I., Solomyak, B.: A spectral cocycle for substitution systems and translation flows. J. Anal. Math. **141**, 165–205 (2020)
15. Bufetov, A.I., Solomyak, B.: Hölder regularity for the spectrum of translation flows. J. Éc. Polytech. Math. **8**, 279–310 (2021)
16. Athreya, J.S., Forni, G.: Deviation of ergodic averages for rational polygonal billiards. Duke Math. J. **144**(2), 285–319 (2008)
17. Forni, G., Matheus, C.: Introduction to Teichmüller theory and its applications to dynamics of interval exchange transformations, flows on surfaces and billiards. J. Mod. Dynam. **8**(3–4), 271–436 (2014). Lectures from the Bedlewo Summer School 2011 (F. Rodriguez Hertz editor)
18. Forni, G.: On the Lyapunov exponents of the Kontsevich–Zorich cocycle. In: Hasselblatt, B., Katok, A. (eds.) Handbook of Dynamical Systems, vol. 1B, pp. 549–580. Elsevier, Amsterdam (2006)
19. Forni, G., Matheus, C., Zorich, A.: Lyapunov spectrum of invariant subbundles of the Hodge bundle. Erg. Theory Dynam. Syst. **34**(2), 353–408 (2012)
20. Avila, A., Forni, G.: Weak mixing for interval exchange transformations and translation flows. Ann. Math. **165**(2), 637–664 (2007)
21. Filip, S.: Translation surfaces: dynamics and hodge theory (2022, preprint). https://math.uchicago.edu/~sfilip/public_files/surf_surv.pdf
22. Masur, H., Tabachnikov, S.: Rational billiards and flat structures. In: Hasselblatt, B., Katok, A. (eds.) Handbook of Dynamical Systems, vol. 1A, pp. 1015–1089. Elsevier, Amsterdam (2002)
23. Hubert, P., Schmidt, T.: An introduction to veech surfaces. In: Hasselblatt, B., Katok, A. (eds.) Handbook of Dynamical Systems, vol. 1B, pp. 501–526. Elsevier, Amsterdam (2006)
24. Zorich, A.: Flat surfaces. In: Cartier, P.E., Bernard Julia, B., Moussa, P., Vanhove, P. (eds.) Frontiers in Number Theory, Physics, and Geometry. I, pp. 437–583. Springer, Berlin (2006)
25. Yoccoz, J.C.: Interval exchange maps and translation surfaces. In: Einsiedler, M.E.A. (ed.) Homogeneous Flows, Moduli Spaces and Arithmetic, vol. 10, pp. 1–69. American Mathematical Society, Providence (2010)
26. Wright, A.: Translation surfaces and their orbit closures: an introduction for a broad audience. EMS Surv. Math. Sci. **2**(1), 63–108 (2015)
27. Katok, A.B., Zemlyakov, A.M.: Topological transitivity of billiards in polygons. Math. Not. Acad. Sci. USSR **18**, 760–764 (1975). Errata **20** (1976), 1051

28. Richens, P.J., Berry, M.V.: Pseudointegrable systems in classical and quantum mechanics. Phys. D Nonlinear Phenom. **2**(3), 495–512 (1981). https://doi.org/10.1016/0167-2789(81)90024-5
29. Fox, R.H., Kershner, R.B.: Concerning the transitive properties of geodesics on a rational polyhedron. Duke Math. J. **2**(1), 147–150 (1936)
30. Masur, H.: The extension of the Weil–Petersson metric to the boundary of Teichmüller space. Duke Math. J. **43**(3), 623–635 (1976)
31. Mirzakhani, M., Wright, A.: The boundary of an affine invariant submanifold. Invent. Math. **209**, 927–984 (2017)
32. Bainbridge, M., Chen, D., Gendron, Q., Grushevsky, S., Möller, M.: Compactification of strata of Abelian differentials. Duke Math. J. **167**(12), 2347–2416 (2018)
33. Masur, H.: Interval exchange transformations and measured foliations. Ann. Math. **115**(1), 169–200 (1982)
34. Masur, H.: Ergodic actions of the mapping class group. Proc. Am. Math. Soc. **94**(3), 455–459 (1985)
35. Veech, W.: Gauss measures for transformations on the space of interval exchange maps. Ann. Math. **115**(2), 201–242 (1982)
36. Eskin, A., Mirzakhani, M., Mohammadi, A.: Isolation, equidistribution, and orbit closures for the $SL(2,\mathbb{R})$ action on moduli space. Ann. Math. **182**(2), 673–721 (2015)
37. Chaika, J., Weiss, B.: The horocycle flow on the moduli space of translation surfaces. In: Paper Presented at the International Congress of Mathematicians (ICM) 2022. Virtual Lecture at https://www.youtube.com/watch?v=6wAUHpG4Qg8 registered at the University of Bologna, Italy on July 1, 2022 (2022)
38. Chaika, J., Smillie, J., Weiss, B.: Tremors and horocycle dynamics on the moduli space of translation surfaces (2020). https://arxiv.org/pdf/2004.04027
39. Forni, G.: Limits of geodesic push-forwards of horocycle invariant measures. Erg. Theory Dynam. Syst. **41**(9), 2782–2804 (2021)
40. Keane, M.: Interval exchange transformations. Math. Z. **141**, 25–32 (1975)
41. Keynes, H.B., Newton, D.: A 'minimal', non-uniquely ergodic interval exchange transformation. Math. Z. **148**, 101–106 (1976)
42. Kerckhoff, S., Masur, H., Smillie, J.: Ergodicity of billiard flows and quadratic differentials. Ann. Math. **124**(2), 293–311 (1986)
43. Vorobets, Y.: Ergodicity of billiards in polygons. Sb. Math. **188**(3), 389–434 (1997). (English translation of Mat. Sb. **188**, no. 3 (1997), 65–112)
44. Athreya, J.S.: Quantitative recurrence and large deviations for Teichmüller geodesic flow. Geom. Ded. **119**, 121–140 (2006)
45. Halmos, P.R.: In general a measure preserving transformation is mixing. Ann. Math. **45**, 786–792 (1944)
46. Rokhlin, V.: A 'general' measure-preserving transformation is not mixing (in Russian). Dokl. Akad. Nauk SSSR **60**, 349–351 (1948)
47. Katok, A.B.: Interval exchange transformations and some special flows are not mixing. Isr. J. Math. **35**(4), 301–310 (1980)
48. Katok, A.B., Stepin, A.M.: Approximations in ergodic theory. Russ. Math. Surv. **22**(5), 77–102 (1967). (English translation of Uspehi Mat. Nauk **22** (1967), no. 5 (137), 81–106)
49. Wright, A.: Cylinder deformations in orbit closures of translation surfaces. Geom. Topol. **19**(1), 413–438 (2015)
50. Avila, A., Delecroix, V.: Weak mixing directions in non-arithmetic Veech surfaces. J. Am. Math. Soc. **29**(4), 1167–1208 (2016)
51. Chaika, J., Forni, G.: Weakly Mixing Polygonal Billiards (2020). https://arxiv.org/pdf/2003.00890.pdf
52. Casati, G., Prosen, T.: Mixing property of triangular billiards. Phys. Rev. Lett. **83**, 4729–4732 (1999)
53. Treviño, R.: On the ergodicity of flat surfaces of finite area. Geom. Funct. Anal. **24**(1), 360–386 (2014)

54. McMullen, C.: Teichmüller dynamics and unique ergodicity via currents and Hodge theory. J. Reine Angew. Math. **768**, 39–54 (2020)
55. McMullen, C.: Billiards and Teichmüller curves. Bull. Am. Math. Soc. **60**(2), 195–250 (2023)
56. Katok, A.B.: Invariant measures of flows on oriented surfaces. Soviet Math. Dokl. **14**(4), 1104–1108 (1973). (English translation of Dokl. Akad. Nauk SSSR **211** (4), 1973)
57. Forni, G.: On the equidistribution of unstable curves for pseudo-Anosov diffeomorphisms of compact surfaces. Erg. Theory Dynam. Syst. **42**(3), 855–880 (2022)
58. Forni, G.: A geometric criterion for the nonuniform hyperbolicity of the Kontsevich–Zorich cocycle. J. Mod. Dynam. **5**(2), 355–395 (2011)
59. Avila, A., Viana, M.: Simplicity of lyapunov spectra: proof of the Zorich–Kontsevich conjecture. Acta Math. **198**(1), 1–56 (2007)
60. Frączek, K., Ulcigrai, C.: On the asymptotic growth of Birkhoff integrals for locally Hamiltonian flows and ergodicity of their extensions (2021). https://arxiv.org/pdf/2112.05939.pdf
61. Frączek, K., Kim, M.: New phenomena in deviation of Birkhoff integrals for locally Hamiltonian flows. J. Reine Angew. Math. **2024**(807), 81–149 (2024). https://doi.org/10.1515/crelle-2023-0090
62. Frankel, I.: Meromorphic l^2 functions on flat surfaces. Geom. Funct. Anal. **32**, 832–860 (2022)
63. Kahn, J., Wright, A.: Hodge and Teichmüller. J. Mod. Dynam. **18**, 149–160 (2022)
64. Forni, G., Matheus, C.: An example of a Teichmüller disk in genus 4 with degenerate Kontsevich–Zorich spectrum (2008). https://arxiv.org/pdf/0810.0023.pdf
65. Forni, G., Matheus, C., Zorich, A.: Square tiled cyclic covers. J. Mod. Dynam. **5**(2), 285–318 (2011)
66. Avila, A., Forni, G., Safaee, P.: Quantitative weak mixing for interval exchange transformations. Geom. Funct. Anal. **33**, 1–56 (2023). https://doi.org/10.1007/s00039-023-00625-y
67. Avila, A., Gouëzel, S., Yoccoz, J.C.: Exponential mixing for the Teichmüller flow. Publ. Math. de l'IHÉS **104**, 143–211 (2006)
68. Avila, A., Bufetov, A.: Exponential decay of correlations for the Rauzy-Veech-Zorich induction map. In: Partially Hyperbolic Dynamics, Laminations, and Teichmüller Flow. Fields Inst. Commun., vol. 51, pp. 203–211. American Mathematical Society, Providence (2007)
69. Yu, F.: Eigenvalues of curvature, Lyapunov exponents and Harder–Narasimhan filtrations. Geom. Topol. **22**(4), 2253–2298 (2018). https://doi.org/10.2140/gt.2018.22.2253
70. Eskin, A., Kontsevich, M., Möller, M., Zorich, A.: Lower bounds for Lyapunov exponents of flat bundles on curves. Geom. Topol. **22**(4), 2299–2338. https://doi.org/10.2140/gt.2018.22.2299
71. Costantini, M.: Lyapunov exponents, holomorphic flat bundles and de Rham moduli space. Isr. J. Math. 240, 345–415 (2020). https://doi.org/10.1007/s11856-020-2060-6

LECTURE NOTES IN MATHEMATICS   Springer

Editors in Chief: J.-M. Morel, B. Teissier;

Editorial Policy

1. Lecture Notes aim to report new developments in all areas of mathematics and their applications – quickly, informally and at a high level. Mathematical texts analysing new developments in modelling and numerical simulation are welcome.

   Manuscripts should be reasonably self-contained and rounded off. Thus they may, and often will, present not only results of the author but also related work by other people. They may be based on specialised lecture courses. Furthermore, the manuscripts should provide sufficient motivation, examples and applications. This clearly distinguishes Lecture Notes from journal articles or technical reports which normally are very concise. Articles intended for a journal but too long to be accepted by most journals, usually do not have this "lecture notes" character. For similar reasons it is unusual for doctoral theses to be accepted for the Lecture Notes series, though habilitation theses may be appropriate.

2. Besides monographs, multi-author manuscripts resulting from SUMMER SCHOOLS or similar INTENSIVE COURSES are welcome, provided their objective was held to present an active mathematical topic to an audience at the beginning or intermediate graduate level (a list of participants should be provided).

   The resulting manuscript should not be just a collection of course notes, but should require advance planning and coordination among the main lecturers. The subject matter should dictate the structure of the book. This structure should be motivated and explained in a scientific introduction, and the notation, references, index and formulation of results should be, if possible, unified by the editors. Each contribution should have an abstract and an introduction referring to the other contributions. In other words, more preparatory work must go into a multi-authored volume than simply assembling a disparate collection of papers, communicated at the event.

3. Manuscripts should be submitted either online at www.editorialmanager.com/lnm to Springer's mathematics editorial in Heidelberg, or electronically to one of the series editors. Authors should be aware that incomplete or insufficiently close-to-final manuscripts almost always result in longer refereeing times and nevertheless unclear referees' recommendations, making further refereeing of a final draft necessary. The strict minimum amount of material that will be considered should include a detailed outline describing the planned contents of each chapter, a bibliography and several sample chapters. Parallel submission of a manuscript to another publisher while under consideration for LNM is not acceptable and can lead to rejection.

4. In general, **monographs** will be sent out to at least 2 external referees for evaluation.

   A final decision to publish can be made only on the basis of the complete manuscript, however a refereeing process leading to a preliminary decision can be based on a pre-final or incomplete manuscript.

   Volume Editors of **multi-author works** are expected to arrange for the refereeing, to the usual scientific standards, of the individual contributions. If the resulting reports can be forwarded to the LNM Editorial Board, this is very helpful. If no reports are forwarded or if other questions remain unclear in respect of homogeneity etc, the series editors may wish to consult external referees for an overall evaluation of the volume.

5. Manuscripts should in general be submitted in English. Final manuscripts should contain at least 100 pages of mathematical text and should always include

   – a table of contents;
   – an informative introduction, with adequate motivation and perhaps some historical remarks: it should be accessible to a reader not intimately familiar with the topic treated;
   – a subject index: as a rule this is genuinely helpful for the reader.
   – For evaluation purposes, manuscripts should be submitted as pdf files.

6. Careful preparation of the manuscripts will help keep production time short besides ensuring satisfactory appearance of the finished book in print and online. After acceptance of the manuscript authors will be asked to prepare the final LaTeX source files (see LaTeX templates online: https://www.springer.com/gb/authors-editors/book-authors-editors/manuscriptpreparation/5636) plus the corresponding pdf- or zipped ps-file. The LaTeX source files are essential for producing the full-text online version of the book, see http://link.springer.com/bookseries/304 for the existing online volumes of LNM). The technical production of a Lecture Notes volume takes approximately 12 weeks. Additional instructions, if necessary, are available on request from lnm@springer.com.

7. Authors receive a total of 30 free copies of their volume and free access to their book on SpringerLink, but no royalties. They are entitled to a discount of 33.3 % on the price of Springer books purchased for their personal use, if ordering directly from Springer.

8. Commitment to publish is made by a *Publishing Agreement*; contributing authors of multiauthor books are requested to sign a *Consent to Publish form*. Springer-Verlag registers the copyright for each volume. Authors are free to reuse material contained in their LNM volumes in later publications: a brief written (or e-mail) request for formal permission is sufficient.

Addresses:
Professor Jean-Michel Morel, CMLA, École Normale Supérieure de Cachan, France
E-mail: moreljeanmichel@gmail.com

Professor Bernard Teissier, Equipe Géométrie et Dynamique,
Institut de Mathématiques de Jussieu – Paris Rive Gauche, Paris, France
E-mail: bernard.teissier@imj-prg.fr

Springer: Ute McCrory, Mathematics, Heidelberg, Germany,
E-mail: lnm@springer.com

SPRINGER NATURE

GPSR Compliance

The European Union's (EU) General Product Safety Regulation (GPSR) is a set of rules that requires consumer products to be safe and our obligations to ensure this.

If you have any concerns about our products, you can contact us on ProductSafety@springernature.com

In case Publisher is established outside the EU, the EU authorized representative is:

Springer Nature Customer Service Center GmbH
Europaplatz 3
69115 Heidelberg, Germany

The manufacturer's authorised representative in the EU is Springer Nature Customer Service Centre GmbH, Europaplatz 3, 69115 Heidelberg, Germany. If you have any concerns regarding our products, please contact ProductSafety@springernature.com

Printed and bound by CPI Group (UK) Ltd, Croydon, CR0 4YY

25/03/2026

02078196-0015